Simulation of Fluid Power Systems
with Simcenter Amesim

Simulation of Fluid Power Systems with Simcenter Amesim

Nicolae Vasiliu

Daniela Vasiliu

Constantin Călinoiu

Radu Puhalschi

CRC Press
Taylor & Francis Group
Boca Raton London New York

CRC Press is an imprint of the
Taylor & Francis Group, an **informa** business

CRC Press
Taylor & Francis Group
6000 Broken Sound Parkway NW, Suite 300
Boca Raton, FL 33487-2742

First issued in paperback 2020

© 2018 by Taylor & Francis Group, LLC
CRC Press is an imprint of Taylor & Francis Group, an Informa business

No claim to original U.S. Government works

ISBN-13: 978-0-367-57201-3 (pbk)
ISBN-13: 978-1-4822-5355-9 (hbk)

Library of Congress Cataloging-in-Publication Data

Names: Vasiliu, Nicolae, 1946- author. | Vasiliu, Daniela, 1958- author. | Calinoiu, Constantin, author. | Puhalschi, Radu, author.
Title: Simulation of fluid power systems with Simcenter Amesim / Nicolae Vasiliu, Daniela Vasiliu, Constantin Calinoiu, Radu Puhalschi.
Description: Boca Raton : Taylor & Francis, CRC Press, 2018. | Includes bibliographical references and index.
Identifiers: LCCN 2017040525| ISBN 9781482253559 (hardback : alk. paper) | ISBN 9781315118888 (ebook)
Subjects: LCSH: Fluid power technology--Computer simulation. | LMS imagine.lab amesim.
Classification: LCC TJ843 .V37 2018 | DDC 620.1/06--dc23
LC record available at https://lccn.loc.gov/2017040525

Visit the Taylor & Francis Web site at
http://www.taylorandfrancis.com

and the CRC Press Web site at
http://www.crcpress.com

Dedication

To Dr. Jan Leuridan, a huge engine of the world scientific technology, and a kind leader anywhere.

Contents

**Chapter 7   Modeling, simulation, and identification of the hydrostatic pumps
                      and motors..305**

Preface

Motto

In order to be useful, a book should be a good impedance match to its audience. Unfortunately, audiences vary widely, and a presentation that attempts to satisfy too wide a range of readers often succeeds in being of little use to any of them. In the case of engineering books, one excellent solution to the problem is to aim the book at the working engineer, and to add material of interest to the student or to the specialist only to the extent dictated by the circumstances of the publication.

John Fitzgerald Blackburn

Fluid Power Systems represents a wide, important, and interesting theoretical and technical field, which can be treated in different ways. The authors of this book regarded the above-mentioned field in close connection with many other scientific and technical fields, trying to create a global image of the realistic mathematical modeling, numerical simulation, experimental identification, and performance tests, as components of any innovation process.

This book contains the results of a systematic research on the steady-state behavior, and the transients occurring in any fluid power systems, trying to draw a straight line between theory and practice through the aid of Simcenter Amesim—Advanced Modeling Environment for Simulation of Engineering Systems (Amesim in short), created by Société IMAGINE from France, and then further developed by LMS Corporation from Belgium. Amesim covers the analysis, synthesis, and design cover flow and pressure valves; fixed and variable displacement pumps and motors; electrohydraulic servovalves, servopumps, and servomotors; hydraulic and electrohydraulic servomechanisms; and some important applications in the field of manufacturing, aerospace, automotive, civil, and power engineering. Most of the theoretical treatises are presented in a practical manner, starting from the needs of the designers.

The classical innovation process in the field of the fluid power systems is built on the following milestone works:

- The fundamental book *Fluid Power Control* published in 1960 by technology press of MIT and Wiley, written by J.F. Blackburn, G. Reethof, and J.L. Shearer from *Dynamic Analysis and Control Laboratory of Massachusetts Institute of Technology*, which remains an almost inexhaustible source of basic information.
- The book *Hydraulic Control Systems* written by Herbert E. Merritt from *Cincinnati Milling Machine Company*, and published by Wiley in 1967; the book is directed more specifically to hydraulic control problems without covering the design phase.

- The work *Analysis, Synthesis and Design of Hydraulic Servosystems and Pipelines* written by Taco J. Viersma from *Delft Institute of Technology,* Delft, Netherlands and published in 1980 by Elsevier Scientific Publishing Company; this textbook draws direct connections to the design of the linear servosystems with constant preload widely used in six DOF dynamic simulators.

The above-mentioned authors set up the theoretical concepts needed for the analysis and synthesis of the fluid power systems by a huge amount of theoretical and applied research, trying to solve the design problems by *paper and pencil* before simulating them on digital computers.

The modern innovation process in the field of the fluid power systems is closely connected with the development of the simulation languages for solving complex sets of nonlinear differential equations. The first simulators used lines of codes in Fortran (similar to ACSL), and needed many running attempts to refine the systems structure and parameters.

In the second stage of evolution, digital simulators such as SIMULINK® gained interactive graphical interfaces using *icons* for many kinds of components and signal libraries. Important progress in the simulation of dynamic systems was produced by Michel Lebrun, professor of mechanical engineering at INSA Lyon, and Claude Richard, a high-level specialist in computer science from Bath, England, using the *multiport approach* in the simulation process. They proposed a new manner of simulating fluid power systems (or any other system) by creating *super components* for each type of component or system. In the historical paper—*How to create good models without writing a single line of code* presented at the Fifth Scandinavian International Conference on Fluid Power at Linköping, Sweden on May 28, 1997, the two authors stated the main reason for creating the Amesim language: The user of the basic element library is relieved of the need to write code and formulate the mathematics: *Understanding of the details of the physics is not needed but decision on assumption are necessary which imply some knowledge of physics. Understanding of the engineering system and an ability to interpret results is still important.*

The incorporation of Société IMAGINE, founded by Michel Lebrun, in LMS International, widened the applications areas of Amesim, especially in the field of Real-Time simulation of the complex engineering systems. The wide approach of the hybrid and electric automotive mechatronic systems dramatically increased the interest of the technical world for the Hardware-in-the-Loop (HiL) simulations by Simcenter Amesim—the new name of Amesim is the wide 1D simulation platform developed by LMS.

This book contains the results of the authors' efforts to explain, refine, or build confident models for the main components of the fluid power systems, and to study the transients that can occur during their operation. The new models were validated and included in home libraries. Different categories of fluid power and complex fluid control systems were represented by super components, which can be combined in a different manner in order to be included in new systems without a major risk.

The present research was carried out at the Fluid Power Laboratory from the Power Engineering Faculty of the University Politehnica of Bucharest (UPB), Bucharest, Romania. This laboratory is recognized by the Romanian Accreditation Association (RENAR) in the field of electrohydraulic servovalves and is operating in close collaboration with the following R&D entities:

- Siemens PLM Software—LMS from Leuven, Belgium, led by Dr. Jan Leuridan
- The Romanian Subsidiary of Siemens Industry Software from Brasov, Romania, led by Petru Cristinel Irimia, PhD

- Fluid Power Laboratory of the Mechanical Engineering Department of INSA. Toulouse, set up by Professor Jean-Charles Mare, PhD
- The Romanian subsidiary from Cluj, Romania of the National Instruments Corporation
- The engineering R&D centers of Brasov and Bucharest supported by Parker Hannifin Corporation from the United States

The introductory chapter, Chapter 1, of this book is devoted to a short review of the main simulation software packages or tools used in industrial applications. The field of applications and the main capabilities of each program or platform are taken from their sites or user's manuals with a positive regard on the performances required by practical applications. Both general purpose languages such as SIMULINK®, ACSL, or MATRIX$_x$, and dedicated modeling and simulation environments such as Simhydraulic and Automation Studio are compared.

Chapter 2 presents the capabilities of Simcenter Amesim for solving engineering problems. The platform overview provides details on some powerful facilities, analysis, or optimization tools, which offer an advanced and easy-to-use environment for 1D system simulation and robust design. Its physically conservative multiport modeling of physical components and its block-diagram approach for control systems enables the coupling of all its libraries together within a built-in and comprehensive workflow. The original engineering solutions offered by LMS Imagine.Lab in almost all technical domains are widely presented.

Chapter 3 is devoted to the numerical simulation of the basic hydraulic components. The mathematical models were built for flow through orifices, three-way flow valves, four-way flow valves, and single-stage and two-stage pressure relief valves, taking into account the real characteristics of all the hydraulic resistances involved. The numerical simulations were performed by SIMULINK and/or Amesim for typical input signals. The complete mathematical models were encapsulated and introduced in the hydraulic library of Amesim.

Chapter 4 of this book presents the mathematical modeling, numerical simulation, and experimental identification of the last generation of both mechanical and electrical feedback electrohydraulic servovalves. The synthesis of a three-stage servovalve by Amesim, the simulation of the power-stage overlap influence on the servovalves performance, the design of the controller for a servovalve by simulation, and the dynamic identification of the electrohydraulic servovalves by simulation with Amesim are other problems studied in this part. The experimental research concerning these problems, performed by the authors, proved that numerical simulations with Amesim can speed the design process of these important boundary components.

Numerical simulation and the experimental identification of the hydraulic servomechanisms are presented in Chapter 5. A wide range of servomechanism types are investigated, in close connection with the application peculiarities: bent axis servopumps, electrohydraulic valves train drive systems, hybrid electrohydraulic flight control actuators, electrohydraulic flight control servomechanisms with increased stability, hydromechanical aerospace servomechanisms supplied by constant pressure sources, electrohydraulic servomechanisms with additional feedbacks, and the synchronous control of the linear hydraulic motors with heavy mass loads. All the models correspond to cases taken from the real technical systems such as aerospace, mobile equipment, hydropower systems, and so on.

Chapter 6 covers the numerical simulation and experimental identification of the automotive hydraulic-steering systems. First, the steady-state automotive hydraulic servomechanisms with open-center flow control valve are studied; then, the dynamics of the same peculiar position control system is studied with Amesim. Finally, the authors present the structure and the performance of a new electrohydraulic-steering system for articulated mobile equipment, which can be remotely controlled via GPS. The experimental results obtained in the laboratory on a forestry tractor are found to be in good agreement with the simulated ones.

Chapter 7 presents the dynamics of the servopumps and servomotors. Both swashplate and bent axis pumps were considered. In the first case, Amesim was used for modeling and simulation of the-real pressure servopumps with a finite number of pistons containing a direct-action pressure compensator. For comparison, the two-stage pressure compensators were studied by SIMULINK for the bent axis pumps. The high level of compression of the Amesim models allowed a deep study of the flow ripple during the transients. Next, the open-circuit swashplate flow servopumps dynamics was studied taking into account the finite number of pistons. The numerical simulations of the step-input response displayed high-speed dynamics, according to the small swashplate inertia. The response of the bent axis flow servopumps for close circuits is much slower. This information is found in good agreement with the experimental information obtained by the authors. The new models were also turned into Amesim super components.

The modern hydrostatic transmissions for mobile equipment including variable displacement axial piston pumps and bent axis axial servomotors controlled by force feedback servomechanisms are investigated by Amesim in Chapter 8 of this book. The mathematical models were assembled in a complete model of a hydrostatic transmission and used for the investigation of different transients in the frame of an automotive system. The overall dynamic model was excited by different input signals applied to the servopump and to the servomotor. The slope of the road was also a way to introduce disturbances in the steady-state behavior of the transmission. Special attention was paid to the conditions of appearance of the cavitation phenomenon into the power lines of the transmission. Finally, the authors have built the super component *Hydrostatic Transmission*, which was included in the Amesim hydraulic library.

The synthesis of the speed governors specific to the hydraulic turbines is presented in Chapter 9. The complete mathematical model of the whole electric, hydraulic, and speed-control system is studied by the Real-Time co-simulation between MATLAB®, Amesim, and LabVIEW. The authors have introduced here new patented concepts for multistage nonlinear servovalves, operating under low-supply pressure. The high-head Francis turbines with pressure compensators are studied in-depth, pointing out the benefits of the high-speed bypass valve regarding the elimination of dangerous overpressure occurring during the turbine's sudden shut down. The simulation with Amesim of a redundant position control system used in the modern speed governors is also presented. The same part offers a complete original example of sizing and tuning the speed governors for hydraulic turbines by Amesim. This part contains a report on a long series of theoretical and experimental activities aiming to create a new type of electrohydraulic digital speed governor for hydraulic Kaplan turbines. Computational methods, control software, design problems, and experimental validation are shortly presented.

The numerical simulation of the modern common rail fuel injection systems (CRFIS) is studied in Chapter 10. Both solenoid and piezoactuated injector dynamics are studied in the same ideal operating conditions (ideal ones). Reduction in the overall computation time by the Discrete Partitioning Technique available in Amesim is also investigated. The principle of this procedure is to divide the complete injection system into smaller subsystems and perform a co-simulation between the master and the slave systems. Finally, the authors discuss the comparative results of the preliminary computer simulations of a mechanically and an electronically controlled injector running on diesel fuel and high oleic sunflower oil.

The numerical simulation and experimental validation of ABS for automotive systems is discussed in Chapter 11. A deep analysis of the typical automotive antilocking systems reveals the possibilities of reducing the degree of complexity of the Amesim models in order to run them in Real Time on different platforms, integrated in virtual car simulations. The reduction of the degree of complexity of the models of different ABS components was validated experimentally on different kinds of road surfaces.

The numerical simulation and experimental tuning of the special purpose electrohydraulic servosystems for mobile equipment are considered in Chapter 12. The theoretical dynamics of the electrohydraulic position servomechanisms with laser beam feedback, set up on leveling machines used for improving the quality of the lands in hydropower dams, agricultural lands, civil engineering works, and so on was validated by experiments in practice. The laboratory simulator built for these kinds of applications is found to be very useful for preliminary tuning of the adequate type of controller.

The last part of this book (Chapter 13) is devoted to an actual design challenge: The use of Amesim for solving multiphysics problems. The Real-Time simulation with HiL of the electric car transmissions is treated by cosimulation between Amesim and LabVIEW. The authors created a modular architecture that includes both hardware and software. To prove the new concept, a patented test bench has been designed. It includes two vector-controlled electric motors, three Real-Time units, and an industrial PC. The electric motors are mechanically coupled by a torque and speed transducer. The Real-Time units perform three individual tasks: (1) acquisition of electrical parameters, (2) electric vehicle command and control, and (3) the simulation of the mathematical Amesim model that includes all the subcomponents of a car. It is integrated using a partially LabVIEW toolkit. A specific fuzzy controller was created using the same software. The test results gathered during a large number of test runs with different scenarios showed that the new control algorithm is a realistic one. This way, an effective tool was developed for tuning the powertrain controller of future cars.

This book is aimed at the researchers, design engineers, hydraulic control systems users, and academic people dealing with this modern kind of concurrent engineering. Any scientific or technical remark concerning this book is welcome through the following e-mail address: nicolae.vasiliu@upb.ro.

MATLAB® is a registered trademark of The MathWorks, Inc. For product information, please contact:

The MathWorks, Inc.
3 Apple Hill Drive
Natick, MA 01760-2098 USA
Tel: 508 647 7000
Fax: 508-647-7001
E-mail: info@mathworks.com
Web: www.mathworks.com

Acknowledgments

Twenty years ago, during one of our numerous professional contacts with Professor *Jean-Charles Mare*, our partner from Mechanical Department of INSA Toulouse, we had the chance to meet Professor *Michel Lebrun* from INSA Lyon. At that time, he was leading Société IMAGINE, a small company he created as a spin-off from the university. Trying to promote engineering innovation, Professor Lebrun imagined a way to encourage the young people to create new products and systems by the aid of numerical simulation without a special mathematical effort. So, Amesim—**A**dvanced **M**odeling **E**nvironment for **Sim**ulation—was *born* in Roanne, France, and soon it was adopted and successfully used by many high-level innovative companies in the field of fluid control systems. The integration of IMAGINE with LMS International Corporation gave to Amesim a new opportunity to be extended for different types of mechatronic systems. Due to the permanent aid of LMS, any Romanian student is able to access LMS Amesim on the national Fluid Power Engineering Platform www.fluidpower.ro built by our research team.

We are particularly grateful to *Dr. Jan Leuridan*, senior vice president in charge of simulation and test solutions from *Siemens PLM Software*, a business unit of the Siemens Digital Factory Division. He also serves as the CEO for Siemens Industry Software NV. From the beginning of our cooperation with LMS, he tried to help us in understanding and developing the scientific and technological achievements of this innovative company in the field of multidomain and mechatronic simulation solutions. Many of our former PhD students, working in automotive fluid control systems, are now deeply integrated in Real-Time simulation innovation teams using the release of LMS Amesim within the framework of some major world-beating corporations, after finishing their research stages in LMS.

The preparation of this book has been encouraged and sustained since the very beginning by Petru-Cristinel Irimia, Director of Siemens Industry Software SRL from Romania. His wide-ranging knowledge in engineering mechatronic systems helped us to keep the right proportion between the scientific information and the technical one. The research stages organized by him in Roanne, Lyon and Leuven for our former PhD students—Daniel Florin Dragne, Ion Guta Dragos, and Catalin Vasiliu—helped them and us in covering some specific parts of the manuscript.

Special thanks are due to Philippe Geril, the General Secretary/Director of European Simulation Society—EUROSIS/ETI Bvba, Ghent, Belgium. He had the kindness and patience to read our *manuscript* and to make a lot of useful semantic and linguistic remarks, helping us to overcome our natural professional and teaching limits.

All along the manuscript *building*, Dr. Gagandeep Singh, editorial manager and senior editor for Engineering & Environmental Sciences of CRC Press/Taylor & Francis Books India Pvt. Ltd., guided us both as editor and manager, kindly filtering our style of presenting a delicate theme. The final printing form of the book was carefully edited by Mrs. Madhuriba Subarayalou, Senior Project Manager at Lumina Datamatics Ltd.

Finally, we express our gratitude to Florin Gheorghe Filip, chair of the *Information Science and Technology* section of The Romanian Academy, for his generous scientific and methodological help.

Nicolae Vasiliu
Daniela Vasiliu

Authors

Nicolae Vasiliu, PhD, graduated in hydropower engineering from the University Politehnica of Bucharest (UPB), Bucharest, Romania in 1969. He earned his PhD in fluid mechanics after a research stage in Ghent State University, Ghent, Belgium and Von Karman Institute from Brussels, Belgium. He became a state professor in 1994, leading the Energy & Environment Research Centre from the UPB. He managed five years in the Innovation Romanian Agency. He always worked for the industry, as project manager or scientific advisor, promoting the numerical simulation as an engineering tool, and has received many national and international distinctions for applied patents, papers, laboratories, and so on. He is a member of the Romanian Technical Science Academy.

Daniela Vasiliu, PhD, graduated in mechanical engineering in 1981 and prepared a PhD thesis in the field of the dynamics of the electrohydraulic servopumps and servomotors for hydrostatic transmissions. She is currently professor in the Department of Hydraulics, Hydraulic Machines and Environmental, head of Fluid Power Laboratory of the UPB, Bucharest, Romania, and director of the Fluid Power Systems Dynamics Master Cycle. She works in the field of modeling, simulation, and experimental identification of the electrohydraulic control systems. She is an active member of The European multidisciplinary society for modelling and simulation technology (EUROSIS), fluid power net international (FPNI), société des ingénieurs de l'automobile (SIA), The American society of mechanical engineers (ASME), and so on.

Constantin Călinoiu, PhD, graduated in power engineering from UPB, Bucharest, Romania in 1976, and in mathematics from the University of Bucharest, Romania in 1981. After completing his studies, he became a scientific researcher in the Hydraulics Laboratory of the Romanian Aerospace Institute. In 1998, he defended his PhD thesis in the field of the experimental identification of the hydraulic servosystems and became an associate professor in the Fluid Power Laboratory from UPB. He is working mainly in modeling, simulation, and identification of the hydraulic and electro-hydraulic control systems.

Radu Puhalschi, PhD, graduated in applied computer science from the UPB, Bucharest, Romania in 2009, and a MD in advanced hydraulic and pneumatic systems in 2011. After a stage of web designer at HP Germany, he performed a PhD thesis on Real-Time simulation of hydraulic control systems at the Fluid Power Laboratory from the Power Engineering Department of the UPB, Bucharest, Romania. Now, he is still contributing to control courses as an associate professor in the same university, and he is working as a supervisor control engineer in the Romanian Division of Honeywell International Inc.

chapter one

Overview on the numerical engineering simulation software

1.1   Introduction

Fifty years ago, Herbert E. Merritt, an eminent fluid power engineer working for the Cincinnati Machines company, Cincinnati, Ohio stated that "Actually a great deal of time and trouble can be saved if a paper and pencil analysis and design of system is made before it is simulated on a computer for final refinements."[1] He was right for that time, but now the pencil and paper have been successfully replaced by a touch screen computer loaded with the proper simulation software. Now, a high-speed *batch* simulation can replace any method of optimization such as random search.

A short and comprehensive definition of simulation in industry, science, and education given in the *Encyclopedia Britannica*[2] is "a research or teaching technique that reproduces actual events and processes under test conditions. Developing a simulation is often a highly complex mathematical process. Initially, a set of rules, relationships, and operating procedures are specified along with other variables. The interaction of these phenomena creates new situations and even new rules, which further evolve as the simulation proceeds."

The need for increasing the productivity in the design phase of manufacturing of any product led to the high-speed development of the simulation languages. First of all, the aerospace competition demanded high-performance fly control simulators. The success of the Apollo missions was facilitated by the use of *Advanced Continuous Simulation Language* (*ACSL*) written in Fortran. A huge extension of the user's circle of this language still promoted by AEgis Technologies Group, Inc. from Alabama as *ACSLX*[3] was demanded by the online simulation of the operation of the classic and nuclear power stations. *nHance company* from the United States[4] has chosen the generic name *Modular Modeling System* (*MMS*) for this software. The project developed under this name was successfully sustained by the Electric Power Research Institute in Palo Alto, California[5], which is an independent and nonprofit organization.

The need of the deep study of complex phenomena described by partial differential equations (PDE) leads to a strong development of the simulation program based on the Finite Element Method. Some important IT companies such as ANSYS[6] offer a comprehensive software suite that spans the entire range of physics, providing access to virtually any field of engineering simulation that a design process requires, that is, from fluids, structures, electronics, and semiconductors to multiphysics and embedded software. Hardware-in-the-Loop (HiL) compatibility with 1D software is also available.

The basic design requirements of the fluid power systems are covered by one-dimensional (1D) simulation programs. The physical interactions with other different kinds of systems can be studied now by cosimulation with different three-dimensional (3D) programs using HiL simulation. The most important step in the field of mathematical computing software for engineers and scientists was taken by the MathWorks company[7] that started from 1984 onward. The enhanced access of matrix laboratory (MATLAB®) and SIMULINK® to control systems, their multidomain simulation, and model-based designed has put this platform of technical computing on a lead position in modeling and design of fluid power systems by the aid of common or custom libraries.

Another important progress in 1D simulation was made by the National Instruments (NI) company, Austin, Texas[8] in the field of innovative hardware and software applicable in many engineering fields. The control hardware using the original software created a valuable platform for the development of technical systems. The NI controllers, LabVIEW simulation, and control software can be used in any fluid power system. The compatibility of LabVIEW with the MathSCRIPT module from NI MATRIXx language widely extended the capability of solving PDE during a Real-Time simulation.

The last remarkable progress in simulation software development was achieved by Société IMAGINE from France, which created Advanced Modeling Environment for Simulation of Engineering Systems (Amesim)[9] in 1986. The main idea of the authors (Michel Lebrun and Claude Richards) was to create very refined libraries for all industrial mechanic, hydraulic, pneumatic, thermal, and hybrid components and systems, validated by high-level manufacturers, in order to automate the design phase of any new mechatronic product without writing equations! This book is devoted to the capabilities of this software.

This introductory part is devoted to a short review of the main simulation software packages or tools used in industrial applications. The field of applications and the main capabilities of each program or platform are taken from their sites or user's manuals with a positive regard on the performances required by the practical applications. The problems encountered by the users of these software packages or tools in finding proper solutions for their problems are not the aim of this short overview.

1.2   Free software capabilities

This short review is devoted to the free software used in the analysis and synthesis of the static and dynamic behavior of the systems from any scientific or technical field. Advanced Simulation Library, ADMB, Chapel, Euler, Fortress, FreeFem++, FreeMat, Genius, Gmsh, GNU Octave, Julia, Maxima, OpenFOAM, R, SageMath, SALOME, ScicosLab, and Scilab®, X10 are presented here.

Advanced Simulation Library (ASL)[10] is a free open-source hardware accelerated multi-physics simulation software. It enables users to write customized numerical solvers in C++ and deploy them on a variety of massively parallel architectures, ranging from inexpensive field-programmable gate arrays (FPGAs), digital signal processor (DSPs), and graphics processing units (GPUs) up to heterogeneous clusters and supercomputers. Its internal computational engine is written in OpenCL and utilizes a variety of advanced numerical methods such as Level set method, Finite Difference, Lattice Boltzmann, and Immersed Boundary. ASL can be used to model various coupled physical and chemical phenomena, especially in the field of Computational Fluid Dynamics. It is distributed under the free GNU Affero General Public License with an optional commercial license.

ADMB or *AD Model Builder*[11] is a free and open-source software suite for nonlinear statistical modeling. It was created by David Fournier and now it is being developed by the ADMB project, a creation of the nonprofit ADMB foundation. The *AD* in AD Model Builder refers to the automatic differentiation capabilities that come from the

AUTODIF Library, a C++ language extension also created by David Fournier, which implements reverse-mode automatic differentiation. A related software package, *ADMB-RE*, provides additional support for modeling random effects.

Chapel[12] is an emerging parallel language being developed at the Cray Inc. with the goal of addressing this issue and making parallel programming far more productive and generally accessible.

Euler (now *Euler Mathematical Toolbox* or *EuMathT*) is a free and open-source numerical software package.[13] It contains a matrix language, a graphical notebook style interface, and a plot window. Euler is designed for higher level math such as calculus, optimization, and statistics. The software can handle real, complex, and interval numbers, vectors, and matrices. It can produce 2D/3D plots, and uses Maxima for symbolic operations. The software is compatible with Windows. The Unix and Linux versions do not contain a computer algebra subsystem.

FreeFem++ is a PDE solver.[14] It has its own language. FreeFem scripts can solve multiphysics nonlinear systems in 2D and 3D. Problems involving PDE (2d, 3d) from several branches of physics such as fluid–structure interactions require interpolations of data on several meshes and their manipulation within one program. FreeFem++ includes a fast $2^{\wedge}$d-tree-based interpolation algorithm and a language for the manipulation of data on multiple meshes. FreeFem++ is written in C++ and the FreeFem++ language is a C++ idiom. It runs on Macs, Windows, and Unix machines. FreeFem++ replaces the older FreeFem and FreeFem+.

FreeMat[15] is a free environment for rapid engineering and scientific prototyping and data processing. It is similar to commercial systems such as MATLAB from MathWorks and IDL from Research Systems, but it is Open Source. FreeMat is available under the GPL license.

The Genius Project Enterprise[16] project management software is designed to adapt to your organization's *business processes*, Genius Project delivers a highly flexible and configurable portfolio and project management software allowing tailored feature sets for a wide array of project teams and project types.

Gmsh[17] is a free 3D finite element grid generator with a built-in CAD engine and postprocessor. Its design goal is to provide a fast, light, and user-friendly meshing tool with parametric input and advanced visualization capabilities. Gmsh is built around four modules: (1) geometry, (2) mesh, (3) solver, and (4) postprocessing. The specification of any input to these modules is done either interactively using the graphical user interface or in ASCII text files using Gmsh's own scripting language.

GNU Octave[18] is a high-level interpreted language, primarily intended for numerical computations. It provides capabilities for the numerical solution of linear and nonlinear problems, and for performing other numerical experiments. It also provides extensive graphics capabilities for data visualization and manipulation. Octave is normally used through its interactive command line interface, but it can also be used to write noninteractive programs. The Octave language is quite similar to MATLAB so that most programs are easily portable. Octave is distributed under the terms of the GNU General Public License.

Julia[19] is a high-level, high-performance dynamic programming language for technical computing with a syntax that is familiar to users of other technical computing environments. It provides a sophisticated compiler, distributed parallel execution, numerical accuracy, and an extensive mathematical function library. Julia's Base library, largely written in Julia itself, also integrates mature, best-of-breed open-source C and Fortran libraries for linear algebra, random number generation, signal processing, and string processing. In addition, the Julia developer community is contributing a number of external packages through Julia's built-in package manager at a rapid pace. IJulia, a collaboration between the IPython and Julia communities, provides a powerful browser-based graphical notebook interface to Julia. Massachusetts Institute of Technology (MIT) licensed this software as free and open source.

Maxima[20] is a system for the manipulation of symbolic and numerical expressions, including differentiation, integration, Taylor series, Laplace transforms, ordinary differential equations, systems of linear equations, polynomials, sets, lists, vectors, matrices, and tensors. Maxima yields high-precision numerical results by using exact fractions, arbitrary precision integers, and variable-precision floating-point numbers. Maxima can plot functions and data in two and three dimensions. The Maxima source code can be compiled on many systems, including Windows, Linux, and MacOS X. The source code for all systems and precompiled binaries for Windows and Linux are available at the SourceForge file manager.

Maxima is a descendant of Macsyma, the legendary computer algebra system developed in the late 1960s at the MIT. It is the only system based on the effort that is still publicly available and with an active user community, thanks to its open-source nature. Macsyma was revolutionary in its days, and many later systems, such as Maple and Mathematica, were inspired by it.

OpenFOAM[21] is a free, open-source CFD software developed primarily by OpenCFD Ltd started in 2004, and distributed by OpenCFD Ltd and the OpenFOAM Foundation since then. It has a large user base across most areas of engineering and science, from both commercial and academic organizations. OpenFOAM has an extensive range of features to solve anything from complex fluid flows involving chemical reactions, turbulence, and heat transfer, to acoustics, solid mechanics, and electromagnetics. OpenFOAM+ uses the OpenFOAM Foundation version as a common code base, and offers wider functionality and platform support. Its purpose is to accelerate the public availability of new features, which are sponsored by OpenCFD's customers and contributed by the OpenFOAM community.

R[22] is a language and environment for *statistical computing and graphics*. It is a GNU project, which is similar to the S language and the environment that was developed at the Bell Laboratories (formerly AT&T, now Lucent Technologies) by John Chambers and colleagues. R can be considered as a different implementation of S. There are some important differences, but much of the code written for S runs unaltered under R.

R provides a wide variety of statistical (linear and nonlinear modeling, classical statistical tests, time-series analysis, classification, clustering, etc.) and graphical techniques, and is highly extensible. The S language is often the vehicle of choice for research in statistical methodology, and R provides an open-source route to participation in that activity. One of R's strengths is the ease with which well-designed publication-quality plots can be produced, including mathematical symbols and formulae, where needed. Great care has been taken over the defaults for the minor design choices in graphics, but the user retains full control.

SageMath[23] is a free open-source mathematics software system licensed under the GPL. It builds on the top of many existing open-source packages: NumPy, SciPy, matplotlib, Sympy, Maxima, GAP, FLINT, R, and many more. It accesses their combined power through a common Python-based language or directly via interfaces or wrappers. The mission of this software is *creating a viable free open-source alternative to Magma, Maple, Mathematica,* and *MATLAB.*

Salome[24] is an open-source software that provides a generic platform for pre- and post-processing of numerical simulation. It is based on an open and flexible architecture made of reusable components. Salome is a cross-platform solution. It is distributed as open-source software under the terms of the GNU LGPL license. Salome can be used as a standalone application for the *generation of a CAD model*, its preparation for numerical calculations, and postprocessing of the calculation results. Salome can also be used as a *platform for integration* of the external third-party numerical codes to produce a new application for the full life-cycle management of CAD models.

ScicosLab[25] is a software package providing a multiplatform environment for scientific computation. It is based on the official Scilab 4.x (BUILD4) distribution, and includes the modeling and simulation tool Scicos and a number of other toolboxes. The latest stable version of ScicosLab is ScicosLab 4.4.1 as of April, 2011. It seems that Scilab/Scicos is currently the most complete alternative to commercial packages for dynamic systems modeling and simulation packages such as MATLAB/SIMULINK and MATRIXx/SystemBuild.

1.3 *Proprietary software capabilities*

Data Analysis and Display (DADiSP)[26] is a numerical computing environment developed by DSP Development Corporation, which allows one to display and manipulate data series, matrices, and images with an interface similar to a spreadsheet. DADiSP is used in the study of signal processing, numerical analysis, statistical, and physiological data processing.

GAUSS[27] is a matrix programming language for mathematics and statistics, developed and marketed by Aptech Systems. Its primary purpose is the solution of numerical problems in statistics, econometrics, time-series, optimization, and 2D- and 3D-visualization. It was first published in 1984 for MS-DOS and is currently also available for Linux, Mac OS X, and Windows.

NI MATRIXx[28] National Instruments Corporation, (short NI), an American company with international operations produces automated test equipment and virtual instrumentation software. One of the most used software packages produced by NI for mechatronics is MATRIX$_x$,[28] which includes four *tools*. These four programs build accurate, high-fidelity models, and perform interactive simulations; analyze system models and build robust control algorithms; automatically document the system model and controller properties; generate readable, traceable, highly optimized C or Ada code; and target a code to a Real-Time platform for control prototyping and HiL test. The MATRIXx suite of software includes the following:

SystemBuild: is an easy-to-use graphical environment for rapid model development and simulation and the development and management of large, complex models. With the SystemBuild intuitive, hierarchical structure, one can segment the model at any level for validation and verification from the top down or the bottom up.

Xmath is a mathematical analysis, visualization, and scripting software environment featuring a high-level, object-oriented programming language called MathScript. With features such as an interactive debugger, programmable graphical user interfaces, and an extensive library of predefined functions, the user development time and effort greatly decreases.

DocumentIt provides a powerful and customizable method to generate formatted documentation of the SystemBuild models. Using template-based technology, you can easily customize the resulting documentation to extract all the desired information and present it in the most useful format, including block and *SuperBlock* names, data types, signal names and flows, and user-entered block comments. The user can create a single template to standardize documentation for the entire enterprise or use different templates to provide input/output interface information among development teams.

AutoCode is a flexible and easily customizable automatic code generation tool. Using template-based generation technology, you can configure the exact format and specification of the resulting code. This makes it easy for you to generate code for rapid control prototyping and HiL testing on one Real-Time platform, and then switch templates to generate code for the final production target.

LabVIEW[29] (short for Laboratory Virtual Instrument Engineering Workbench) is a system-design platform and development environment for a visual programming language from NI. The graphical language is named *G*. Originally released for the Apple Macintosh in 1986, LabVIEW is commonly used for data acquisition, instrument control, and industrial automation on a variety of platforms, including Microsoft Windows, various versions of UNIX, Linux, and OS X. The latest version of LabVIEW was released in August 2015 under the name *LabVIEW System Design Software.*

LabVIEW is a development environment designed specifically to accelerate the productivity of engineers and scientists. With a graphical programming syntax that makes it simple to visualize, create, and code engineering systems, LabVIEW is helping engineers to translate their ideas into reality, reduce test times, and deliver business insights based on collected data. From building smart machines to ensuring the quality of connected devices, LabVIEW has been the preferred solution to create, deploy, and test the Internet of Things for decades.

Maple[30] is a commercial computer algebra system developed and sold commercially by *Maplesoft*, a software company based in Waterloo, Ontario, Canada. The current major version was released in March 2015. It was first developed in 1980 by the Symbolic Computation Group at the University of Waterloo, Ontario, Canada. In 1988, Maplesoft (then known as Waterloo Maple Inc.) was founded to commercialize the technology. Maplesoft Engineering Solutions are uniquely positioned to offer cutting-edge software tools and expertise for design, modeling, and high-performance simulation that will help you meet the challenges of engineering design projects. With experts in a variety of engineering fields, extensive experience in model-based design and the superior system-level modeling and analysis tools, MapleSim and Maple, and Maplesoft can help you reduce developmental risk and bring high-quality products to market faster. Maplesoft Engineering Solutions support the following: System functional verification, Design parameter optimization, Design trouble-shooting and improvement, System-level modeling for component sizing, and Optimized model code generation for HiL and Software-in-the-Loop (SiL) testing.

Mathcad[31] is a computer software primarily intended for the verification, validation, documentation, and the reuse of engineering calculations. First introduced in 1986 on DOS, it was the first to introduce live editing of typeset mathematical notation combined with its automatic computations. Calculations are the heart of engineering information. The design team must be able to find, reuse, and share this important intellectual property. PTC Mathcad has all your engineering notebook's ease-of-use and familiarity—combined with live mathematical notation, unit's intelligence, and powerful calculation capabilities. This engineering math software allows presenting the calculations with plots, graphs, text, and images in a single document. Nobody needs specialized skills to understand

PTC Mathcad data, and now that your intellectual property has been preserved, anyone can leverage it for other projects.

Mathematica[32] is a symbolic mathematical computation program, sometimes called *a computer algebra program*, used in many scientific, engineering, mathematical, and computing fields. It was conceived by Stephen Wolfram and is developed by Wolfram Research of Champaign, Illinois. The Wolfram language is the programming language used in Mathematica. This software has nearly 5000 built-in functions covering all areas of technical computing—all carefully integrated, so they work perfectly together, and all included in a fully integrated system. Mathematica builds in unprecedentedly powerful algorithms across all areas—many of them created at Wolfram using unique development methodologies and the unique capabilities of the Wolfram language. Mathematica is built to provide industrial-strength capabilities—with robust, efficient algorithms across all areas, capable of handling large-scale problems, with parallelism, GPU computing, and more. Mathematica uses the Wolfram Notebook Interface, which allows you to organize everything you do in rich documents that include text. Wolfram System Modeler is an easy-to-use, next-generation modeling, and simulation environment for cyber-physical systems. Using drag-and-drop from the large selection of built-in and expandable modeling libraries, you can build industrial strength, multidomain models of your complete system. Adding the power of Mathematica the designer obtains a fully integrated environment for analyzing, understanding, and quickly iterating system designs.

MATLAB[33] is a multiparadigm numerical computing environment and fourth-generation programming language. A proprietary programming language developed by MathWorks, MATLAB allows matrix manipulations, plotting of functions and data, implementation of algorithms, creation of user interfaces, and interfacing with programs written in other languages, including C, C++, Java, Fortran, and Python. Although MATLAB is intended primarily for numerical computing, an optional toolbox uses the MuPAD symbolic engine, allowing access to symbolic computing capabilities. An additional package, SIMULINK, adds graphical multidomain simulation and model-based design for dynamic and embedded systems. MATLAB has some million users across the industry and academia.

SIMULINK is the toolbox of the MATLAB-programming environment dedicated to numerical simulation in all fields. It is a graphical environment, in which models are built as block diagrams, chosen from a library that includes general elements (mathematical operations, transfer functions, and common signal generators such as square, sine, or step functions) and blocks that represent specific components from various fields of application (mechanical, electrical, hydraulic, etc.).

In SIMULINK, the whole modeling process is based on building the equations that make up the mathematical model of the analyzed process. Even the standard components from the library use the same approach, the explicit mathematical model of each one accessible by the *Look under mask* option from the right-click menu. As a MATLAB toolbox, all the data used while running a SIMULINK model are stored within the MATLAB workspace, which makes it very easy to use the excellent data analysis functions (graphs and charts, statistical analysis, stability analysis with various criteria, transfer function calculation, etc.) included in MATLAB on the result of any simulation of a SIMULINK model. For Real-Time and HiL simulation, SIMULINK uses an additional module called SIMULINK Coder™ (formerly known as Real-Time Workshop). This module generates C or C++ code that can be used for Real-Time simulation. The C code can be run on a dedicated Real-Time simulation platform (using xPC Target), or on any Windows machine (using Real-Time

Windows Target). xPC Target is a SIMULINK module that allows running the C code generated from the SIMULINK model on any compatible Real-Time hardware. The list of compatible hardware platforms includes many of the leading products in the field, such as ADWin or dSPACE.

Real-Time Windows Target allows SIMULINK to run a Real-Time simulation on any Windows PC (only 32 bit version currently, which limits its usefulness significantly). It is an accessible and quick solution, but limited by the operating system's own limitations. Since Microsoft Windows is not a dedicated Real-Time operating system, direct control and rigorous prioritization of the simulation loop's access to the hardware resources (CPU, RAM, etc.) is not possible. This shortcoming can lead to random delays in the simulation loop's execution of up to 10–20 ms. The problem is not very relevant for the simulation of slow processes (chemical, power, etc.) but *can significantly alter* the results when simulating rapid processes (electrical, optical, electronics, etc.). The authors of this book have successfully employed Real-Time Windows Target for HiL simulation of turbines speed governors.

VisSim[34] is a visual block diagram language for the simulation of dynamical systems and model-based design of embedded systems. It was developed by Visual Solutions of Westford, Massachusetts. It uses a graphical data flow paradigm to implement dynamic systems based on differential equations. Version 8 adds interactive UML OMG 2 compliant state chart graphs that are placed in VisSim diagrams. This allows easy modeling of state-based systems such as start-up sequencing of process plants or serial protocol decoding. The highly efficient fixed point code generator allows targeting of low cost fixed-point embedded processors. VisSim is widely used in control system design and digital signal processing for multidomain simulation and design. It includes blocks for arithmetic, Boolean, and transcendental functions, and digital filters, transfer functions, numerical integration, and interactive plotting. The most commonly modeled systems are aeronautical, biological/medical, digital power, electrical, hydraulic, mechanical, process, thermal/HVAC, and econometric. The *VisSim/C-Code* add-on generates ANSI C code for the model, and generates target-specific code for on-chip devices such as pulse-width modulation (PWM), analog-to-digital converter (ADC), encoder, generic pin on an integrated circuit or computer board (GPIO), multi-master bus (I2C), and so on. This is useful for the development of embedded systems. After the behavior of the controller has been simulated, C-code can be generated, compiled, and run on the target.

Dymola[35] is a physical modeling and simulation tool used for model-based design of complex engineering systems. Multidomain libraries covering the mechanical, electrical, control, thermal, pneumatic, hydraulic, powertrain, thermodynamics, vehicle dynamics, and air-conditioning domains can be coupled together to form a single complete model of the system. Dymola is used by companies operating in many industries, including automotive, aerospace, motorsport, energy, and high tech.

Dymola uses the *Modelica*[36] modeling language to define models and provides the user with open access to the language. This means that users are free to create their own model libraries or extend from the existing libraries to accelerate development times, reduce maintenance efforts, and improve the level of reuse across projects. Modelica is a nonproprietary, object-oriented, and equation-based language to conveniently model complex physical systems containing, for example, mechanical, electrical, electronic, hydraulic, thermal, control, electric power, or process-oriented subcomponents. The language is developed and maintained by The Modelica Association, a nonprofit organization with members from Europe, United States, Canada, and Asia. Since 1996, its simulation experts have been working to develop the open-standard Modelica and the open-source Modelica Standard Library.

The industry is increasingly using Modelica for model-based development. Especially, many automotive companies, such as Audi, BMW, Daimler, Ford, Toyota, and VW use Modelica to design energy-efficient vehicles and/or improved air-conditioning systems. In addition, power plant providers, such as ABB, EDF, and Siemens use Modelica, and many other companies. Models in Modelica are mathematically described by differential, algebraic, and discrete equations. These equations are then typically manipulated using symbolic manipulation to generate efficient simulation code directly from the models. The object-oriented modeling language Modelica is designed to allow convenient, component-oriented modeling of complex physical systems, for example, systems containing mechanical, electrical, electronic, hydraulic, thermal, control, electric power, or process-oriented subcomponents. The development and promotion of Modelica is organized by the nonprofit Modelica association.

ANSYS[37] offers a comprehensive software suite that spans the entire range of four-dimensional (4D) physics, providing access to virtually any field of engineering simulation that a design process requires. Simulation-Driven Product Development takes engineering simulation to a higher level, applying engineered scalability and adaptive architecture to comprehensive multiphysics foundation. By partnering with key hardware and independent software vendors, the company extended software product functionality to ensure the most accurate solution in the least amount of time.

From 3D unsteady viscous fluids motion, to structures, multiphysics problems, and semiconductors to hybrid-system dynamics, very accurate solutions are offered both for scientific and industrial problems. Fluid forces, thermal effects, structural integrity, and electromagnetic radiation can all impact performance of products and industrial processes. If the engineer is trying to isolate the multiple forces in play, he is not getting an accurate prediction of systems behavior. ANSYS multiphysics solutions can help engineers examine these effects in combination and isolation, achieving the highest fidelity solution when it is needed. ANSYS provides even a model-based embedded software development and simulation environment with a built-in automatic code generator to accelerate embedded software development projects. The need of improving the dynamic systems performances needs the co-simulation with 1D control software package.

Simcenter Amesim This *mechatronic system-simulation software of physical multidomain systems* was invented by the French company IMAGINE[38] in 1986. It has a broad range of application and physical domains: automotive, aerospace, and off-highway-specific solutions. Both steady-state and transient analyzes are possible. The LMS Amesim plant model can be coupled with the SIMULINK control system model. The interface to MATLAB allows performing complex and automated pre- and postprocessing. Simcenter Amesim™ offers a complete system-simulation platform to model and analyze multidomain intelligent systems. It provides an extensive set of application-specific solutions that comprise a dedicated set of application libraries and focus on delivering simulation capabilities to assess the behavior of specific subsystems: internal combustion engines, transmissions, thermal management systems, vehicle systems dynamics, fluid systems, aircraft ground loads, flight controls, and electrical systems.

The interface between LMS Amesim and SIMULINK software enables the user to couple an LMS Amesim plant model with a SIMULINK control system model. This one can be exported from LMS Amesim into SIMULINK and also into SIMULINK Coder. Co-simulation between LMS Amesim and SIMULINK, or import from SIMULINK into LMS Amesim is also possible. LMS Amesim also offers a scripting facility with MATLAB to perform complex and automated pre- and postprocessing or to build customized graphical interfaces. The development of the platform Simcenter Amesim (shortly—Amesim) is continuously developed by Siemens Industry Software.[39]

Bibliography

1. Merritt, H.E. *Hydraulic Control Systems.* John Wiley and Sons Inc., New York, 1967.
2. http://www.britannica.com/
3. http://www.acslX.com/
4. http://www.nhancetech.com/
5. http://www.epri.com/
6. http://www.ansys.com/
7. http://www.mathworks.com/products/matlab/
8. http://ni.com
9. http://www.industry.siemens.com/
10. http://asl.org.il/
11. http://admb-project.org/
12. http://www.cray.com/blog/chapel-productive-parallel-programming/
13. http://euler.rene-grothmann.de/
14. http://www.freefem.org/ff++/
15. http://freemat.sourceforge.net/
16. http://www.geniusproject.com/software/
17. http://gmsh.info/
18. https://www.gnu.org/software/octave/
19. http://julialang.org/
20. http://www.maxima.sourceforge.net/
21. http://www.openfoam.com/
22. https://www.r-project.org/about.html
23. http://www.sagemath.org/
24. http://www.salome-platform.org/
25. http://www.scicoslab.org/
26. https://www.dadisp.com/
27. http://www.aptech.com/products/gauss-mathematical-and-statistical-system/
28. http://sine.ni.com/nips/cds/view/p/lang/ro/nid/12153/
29. http://www.ni.com/labview/
30. http://www.maplesoft.com/products/
31. http://www.ptc.com/engineering-math-software/mathcad/
32. http://www.wolfram.com/mathematica/
33. http://www.mathworks.com/
34. http://www.vissim.com/
35. https://www.modelica.org/
36. http://www.claytex.com/products/dymola/
37. http://www.ansys.com/
38. Lebrun, M. and Richard, C.W. How to create good models without writing a single line of code. *The Fifth Scandinavian International Conference on Fluid Power at Linköping*, Sweden, 1997.
39. http://www.plm.automation.siemens.com/en_us/products/lms/imagine-lab/index.shtml

chapter two

Capabilities of Simcenter Amesim platform for solving engineering problems

2.1 Platform overview

The high-speed development of mechatronic systems needs an innovative model-based system-engineering approach. The increasing product complexity, including more controllers and software, is driving the market toward *systems engineering*. Processes are transforming into model-based system-engineering approaches to support mechatronic product development. Multiphysics system simulation in combination with controls, also known as *mechatronic system simulation*, is essential to validate alignment to performance requirements during all the development phases. With the new developments in the Leuven Measurement Society (LMS) Imagine.Lab platform [1], LMS answers the trend toward model-based system-engineering by making multiphysics subsystem and system models available as *plant* models to frontload control engineering based on Model-in-the-Loop (MiL), Software-in-the-Loop (SiL), and Hardware-in-the-Loop (HiL) development approaches.

LMS Imagine.Lab is a productive environment for model-based system engineering. Directed toward mechatronic system simulation, the LMS Imagine.Lab platform offers an open approach starting from functional requirements to physical modeling and simulation, through its following three components:

1. *Simcenter Amesim*: Creates and runs multiphysics simulation models to analyze complex system behavior and support the design of controlled system from early specification to subsystem testing.
2. *LMS Imagine.Lab SysDM*: Stores and organizes system mechanical and controls models and data across the organization
3. *LMS Imagine.Lab System Synthesis*: Synthesizes even more complex systems and creates product architectures based on performance requirements, solving so-called *inverse problems*.

Simcenter Amesim simplifies multidomain integration, thanks to its easy-to-use simulation platform. All an engineer needs to do is connect various validated components to simply and accurately predict multidisciplinary system performance. With extensive dedicated libraries, Simcenter Amesim actually saves enormous amounts of time by eliminating the need for extensive modeling. Thanks to application-specific simulation, engineers can assess a variety of subsystems in multiple physical domains. This way, design and engineering teams can carefully balance product performance according to various brand-critical attributes to achieve the best possible design way before committing to expensive and time-consuming prototype testing. Since Simcenter Amesim actually frontloads system simulation early in the development cycle, it truly allows mission-critical design functionality to drive new product development.

2.2   Amesim platform capabilities

The Amesim platform analyzes the functional performance of intelligent, mechatronic systems from the early development stages onward. Through its powerful facilities, analysis, or optimization tools, it offers an advanced and easy-to-use environment for 1D system simulation and robust design. Its physically conservative *multiport modeling* [1] of physical components and its block-diagram approach for control systems enables the coupling of all its libraries together, within a built-in and comprehensive workflow. Various scripting and customization capabilities provide seamless integration of Amesim in the customers' existing design processes. Open and flexible, the Amesim platform efficiently interfaces with many 1D/3D CAE software and helps users to quickly derive and export models for standard Real-Time targets by providing a consistent and continuous MiL/SiL/HiL capable framework.

The Amesim platform offers numerous facilities in terms of usability and increased model scalability. With supercomponents, users have the possibility to *shrink–wrap* a group of components, so as to create subsystems. This feature is adapted to complex systems in which topological subsets can be identified.

The Interactive Help is always easily accessible. It provides full-text search, all platform and library user manuals in HTML and PDF formats, and full HTML documentation for every submodel, utility, and demo.

Contextual windows allow users to trace any parameter or variable, with no predefined declaration. Powerful postprocessing features offer a convenient way to define performance criteria.

For variant analysis and parameter sensitivity studies, the user can launch *batch runs* and experiments. It is also possible to trace the changes between system versions with the Compare Systems feature.

2.2.1   Personalization features

The Amesim platform finally offers many personalization features to tune the everyday design environment by defining *favorites* for usual libraries and components.

1. *Supercomponent facility*: As a model gets bigger it becomes more and more difficult to find a particular constituent component in it and a single snapshot of the whole system cannot be taken easily. The supercomponent facility overcomes these problems. The principle is to *select* a group of components and *shrink–wrap* them into a single icon. For all libraries, there are big advantages in using this approach. Users can simply explore and modify their supercomponents by double-clicking on them, when saved locally in their sketches or when saved in their user libraries.

2. *Batch Runs, Experiment Manager, and Run Monitor*: Batch runs help optimizing the parameters of your system. They can be monitored during the simulation by the Run Monitor facility. More advanced simulation combinations can be organized by the Experiment Manager within a convenient hierarchical representation. This provides better management of the simulation runs because all the sets of use cases can be easily examined, saved, and applied back to the system.

3. *Postprocessed variables*: Postprocessing is a powerful facility that is accessible to users by means of a simple drag-and-drop of any variable into a dedicated post-processing window. The chosen variables can be combined to define expressions of relevant indexes to evaluate the performance of your system. For these expressions, a wide range of mathematical functions and operations (Sum, Subtract, Root Mean Squares, Minimal, or Maximal value, etc.) are available within a built-in and convenient Expression Editor. The postprocessed variables are reusable in the Animation module.

4. *Interactive Help*: Amesim comes with integrated interactive help that is always available and provides the user with useful functionalities such as full-text search, index, and the possibility to add bookmarks. It gives access to all platform and library manuals both in HTML and PDF formats and to the HTML documentations of all the submodels and utilities available. Tutorial examples, demos, and videos are also provided to ensure a quick start for newcomers and efficient productive work for end users.

2.2.2   Analysis tools

Analysis tools, such as Fast Fourier Transform (FFT), spectral map, linear analysis order tracking, and activity index, help in explaining system behavior, highlighting main dynamics so that the user can adapt the required level of modeling to get the best accuracy in time and frequency domains. Amesim provides a comprehensive set of methods, representations, or animation capabilities that help users to analyze their system.

1. *Plotting Facilities*: To facilitate the visualization of simulation results, a large number of advanced, specialized, and convenient types of plots exist. Use thermodynamic plots, efficiency maps, or contour levels for fuel consumption analysis and various plot renderers (3D surfaces, 2D color maps) for everyday result analysis. Order tracking and spectral maps are proposed for vibration analysis. Use Bode, Nichols, Nyquist, or Modal Shapes for frequency domain exploration.

2. *Dashboard*: The dashboard is a postprocessing capability that can be used to create 2D animations of simulation results under the form of gauges, sliders, buttons, and switches. A predefined generic set of basic and fully customizable shapes are available and are allowed to quickly generate a dashboard within an Amesim or AMERun environment. Predefined toolboxes are by default proposed with possible user toolboxes extensions.

3. *Animation*: Animation is a handy way to create a 3D animation of any Amesim simulation. Users can simply create and link objects in Animation to the simulation itself. Users can easily visualize the physical component behavior according to set parameters to demonstrate the final results of the simulation.

4. *Table Editor*: The Table Editor is a powerful tool for working with data files that are often needed for a large variety of purposes. Data files can be created, loaded, or modified, either manually or by applying mathematical transformations (interpolation, extrapolation, derivation, and parametric variation) to the data. A large number of formats are supported and adapted graphical previews can be produced to prepare or verify the data.

5. *Linear Analysis*: The Linear analysis tools (eigenvalues, modal shapes, transfer functions, and root locus) enable powerful and clear analysis of the structural properties of dynamic systems with a frequency domain view. In a very limited CPU-time,

meaningful conclusions on the dynamic behavior of systems are obtained and can be judiciously exploited to perform model reduction or to study potential mechanical couplings.

6. *Activity Index*: The Activity Index is an energy-based model reduction metric that identifies the contributions of individual components in the overall dynamics of the system. It therefore provides valuable information on ways to significantly accelerate simulations by reducing the system's complexity while preserving its relevance with respect to physics.

7. *Replay*: To evaluate the behavior of a system, it is not sufficient to check the result of a specific variable with respect to time. A more global view is often necessary. Typically, engineers need to know at first sight how the variables evolve in the different parts of the sketch and which values they reach. The Replay facility proposes an easy-to-understand graphical representation for the evolution of variable results throughout the model sketch.

2.2.3   Optimization, Robustness, and Design of Experiments

In the world of multidomain system simulation, the realm of 1D is probably the place where design exploration techniques are the most effective. Whether used for design or validation, system models can provide global parameter access that directly influences design decisions. Simcenter Amesim integrates tools for design exploration, optimization, and robustness analysis and can also be interfaced with Noesis OPTIMUS and other well-established third-party optimization software for advanced studies or better process integration.

1. *Optimization*: Simcenter Amesim integrates techniques for local (NLPQL) and global (Genetic Algorithm) single or multiobjective optimization that cover the needs of everyday engineering studies. For user convenience, it can also be interfaced with almost all third-party optimization software. Finally, thanks to the multiobjective state-of-the-art optimization methods adapted from Noesis OPTIMUS, you can enhance Amesim capabilities and find the best-possible set of design variables.

2. *Robustness*: To assess and optimize design responses taking into account the variability present in the design input parameters, Simcenter Amesim provides users with a built-in tool for Monte Carlo studies and with various possibilities of interfacing Amesim with third-party software to perform robustness analysis.

3. *Design of Experiments*: Design of Experiments (DOE) allows quick exploration of the design space and parameter screening. It helps users to define an optimal set of experiments to obtain maximum information with the highest accuracy level at the lowest cost. DOE also allows users to perform sensitivity analysis. Associated with Amesim's Response Surface Methodology (RSM) techniques, it provides users with critical insight into possible design alternatives.

2.2.4   Amesim Simulator Scripting

A time-saving feature, Simcenter Amesim provides a comprehensive set of scripts that support programming in higher abstraction-level languages such as MATLAB®, Python, Scilab®, or Microsoft Excel®, and Visual Basic Application® (VBA) to automate model interaction for batch runs, perform complex or automated preprocessing, or integrate an Amesim model within an external application. In particular, the Amesim Circuit

Application Programming Interface (API) enables the writing of powerful user-defined Amesim-based applications to meet specific customer needs. For instance, it is possible to create an Amesim-powered application where a Graphic User Interface (GUI) is designed for your endusers or presales engineers to facilitate its use in their daily work.

1. *Amesim Scripting*: With Simcenter Amesim Simulator Scripting, the user is able to set or get parameters within a Microsoft Excel, MATLAB, Scilab, or Python environment and to pilot temporal simulations or linear analyses in an automated way. Amesim models can thus be easily included into existing design processes.
2. *Amesim Circuit API*: The Circuit API allows the building of Amesim-based applications and is available in three versions: (1) C, (2) Python, and (3) VBA. This powerful API contains all the necessary functions to add or remove components, to set submodels and connect them, to handle parameters or variables, and to manage simulations.

2.2.5   Amesim Customization

Simcenter Amesim offers a wide range of customization capabilities. Users can create their own customized GUIs for pre- and postprocessing directly attached to models by means of convenient Application Assistants.

Simcenter Amesim supports Python as the User Interface framework for creating a customized component or model interface. Customization is also possible via the Model Properties. For instance, it is possible to attach metadata to Amesim models that come from other software environments involved in the typical workflow. Users can thus use their specific classifications or terminology. This metadata can then be accessed via custom scripting or GUIs, allowing managing component properties in the exact same way as engineers are used to.

1. *Customized Graphical User Interfaces*: A script caller connected to AMESet helps users create customized GUIs for the preprocessing of parameters. This functionality is useful for expert users who want to create and share user-friendly models that meet application-specific needs. As a simpler alternative to classic Python technology, Simcenter Amesim supports usage of common User Interface files for the GUI creation of dedicated applications. With these two technologies, you can either quickly design simple GUIs or create more advanced GUIs with full flexibility calling for some high-level Python scripts.
2. *Model Properties with Metadata*: Amesim enables to associate any type of metadata with sketches or components: numbers, simple or rich text, dates, files, or images. It is possible to access or modify these documents easily by clicking on the corresponding models. For custom applications, this information attached to components can be handled by scripting or GUI filters. For easier management of these properties, Amesim offers both a contextual and a global view. These Model Properties provide a better link between models and the corresponding real systems.

2.2.6   Solvers and Numerics

Based on the most advanced numerical integrators, the Amesim integration algorithms support ordinary differential equation (ODE) and differential algebraic equations (DAE). The solver automatically and dynamically selects the best-adapted calculation method, depending on the dynamics of the system, among 17 available algorithms. In addition, the Performance Analyzer provides users with a comprehensive and easy-to-use set of graphical utilities for in-depth monitoring of the simulation performance. Amesim also comes along with Parallel Processing and the Discrete Partitioning facilities, to take advantage of multiprocessor, multi-thread, or multicore machines and to reduce computation times on very large systems.

1. *Performance Analyzer*: The Performance Analyzer offers a set of facilities for in-depth monitoring of the simulation performance. It provides users with valuable statistics on the simulation performance (CPU time, integration steps, and number of Jacobians, function evaluations or discontinuities processed). It allows to clearly identify which state of variables can be looked after to speed-up your simulation.
2. *Parallel Processing*: *CPU maximization* Thanks to new technologies appearing at highly reduced costs (multicore computers, clusters, etc.), Parallel Processing can succeed in performing distributed computing with a reasonable investment. This facility can also be used to take advantage of multiprocessor or multicore machines. This makes it possible to solve long-running and computationally intensive problems, especially when large independent data sets are involved. This results in a significant speed-up of simulation runs.
3. *Discrete Partitioning*: Discrete Partitioning is a technique that can lead to drastic run-time reductions for certain types of hydraulic systems, such as fuel injection, automatic gearbox command, and anti-lock braking system (ABS). Discrete Partitioning uses genuine physical discrete communication to produce a model that is suitable for co-simulation without accuracy loss. It takes advantage of wave propagation physics in hydraulic lines.

2.2.7   Model-in-the-Loop, Software-in-the-Loop, Hardware-in-the-Loop, and Real Time

Design of controlled systems requires the integration of both mechanical and control systems during the simulation phase in order to efficiently ensure proper results. This integration is done at different steps of the process, first with MiL or SiL and then in HiL. The integration of a plant model with a controls' model or code will ensure the required accuracy, and the accessibility of the variables needed for controls.

With Simcenter Amesim, LMS proposes a unique integrated platform that provides realistic component and system models for every stage of the development cycle, enabling both system and control engineers to start evaluation and validation phases early in the design cycle. Development cycle uncertainty, resulting from late design process integration, is practically eliminated.

Simcenter Amesim supports dSPACE®, xPC Target®, NI LabVIEW RT®, OPAL RT-LAB Real-Time®, ETAS LABCAR®, and HWA Create® products. Users of these control systems can export Simcenter Amesim models to one of these Real-Time platforms.

1. *Plant/Control modeling*: There is a missing link between the engineers in charge of physical system detail design and the ones in charge of the control systems. How do you validate the control system before hardware prototypes become available? The answer is to build validated Real-Time simulation of the behavior of the physical system and integrate the control-system hardware and software combined with the simulation. The Amesim platform covers this missing link and provides unique capabilities from high-fidelity physical system modeling to validated Real-Time

simulation. Amesim provides three main features: (1) model reduction methods, (2) suitable integration algorithms, and (3) automatic code generation for standard Real-Time targets.

2. *Amesim Model-in-the-Loop*: With MiL in Amesim, users can design the control strategies with the help of a representative model of the system to pilot. Virtual models for the control (SIMULINK® or equivalent) and the system (Amesim) can be used and co-simulated for function specification or investigation on concepts. Plant models can be used by both the design and the control team. Several levels of models can be evaluated for a good accuracy/CPU compromise.

3. *Amesim Software-in-the-Loop*: During the implementation stage, users can test control strategies (software) using a plant model, including CAN or FlexRay protocol. The generated C code for the control or a SIMULINK model can be used in a *virtual* Real-Time environment. Amesim model's export within SIMULINK is supported (SiL within SIMULINK) and the C code import for the control of a SIMULINK model with Amesim (SiL within Amesim). The system model can be adapted and tested with Real-Time simulation. For calibration with the detailed model, the software is connected to the plant model while the operating environment of the controller is simulated.

4. *Amesim Hardware-in-the-Loop*: For ECU testing and validation phases, control strategies implemented on the real ECU can be evaluated. Regulation, security, and failure tests can be performed without any risk and the interaction between several ECU/TCUs can be investigated, thus ensuring a high level of robustness and quality. The methodology provides an efficient and secure link with the model used for system design and complies with major hardware manufacturers.

5. *Precalibration stage*: Precalibration can then be achieved while keeping costs down on a virtual test cell that simulates an environment in which the ECU and the real system are integrated: to validate the ECU strategy; the real prototype of the system is replaced by accurate models.

2.2.8   Amesim Software Interfaces

Simcenter Amesim provides a Generic co-simulation Interface that allows users to interface Amesim with third-party software, via the co-simulation API, or to perform Amesim–Amesim co-simulations within a master–slave relationship. In this case, depending on the coupling strength of the models considered and on the available hardware capabilities, computation speedups are obtained via parallelization. Three different protocols are supported: (1) network co-simulation (two computers, each one running on Amesim), (2) shared memory co-simulation (two Amesim running on a multi-CPU computer), and (3) import co-simulation (two Amesim running on a single-CPU computer). Simcenter Amesim also supports Functional Mockup Interface (FMI): export of submodels to the Functional Mockup Unit (FMU) and import of a FMU as external models toward the FMI co-simulation.

1. *Generic co-simulation*: Co-simulation means that two models are simulated within their own simulation environments (including solvers) while communicating at regular intervals. The Generic co-simulation Interface is primarily intended for establishing co-simulation with third-party software, such as Computational Fluid Dynamics (CFD) software or any proprietary/in-house code, but it can also be used for Amesim–Amesim co-simulation.

2. *Functional Mockup Interface*: The FMI offers a standardized interface for model exchange and tools coupling. It is implemented by simulation models as an FMU and provides a standard definition for model coupling.

2.2.9   1D/3D CAE

Physical systems are often composed of different elements all working together, such as pneumatics, mechanics, hydraulics, electrics, or control systems. Interactions between multidisciplinary systems and complex 3D systems can be difficult to manage in a single modeling software package. This is why users can connect Amesim to specialized 1D–3D CAE Simulators.

For different purposes and applications, Simcenter Amesim can be coupled with various external software, such as computer-aided engineering (CAE), computer-aided design (CAD), computer-aided manufacturing (CAM), finite elements analysis/finite elements method (FEA/FEM), or computational fluid dynamics (CFD). Co-simulation provides a link between Simcenter Amesim and CAE tools with predefined setups. This coupling ensures a good dialog between the tools and the simulation software. Note that both time and frequency domains are supported by this combined approach, with the 1D/3D CAE software simultaneously involved. All analysis tools for a structural and system analysis are still compatible when the Amesim models are connected to the CAE software. It is still possible to easily perform user-defined batch runs, parameter studies, or optimizations.

1. *Finite Element Analysis (FEA)*: With the FEA Import or co-simulation, Amesim users can connect model description of their mechanical structures from their usual FEA software to Amesim. With this functionality, Amesim allows users to check coupling between the mechanical structures and any kind of connected nonlinear actuator (hydraulic, pneumatic, electric, magnetic, piezoelectric, etc.). All this can be, of course, combined with control strategies included in Amesim.
2. *Multi-Body Systems (MBS)*: Amesim can be connected to specialized multibody packages, such as LMS Virtual. Lab Motion. The multibody interfaces provided by LMS Imagine make it simple to model each subsystem directly in the appropriate environment and perform combined simulation using either model export facilities or co-simulation.
3. *Computational Fluid Dynamics* : Amesim can be connected to third-party CFD software, such as Fluent®, Ansys®, CFX®, StarCD®, or Eole® to refine models or obtain more realistic boundary conditions, especially for transient dynamics. Coupling Amesim with CFD can also be a good way of calibrating 1D models.

2.2.10   Amesim Libraries

Amesim Libraries exploit a large library of predefined and validated components from different physical domains, such as fluid, thermal, mechanical, electromechanical, and others.

1. *Accelerating model creation*: To create a system simulation model in Simcenter Amesim, users simply access one of the numerous libraries of predefined and validated components from different physical domains, such as fluid, thermal, mechanical, electromechanical, powertrain, and many others. All library components are completely validated to guarantee the accuracy and reliability of the simulation. By selecting the required validated component from the related library, users completely avoid the step of creating their own complicated code. Not only does this save enormous

amounts of time, it also allows teams to easily create complex system models covering multiple domains. Rather than spending time simply building a functional model, engineers can focus on design-critical tasks such as optimizing the design for best-in-class product behavior early in the process.

2. *Easy and accurate model creation*: By combining library components, users create an easy-to-understand working sketch of the system model. To aid investigation, varying model complexity can be selected for each component. Parameters and measuring units can be set in an easy and interactive way. Thanks to this completely transparent concept with easy-to-access embedded model information, users can capture, reuse, and share engineering knowledge. Engineers can also start from a simplified model representation in the early development stage, and gradually add more detail to the model as design information becomes available.

3. *A broad range of applications and physical domains*: With more than 38 validated libraries filled with more than 5000 dedicated models, Simcenter Amesim enables to cover a multitude of physical domains and engineering applications:
   - Control: Signal, Control, and Observers
   - Electromechanical: Electrical Basics, Electromechanical, and Electric Motors and Drives
   - Fluid: Hydraulic, Hydraulic Component Design, Hydraulic Resistance, Filling, Pneumatic, Pneumatic Component Design, Gas Mixture, and Moist Air
   - Internal Combustion Engine: IFP-Drive, IFP-Engine, IFP-Exhaust, and IFP-Combustion 3D
   - Mechanical: Mechanical, Planar Mechanical, Powertrain, and Vehicle Dynamics;
   - Thermal: Thermal, Thermal–Hydraulic, Thermal–Hydraulic Component Design, Thermal–Pneumatic, Two-Phase Flow, Air-Conditioning (AC), Cooling System, and Heat Exchanger Assembly Tool

2.3   LMS Imagine.Lab Solutions

Amesim Solutions focus on critical design and engineering issues with an extensive set of application-specific solutions for automotive, aerospace, and mechanical industries. Simcenter Amesim provides users direct access to multidomain simulation solutions for powertrain, transmission and internal combustion engines, vehicle thermal management, fluid systems, ground loads and flight controls, vehicle system dynamics, and energy and electromechanical systems. Each of these solutions comprises a specific set of tools and application libraries and focuses on delivering simulation capabilities to assess the behavior of specific subsystems. For example, engineering teams can analyze and optimize the Noise–Vibration–Harshness (NVH) performance of transmission systems, or fine-tune air management and ECU calibration for combustion engines. With Simcenter Amesim, systems can be studied independently, but what makes LMS Imagine.Lab truly unique is the possibility to integrate subsystems into a unique environment to evaluate their interaction.[*]

2.3.1   LMS Imagine.Lab Powertrain Transmission

LMS Imagine.Lab Powertrain Transmission provides a generic platform for analyzing and designing optimal transmission systems. LMS Imagine.Lab Powertrain Transmission gives access to driveline, engine and transmission models and components, and focuses

---

[*] The names in this section come from the previous release of the software.

on comfort, performance, and losses, and NVH and Hybrid issues. These solutions help users to study the global behavior of the entire powertrain architecture from low (<20 Hz) to high frequencies.

LMS Imagine.Lab Powertrain Transmission facilitates the development of new concepts such as hybrid architectures and solves powertrain transmission challenges such as high-shift quality and low noise level of drivelines. It gives access to robust and effective modeling of nonlinear phenomena as one may find electric or hydraulic actuators, dry or wet clutches and also in dampers, dual-mass flywheel, universal joints, and gears backlash in all transmission. The development time of powertrain systems can be significantly reduced from months to weeks, and the maintainability of models is greatly facilitated, thus increasing life length while reducing costs of systems development. The constant evolution of application libraries ensures applicable models in an ever-changing industrial world.

2.3.1.1   LMS Imagine.Lab Drivability

The main challenge for today's car and truck manufacturers is to increase their vehicles' performance while reducing fuel consumption pollutant emissions. At the same time, the vehicle's comfort needs to be increased. LMS Imagine.Lab Drivability helps to accelerate the complete powertrain design. It guarantees maximum driver comfort through optimal shift transmission quality and provides a good torque applied to vehicles from engine through driveline. With LMS Imagine.Lab Drivability, users can comprehensively study the entire physics and control strategies of gear shifting for every kind of vehicle architecture (gearbox, driveline, and engines) to improve comfort and avoid bad oscillations (0–20 Hz). Every classic or new hybrid vehicle's architectures can be computed with a flexible toolset that is included in the various structured libraries of physical models and consists of different complexity levels (from complex models, including actuators coupling and their controllers to Real-Time models).

LMS Imagine.Lab Drivability provides a comprehensive, flexible development framework ranging from design to validation and control. Thanks to off-the-shelf component libraries, users have access to graphical multiphysics system design, simulation, and analysis in a single environment. The solution handles various levels of engine behavior (cold start, starter, etc.) and any kind of transmission (hybrid, DCT, IVT, AT, MT, CVT, etc.). Users can combine all possible components in a gearbox, driveline, or engine and subsequently analyze the total system design, thereby focusing on the comfort impact of defined strategies. The solution further offers to connect engine acyclism for damper analysis and complex vehicle driveline for torque vectoring. Frequency analysis tools (modal shapes, eigenvalues, FFT, etc.) provide an accurate environment to represent physical details of a specific component design. It also provides efficient simulations and quality results for optimized engine (ECU) and gearbox (TCU) control design and validation.

LMS Imagine.Lab Drivability makes it possible to run off-line test procedure validation (HiL and/or SiL) and fully interface with MATLAB/SIMULINK and common Real-Time platforms (dSPACE, Opal RT, and xPC Target), in this way integrating the design process from simulation to test bench. Finally, the solution can run performance testing and comparisons with customer's requirements, and evaluate fuel consumption and pollutant emissions according to ISO requirements (cumulated raw emission calculation) with improved drivability and quality levels.

2.3.1.2   Noise, Vibration, and Harshness

LMS Imagine.Lab Noise, Vibration, and Harshness (NVH) gives users an in-depth understanding of the NVH powertrain system performance. The solution provides all the required information on the root causes of noise and vibration problems related to hydraulic dynamics, mechanical contacts, or slip control. These can potentially generate a negative quality perception or key component durability problems. Moreover, with LMS Imagine.Lab NVH, mechanical parts and overall system architecture can be optimized. Users can focus on NVH sources and related corrective component efficiency: engine torsional harmonics, driveline vibration analysis, dual mass flywheel, clutch dampers, and rattle noise (<200 Hz Max), and whining noise (>1 kHz) with Finite Element (FE) coupling, if necessary.

The solution provides a better physical understanding of driveline vibrations due to a combination of linear and nonlinear systems (dry frictions, variable stiffness, end stops, bearings, joints, and gear backlash). The FE import interface is able to include any FE mesh (modal base or condensed) in an Amesim sketch to study the coupling between mechanical 3D structures or 3D shafts with actuators (electrics or hydraulics).

Engineers can considerably reduce the number of vehicle dynamics test benches, using Amesim's off-line test procedure validation on any vehicle powertrain architecture:

- Transmission part: Mechanical, automated, dual clutch, infinitely variable, and continuously variable
- Driveline part: Universal joints, clutch dampers, dual mass flywheel, and chassis
- Engine part: Crankshaft, camshaft, valves, and rocker arms

2.3.1.3   Performance and Losses

With LMS Imagine.Lab Performance and Losses, users can define vehicle powertrain architectures for dedicated studies on performance and consumption. This helps engineers to design optimal strategies to reduce fuel consumption while providing a consistent output power curve within the engine's best operating range, which reduces mechanical losses and optimizes controls. LMS Imagine.Lab Performance and Losses give engineers the required insight into taking key design decisions for optimal customer satisfaction and comfort.

LMS Imagine.Lab Performance and Losses helps to accelerate the design of any kind of powertrain, drivelines, and gearboxes: gasoline/diesel vehicles (sedan cars, utilities, and trucks), hybrids (light, mild, series, parallel, and other), automatic gearboxes, CVT, IVT, DCT transmissions, and accessories (AC, power steering). Users can perform a detailed study of the various power consumptions in chosen car architectures and are able to meet specific requirements with a system-level analysis. LMS Imagine.Lab Performance and Losses comes with a set of state-of-the-art physical models and libraries to study the couplings between thermal, hydraulics, electrical, and mechanical domains. It also provides efficient simulations and quality results for optimized engine (ECU) and gearbox (TCU) control design and validation.

LMS Imagine.Lab Performances and Losses make it possible to run off-line test procedure validation (HiL and/or SiL) and fully interface with MATLAB, SIMULINK, and common Real-Time platforms (dSPACE, Opal RT, and xPC Target), in this way integrating the design process from simulation to test bench. Finally, the solution can run performance testing and comparisons with customer's requirements, and evaluate fuel consumption/emission according to ISO requirements (cumulated raw emission calculation).

2.3.1.4   LMS Imagine.Lab Hybrid Vehicle

Hybridization and full electrification are one of the most effective measures to reach challenging emission and fuel consumption targets imposed by regulations.

Simultaneously, vehicles have to keep providing a high level of driving comfort. With LMS Imagine.Lab Hybrid Vehicle, engineers can size components. They can design, analyze, validate, and optimize architectures and power-management strategies. It helps to define and specify the best powertrain architecture, taking vehicle thermal and power management into account.

LMS Imagine.Lab Hybrid Vehicle is based on the Amesim multidomain system simulation approach and provides dedicated tools that help to model and design hybrid engine architectures. With the Amesim models, a conventional propulsion system with an onboard rechargeable energy storage system can be designed through a set of specific multidomain libraries (IFP-Drive, Electric Motors and Drives, and Powertrain) and, subsequently, the behavior of the entire system and its individual components can be studied. The Amesim platform also takes into account control systems, thanks to advanced real-time interfaces with most commonly used Real-Time targets.

Using the facilities of the Powertrain Transmission platform, a lot of high-level developments were achieved: Automatic transmission development and HiL testing; Modeling for electrohydraulic subsystems in automatic transmissions; Complete manual, automatic hybrid gearbox design; Transmission modeling for dynamic behavior, axial, and torsional vibrations; Simulation of the hydraulic control unit of the six-speed automatic transmission for a passenger car, and so on [2–9].

2.3.2   *LMS Imagine.Lab Internal Combustion Engine*

LMS Imagine.Lab Internal Combustion Engine helps users evaluate, design, and optimize comprehensive engine systems from air management and combustion to fuel injection and engine control by providing accurate physical engine models and components.

LMS's Imagine.Lab Internal Combustion Engine has been developed in close collaboration with IFP Energies nouvelles. It offers a cutting-edge, flexible environment for designing and optimizing *virtual* engine and automotive subsystem concepts. Users have the ability to study couplings with fuel injection subsystems, vehicle thermal management, powertrain, and any other components, and can adapt model definitions to a wide range of usage scenarios. This solution enables to analyze the advanced technology choices and gives engineers a powerful toolset to investigate alternative architectures.

2.3.2.1   *Engine Control*

LMS Imagine.Lab Engine Control is a complete and integrated solution to design and set up robust engine controls. From design to validation, it provides a relevant toolset to manage the growing complexity of engine-control strategies and drive assistance technologies directly linked to the engine (speed regulation, gearbox). LMS Imagine.Lab Engine Control goes further than the classical *automatic*-control approach, and provides a unique methodology for control design. The solution is mainly based on a strong modeling competence and a thorough understanding of the system physics.

OEMs and suppliers, engine manufacturers, engine control designers, and testers (HiL test-benches) will be able to use state-of-the-art engine models suitable for every stage from design to validation, and in real time. By using Simcenter Amesim throughout the development process, project engineers can implement and capitalize on their own modeling know-how, and share models to gain time and efficiency. Model accuracy and Real-Time analysis are no longer incompatible.

LMS Imagine.Lab Engine Control comes with a set of cutting-edge components: The Simcenter Amesim core platform; Control-specific interfaces (SIMULINK and Real-Time);

Links with market reference software: SIMULINK (MathWorks), Morphee2 (D2T), LabView RT (NI); Standard and specific libraries and components/models; analysis tools and methodologies for model reduction.

For more specific usage, detailed models of engine subsystems are available (injection, air loop, driveline, thermal, electrical actuators, and motors) and analysis tools to study the combustion process in real time.

Implementing LMS Imagine.Lab Engine Control helps engineers reduce time to market (gain in efficiency) and design/validation costs (mainly a reduction in testing needs) while improving quality and robustness (better physical understanding and increased validation capacity). It offers unrivalled model-based, multidomain system simulation within a single-simulation platform, both in the design and validation process (HiL and SiL).

LMS Imagine.Lab Engine Control has been developed in close collaboration with IFP, ensuring delivery of cutting-edge, innovative, and robust solutions.

2.3.2.2 Air Path Management

LMS Imagine.Lab Air Path Management helps users design different air-path architectures and management strategies. It provides detailed modeling of air-path actuators and controls for the air mass and burned gas ratio in the combustion chamber. With the phenomenological combustion model, the interaction between air-path and other engine subsystems can be examined in detail.

LMS Imagine.Lab Air Path Management helps users to design all kinds of air-path architectures—including exhaust systems—for any kind of technical choice. The solution permits a quick analysis of the impact of selected technical architectures (variable valve timing and lift, charging systems, exhaust gas recirculation, and coolers) on the gas management and hence the combustion.

LMS Imagine.Lab Air Path Management has been developed in close collaboration with IFP Energies nouvelles, ensuring the delivery of cutting-edge, innovative solutions.

LMS Imagine.Lab Air Path Management is based on the Amesim multidomain system simulation approach and on a set of libraries and component models, which provide the necessary tools to design advanced air-path management systems: mechanical, IFP-engine, CFD 1D and thermal libraries, powerful analysis tools, turbo map preprocessing, linear analysis, SIMULINK interface, and generic co-simulation to allow coupling with CFD tools…

LMS Imagine.Lab Air Path Management helps to design all kinds of configurations and provides detailed modeling of air-path actuators such as valves, variable valve timing (VVT), and varible valve for air-path (VVA). Moreover, with the phenomenological combustion models, the impact on torque, emissions, and consumption can be analyzed. Furthermore, the interaction between air path and other engine subsystems (cooling system, electrical devices, etc.) can be detailed and examined.

2.3.2.3 Combustion

LMS Imagine.Lab Combustion supports the design and the optimization of new combustion processes and adaptation of engines to alternative fuels. It helps users to optimize the cylinder geometry (piston shape, head, location of injectors, and plug) to optimize engine parameters (such as advance, turbulence: swirl/tumble/squish, lambda control, or injection split. etc.) for different fuel types, and study advanced combustion processes, including homogeneous charge compression ignition technology (HCCI) and controlled auto-ignition technology (CAI). Combustion processes get increasingly complex in order to comply with conflicting demands. More stringent emission standards have to be met while customers desire higher engine response and lower noise levels. Engine designers are therefore looking for interesting alternatives beyond the continuously evolving conventional gasoline Spark Ignition (SI) and diesel Compression Ignition (CI) combustions.

LMS Imagine.Lab Combustion provides an efficient way to investigate the transient engine behavior by using an unparalleled numerical multidomain system simulation approach. The IFP-Engine library provides dedicated components for combustion simulations and can be seamlessly coupled with 3D combustion codes and models (such as IFP-C3D).

LMS Imagine.Lab Combustion provides the ideal toolset to analyze the dynamic behavior of any combustion process, and enables flawless 3D calculation integration in a system approach. Dedicated libraries come with state-of-the-art 1D and 3D models suitable for various applications from large diesel to high-revving gasoline engines.

2.3.2.4 Emissions

LMS Imagine.Lab Emissions focuses on overall engine optimization according to the required emission standards. The ready-to-use models enable users to test exhaust configurations and associated control strategies. LMS Imagine.Lab Emissions assists in meeting the most stringent emission standards (Euro V and VI, Tier III) and helps to accurately model and analyze NOx, CO, HC, and particle emission levels depending on technological choices.

LMS Imagine.Lab Emissions is based on the Amesim multidomain system simulation approach and comes with a set of advanced tools and libraries for modeling the exhaust system architecture and its environment.

LMS Imagine.Lab Emissions is especially useful for engineers in charge of exhaust components and system development and calibration. It offers new opportunities in terms of system optimization and focuses on the possibility to couple the engine and exhaust system. It helps to investigate the effect of engine design parameters and control strategies on the engine exit emissions taking into account the running conditions, the combustion process itself, and the influence on the exhaust process and the tailpipe emissions. LMS Imagine.Lab Emissions presents a new method to explore and optimize engine emissions during cold start with engine warm-up and catalytic converter light-off.

2.3.2.5 Internal Combustion Engine Related Hydraulics

LMS Imagine.Lab Internal Combustion Engine Related Hydraulics helps design and optimize fuel systems and components from the tank to the injector. It further assists in designing valve actuation systems in relation to the engine cylinder: Injection Systems (Gasoline, diesel, and alternative fuels), low- and high-pressure injection systems (Indirect/direct injection, common rail, unit injector, in-line pump, Solenoid, piezo, electrohydraulic valve, or mechanical actuation); Valve train (Variable valve timing and cam phasing, Variable valve actuation with mechanical [MVT], electromechanical [EMVT], or electrohydraulic [EHVT] systems, Engine compression brake, and Camless systems).

LMS Imagine.Lab Internal Combustion Engine Related Hydraulics is based on the Simcenter Amesim multidomain system-simulation approach and helps develop new concepts to confront challenges posed in systems such as high-pressure multiple injections, gasoline direct injection, return-less low-pressure gasoline systems, variable valvetrain, and engine compression brake systems. The solution comes with a large and flexible set of models, addressing different complexity levels (leakage, compressibility, stiffness, inertias, friction, etc.).

The accessibility, solver efficiency, and accuracy of the models help both specialist and nonspecialist engineers to design robust systems while keeping costs and delays under control. These qualities allowed innovative companies to solve problems such as gasoline engine ECU validation on HiL, diesel engine ECU validation on HiL, test of new warm-up strategies for fuel consumption reduction, development of VVA control strategies, new combustion concepts to satisfy future emission legislation, and observer design for down-sized gasoline engine control [10–17].

2.3.3   LMS Imagine.Lab Vehicle System Dynamics

LMS Imagine.Lab Vehicle System Dynamics offers an application-oriented environment and dedicated capabilities to specify, design, and validate the full vehicle and chassis subsystem components (brakes, suspension, steering, and antiroll system). The solution allows easy and seamless integration in a single system to simulate, analyze, assess physical interactions, and validate the entire vehicle system and global chassis control strategies.

LMS Imagine.Lab Vehicle System Dynamics presents a unique open-integration platform to model and simulate cars and light trucks with their actuators in a straightforward and continuous integration process (from MiL and SiL to HiL). The solution includes an extensive vehicle dynamics application, physical modeling libraries (hydraulic, pneumatic, electrical, and mechanical), ready-to-use template, and solution demonstrators and interfaces with simulation suites such as LMS Virtual. Lab Motion, MSC.ADAMS, and SIMULINK.

2.3.3.1   Vehicle Dynamics

LMS Imagine.Lab Vehicle Dynamics solution offers an application-oriented environment with a complete and dedicated toolset to test and optimize vehicle comfort and riding and handling behavior from the early concept stage onward. The solution enables to assess the interaction between the vehicle, its key subsystems (steering, brakes, and suspension), and their controllers to ensure optimal ride and handling performances, driving pleasure, comfort, and enhanced safety.

To meet the increasingly shorter full-vehicle development process, LMS Imagine.Lab Solution presents a dedicated and easy-to-use environment for chassis and subsystems specification, design, and validation with functional to high-fidelity network level models for vehicle dynamics studies.

With an application-oriented GUI *iCAR* fully embedded in Simcenter Amesim, the user can use predefined chassis subsystem templates and also benefits from the solution modularity to create functional models for vehicle dynamics and carefully evaluate the full vehicle behavior. The chassis design-oriented environment takes advantages of the Amesim multidomain platform as modeling modularity, excellent solver capabilities, and the specific physical libraries then open up integration possibilities for OEMs and suppliers.

The solution comes with a large set of dedicated tools especially useful for chassis and subsystems design, including data management, library of subsystem templates, ISO and NHTSA preset maneuvers and user define maneuver possibilities, parametric functions on kinematics tables for axle specification, and specific chassis pre/postprocessing and optimization tools. The strong integration capabilities of the vehicle dynamics engineering process provide a detailed insight into the numerous component and system interactions that determine a vehicle ride and handling profile.

LMS Imagine.Lab Vehicle Dynamics solution creates the possibility to share datasets, libraries, and subsystem models not only between internal departments but also with external supplier companies. The many possibilities of the platform interfaces help to effortlessly link 1D Amesim simulation studies with in-depth 3D multibody analysis tools and controller design tools.

2.3.3.2   *LMS Imagine.Lab Vehicle Dynamics Control*

With LMS Imagine.Lab Vehicle Dynamics Control, chassis designers and engineers can simulate vehicles, sensors, actuators, and the control strategy in a single comprehensive platform, which can be extended to SIMULINK/RTW for Real-Time simulation purposes. Through the SIMULINK interface, the solution can handle high-quality control strategies on detailed models, and tests on safety related electronic systems (ECU design, testing, robustness, and fault diagnosis).

Users can explore functional architectures for vehicle dynamics control to enhance vehicle ride comfort, handle the best compromises, and integrate new concepts, all in one single platform. It offers the possibility to explore overlapping functions (interactions of tire, driveline, brake, suspension, and steering) and analyses couplings in critical situations in which actuator dynamics are relevant. The platform's open-ended architecture makes it possible to include customer models/libraries and seamlessly run complex analyses, model reduction, and Real-Time tests.

LMS Imagine.Lab Vehicle Dynamics Control ensures significant productivity gains, maximizes knowledge, and minimizes time by using reliable libraries and advanced interfaces with most commonly used third-party technologies for controls. Engineers can identify the optimal configuration between vehicles, actuators, and controllers, and predict vehicle-system response to driver steering, braking, throttling and shifting, road inputs, and wind perturbations.

2.3.3.3   *LMS Imagine.Lab Braking System*

LMS Imagine.Lab Power Steering provides models to design robust power-steering systems while reducing the time-to-market. The solution involves Amesim tools and libraries that focus on a real multidomain approach required to deal with hydraulic, electrohydraulic, and electric power-steering systems. It helps engineers examine functionality and performance in any power-steering system to assess risks.

With LMS Imagine.Lab Braking System solution, users can design and optimize individual braking components (booster, master cylinder, and ESP hydraulic modulator valves) and the complete hydraulic braking circuit (ESP hydraulic modulator, piping system, and calipers). These elements can be used early in the process and for testing control strategies using HiL. It facilitates the development of standard, ABS, and ESP braking systems. It gives further understanding/insights into the noise and vibration behavior and control of the braking system.

LMS Imagine.Lab Braking System gives access to braking circuit dynamics and helps to analyze and compare different hydraulic architectures (X, H, and I) to evaluate braking distance or vehicle stability if connected to a vehicle model. Moreover, the solution seamlessly handles both hydraulic and/or pneumatic systems for cars, trucks, and buses and trains.

In more detailed studies, the LMS Imagine.Lab Braking System solution has the capability to perform global system sizing and handle behavior analyses of typical braking circuit components (both for low and high frequency). It has the ability to study pressure valve's stability of ABS and ESP systems, run analyses of the ABS and ESP, and complete hydraulic circuit and cavitation (noise) in the master cylinder when releasing pressure. The complete braking circuit including the booster, master cylinder, and valves can be assessed in a detailed or in a more basic way while being able to run real time.

2.3.3.4   LMS Imagine.Lab Power Steering

LMS Imagine.Lab Power Steering solution provides reliable and accurate models to design robust power-steering systems (stability, vibration, and couplings analysis) while reducing the time-to-market. The solution involves different Amesim tools and libraries focusing on the reality of the multidomain approach, which is required when dealing with hydraulic (HPS), electrohydraulic (EHPS), and electric (EPS) power-steering systems.

Power-steering solution focuses on the functional design (mainly stability or vibration issues) and enables engineers to examine functionalities and performances of any power-steering system to assess technology risks and examine and test the related Electronic Control Units.

Main EPS structure types can easily be modeled (belt or direct drive, rack assist, dual pinion, pinion assist, and column assist) and advanced power-steering features such as active front steering (AFS, ESAS). Concerning HPS and EHPS, detailed views of the rotary valve, pump, and the hydraulic circuit can be generated for more in-depth studies and for the analysis of vibrations, couplings, and stability in car park maneuvers or normal driving conditions. It is even possible to analyze the shimmy phenomenon and display its contributions on the steering system through interfaces of multibody software such as LMS Virtual. Lab Motion. The power-steering solution could also be enhanced with the complete LMS vehicle dynamics solution for full integration of the steering system in the chassis system, assessing steering fell and safety issues. and second-generation ESC interaction with EPS.

LMS Imagine.Lab Power Steering will significantly help to define optimal-steering performance, present trade-offs between technologies, identify the optimal-steering feel, and assess the impact of driver assistance on the driver's perception.

2.3.3.5   LMS Imagine.Lab Suspension and Anti-Roll

LMS Imagine.Lab Suspension and Anti-Roll helps users analyze circuits and components in suspension and anti-roll systems. It can optimize the dynamic contribution of these systems to passenger comfort and vehicle handling, improve damper design, and examine causes of vibration and suspension control issues. It can help model passive and active suspensions for both hydraulic and pneumatic technologies.

Suspension and anti-roll solution helps users to size and analyze circuits and components involved in suspension and antiroll systems. It can optimize the dynamic contribution of these systems to passenger comfort and vehicle handling, improve damper design versus cavitation, model air-spring systems, and understand causes of vibration whatever technology is involved.

Suspension and antiroll solution helps to model and analyze passive and active hydraulic suspensions and analyze their control issues. Magneto–rheological actuators can be modeled on demand.

The different levels of modeling involved in the Amesim Hydraulic and Pneumatics libraries allow functional design and detailed analyses of such components in order to assess performance advantages (dynamics) and design/test those components according to control strategies. Finally, it is possible to accurately design control strategies regarding the technology and the dynamics involved. Moreover, active roll systems can be modeled focusing on the component and/or circuit design, for example, pressure valves stability and priority valve testing, couplings between the steering and roll bar circuits, and cavitation in cylinders.

The above-mentioned platform was successfully used for Vehicle dynamics engineering, ABS pump simulation, design of the antiroll system, simulation of the power-steering systems, analysis of a F1 car braking system, steering system of a wheel loader, Chassis applications, modeling of a brake actuation system, electrohydraulic braking, and brake pedal feel prediction.

2.3.4   LMS Imagine.Lab Vehicle Thermal Management

LMS Imagine.Lab Vehicle Thermal Management solutions provide dedicated tools to build and analyze complete vehicle thermal management models in a single environment. The solutions let users to: model, size, and analyze components, subsystems, subsystem interaction, run steady-state and real transient multidomain simulations, and handle strategic heat management scenarios and their impact on fuel consumption and pollutant emissions and passenger comfort and engine performance.

Vehicle thermal management solutions give engineers the possibility to work on detailed models of vehicle thermal management subsystems such as cooling, lubrication, thermal engines, AC, and in-cabin systems.

LMS Imagine.Lab Vehicle Thermal Management provides a dedicated set of tools and libraries to study the energy flows under the hood and within the cabin, which are directly or indirectly contributing to pollutant emissions, fuel consumption, engine performance, and passenger comfort. It helps engineers to focus on the interactions between different subsystems (lubrication, cooling, exhaust line, combustion chamber, engine thermal masses, AC, electric auxiliaries, cabin, and vehicle) to find the best compromises for energy flow management (thermal, thermal fluids, electrical, and mechanical).

LMS Imagine.Lab Vehicle Thermal Management, based on multiphysical component libraries, offers a decisive approach to control the energy flows.

Furthermore, LMS Imagine.Lab Vehicle Thermal Management helps analyze the influence of new-generation subsystems (heat pump systems, energy storage systems, and heat recovery system) or components (electric pumps, controlled valves, and immersion heaters) on the overall system behavior and especially on fuel consumption, pollutant emissions, and passenger comfort.

LMS Imagine.Lab Vehicle Thermal Management provides a cost-effective alternative to prototypes so that virtual vehicle subsystems integration can be easily and rapidly studied. It is fully integrated into the Simcenter Amesim platform, and therefore takes advantage of its flexibility and its multidomain modeling capabilities.

LMS Imagine.Lab Engine Thermal Management helps model the overall energy balance of an engine in a single platform by considering thermal interactions between the lubrication and cooling systems, the thermal engine, the combustion chamber, and the intake and exhaust pipes.

System integration is one of the current challenges faced by car manufacturers and suppliers, especially in the field of engine thermal management. To ensure engine performance and design low-consumption and low-emission engines while providing the best possible comfort to passengers, engineers have to be able to optimize the engine warm-up.

With LMS Imagine.Lab tools, users can model engine warm-up and associated key-criteria (consumption, cabin heating) in a specific normalized cycle and rapidly study the influence of topological modifications of each subsystem (split cooling, energy storage tank, engine material changes, and electric pump). Engine Thermal Management solution comes as a set of physics-based libraries such as Thermal, Thermal-Fluid, and application libraries such as Cooling Systems, IFP-Engine (combustion chamber thermal losses). Each product assists the user in building individual subsystems, analyzing their independent behavior, and defining interaction ports with the other subsystems.

Finally, all the modules are connected together to create the Engine Thermal Management model in a single environment.

2.3.4.1 LMS Imagine.Lab Engine Cooling System

With LMS Imagine.Lab Engine Cooling System, engineers can model the complete cooling system of a vehicle, including all the components (pump, thermostat, and radiator) with the associated heat exchanges and interactions with the other under the hood subsystems. It comes with a set of physics-based elements, advanced component libraries, and a specific heat exchanger stacking tool (vehicle front end) for thorough system analysis (isothermal or thermal) in a steady-state and or transient configuration.

Innovations in the engine cooling system field are clearly influenced by restrictive emissions requirements for reduced fuel consumption.

With LMS Imagine.Lab Engine Cooling System, engineers can calculate the coolant flow-rate distribution and predict pressure and temperature levels throughout the circuit to study individual component and global system performances and behavior. Moreover, component's thermal interactions and system architecture modifications (front end heat exchanger stacks) can be studied to perform drive cycle analysis and test new control strategies (fans, blowers, and pumps).

2.3.4.2 LMS Imagine.Lab Refrigerant Loop

With LMS Imagine.Lab Refrigerant Loop, the design of AC loops, heat pumps systems, and Rankine cycle can be studied and performed. The climate control system has become one of the most important features in automotive comfort, and the refrigerant loop significantly influences fuel consumption and pollutant emissions.

Refrigerant Loop comes with a dedicated set of tools and libraries for pressures, temperatures, flow rates, prediction, heat exchangers characterizations, component performance optimization, coefficient of performance calculation, and drive-cycle analysis.

A new step in the understanding of the transient behavior of refrigerant loops has been achieved by making it possible to efficiently size components (heat exchangers, compressor, and thermal expansion value), test alternative fluids (e.g., CO_2 or 1234yf), and optimize/evaluate new strategies or architectures (stop and start strategy, several stage compressors, multievaporator systems, and heat pump systems).

Moreover, engineers can further investigate more global thermal issues associated with refrigerant loop behavior, such as developing and testing control strategies and studying the impact of the system on the engine thermal management. LMS Imagine.Lab Refrigerant Loop gives engineers the ability to ascertain that the system performs to the highest standards for optimal passenger comfort regardless of operating conditions.

2.3.4.3 LMS Imagine.Lab Passenger Comfort

LMS Imagine.Lab Passenger Comfort helps to study thermal interactions between the (AC) system, the cooling system, and the cabin or vehicle interior. Engineers can evaluate and control the cabin refrigeration or heating processes (air temperature and humidity), and study the integration of additional heaters and their influence on passenger comfort, especially when using high-efficiency engines with low thermal losses.

Engineers can study the impact of exterior conditions and technological choices on the air temperatures and humidity within the cabin.

The different analysis capabilities of the platform help the user thoroughly study the behavior of the entire system according to heating strategies and drive cycles under specific operating conditions. With LMS Imagine.Lab Passenger Comfort, it is possible to accurately size components (heater core, evaporators, ducts, and fans/blowers), test control strategies (AC compressor displacement, recycling modes, automatic AC control, control of blowers, and fan rotary speeds), and analyze innovative AC architecture.

2.3.4.4 LMS Imagine.Lab Lubrication

LMS Imagine.Lab Lubrication solution offers the required tools to model and design the entire engine lubrication system with all the associated components (pump, valves, and bearings) for performance validation, system optimization, failures investigation, and evaluation of new architectures. With Lubrication solution, users can perform steady-state and/or transient analyses, and isothermal or thermal analyses.

The complete oil path through the engine model can be used to make sure that the required amount of oil is delivered to the different components. The aim is to optimize pump sizing to develop intelligent systems (piloted pump), thereby reducing power absorbed by these components and systems using oil pressure (VVT, VVA, and more). In this way, the integration of new components (oil cooler, VVT, VVA, and piston cooling jets) can be evaluated. A total synchronization of components (pump) for different engines and platforms can thus be achieved and a detailed validation of production-driven architecture changes.

LMS Imagine.Lab Lubrication is able to run steady-state and transient analyses, which take thermal effects into account. The user can integrate frictional heat sources in pumps, bearings, and contacts between piston rings and cylinder liner to evaluate the oil temperature increase during warm-up. Moreover, it is possible to assess the thermal interactions between components and develop related heat management strategies (oil cooler and piston cooling jets) [18–27].

2.3.5 LMS Imagine.Lab Aerospace Systems

Aerospace manufacturers are faced with the challenge of designing systems and components that have to be safer, more reliable, and cheaper to operate, deliver better passenger comfort, and have less environmental impact than their competitors. In addition, they have to systematically reduce development times in order to get new products to market earlier.

LMS Aerospace Systems solutions enable aviation and aerospace manufacturers and suppliers to ensure the structural integrity of the aircraft and its critical components, while applying lighter materials. The Solutions help engineers to develop robust landing gear, steering and anti-skid systems, and supports the design and optimization of flight control systems. Moreover, it enables to assist engineers in designing and industrializing fuel systems (metering units, pumps, nozzle, starters, and heat exchangers) and their controls, and engine control actuators. Finally, the Aerospace Systems Solutions take care of the design of optimal environmental control system that makes air breathable and comfortable—in terms of pressure, temperature, flow, and humidity.

2.3.5.1 LMS Imagine.Lab Landing Gear

LMS Imagine.Lab Landing Gear delivers tools to assess the whole landing gear system: performance analysis over all its required functionalities: landing, extension, and retraction, braking and steering systems. The solution supports the multidisciplinary nature of the task (hydraulics, electrics, and thermal) and takes into account system structure/actuator coupling. It addresses the challenges posed by physical tests and maintenance costs reduction, sophisticated technologies and materials, and safety and certification requirements.

For the landing gear system and suspension, LMS Imagine.Lab Landing Gear provides a complete environment for multilevel modeling and simulation with its multiphysical domain capability. Thanks to its full integration with multibody dynamics, structural dynamics, and optimization, LMS Imagine.Lab Landing Gear allows for assessing the full aircraft ground load analysis. This makes it possible to analyze earlier in the design process in order to diminish risks and uncover problems. With LMS Imagine.Lab Landing Gear, users can develop validated Real-Time prototypes from high-fidelity simulation models using the same platform and run fewer physical drop tests.

For anti-skid systems, LMS Imagine.Lab Landing Gear helps design and validate the complete braking function with a combination of electrical, mechanical, hydraulic, and control equipment for the modeling of each single component, such as pumps and electrohydraulic valves. Engineers have access to a complete platform to study multiple technologies: steel, carbon, or ceramic brake technology and hydraulic or electric actuation. It accelerates and enhances the design and validation of these systems to ensure a good compromise between performance and risk.

LMS Imagine.Lab Landing Gear also handles the design and validation of the complete steering function with a combination of electrical, mechanical, hydraulic, and control equipment design of single components such as servocontrols and actuators, validation of component integration along with design, and validation of control strategies.

2.3.5.2   LMS Imagine.Lab Flight Controls Actuations

LMS Imagine.Lab Flight Controls supports the design and optimization of high-lift devices actuation systems by using an unparalleled multidomain system simulation approach. It easily handles the combinations of hydraulics, 2D mechanics, and electrics in a unique modeling environment.

As high-lift devices actuation systems become more and more complex, the Amesim multidomain approach is essential to understand the interactions between diverse physical domains, including electrical (elevators, ailerons, roll spoilers, tail plane trim, slats and flaps, speed brakes/lift dumpers, and trims), mechanical (rudder, tail plane trim), and hydraulic control systems. Moreover, the solution makes it possible to handle the development of various high-lift devices actuation systems (mechanical, direct drive, electromechanical, and electrohydrostatic) by integrating it into different CAE attribute analyses such as flexible multibody and stress analysis and fatigue damage prediction.

The unique Component Design libraries concept (Hydraulic, Thermal Hydraulic, and Pneumatic), in association with mechanical and electromechanical libraries can execute a detailed modeling in order to design high dynamic components such as nozzle flapper flow control servovalves, pressures valves, and any kind of piloted fluid component.

With LMS Imagine.Lab Flight Controls, one can explore various technological solutions (electrical, mechanical, and hydraulic) and assess the evaluations of their functional performance attributes (multibody, stress, and durability) in a common environment. LMS Imagine.Lab Flight Controls supports the system validation process and explores the flexible bodies and mechanisms of the complete system through its co-simulation capabilities with most common mechanical FEA software.

2.3.5.3   LMS Imagine.Lab Aerospace Engine Equipment

Aerospace Engine Equipment solution supports the design and industrialization of fuel systems (metering units, pumps, nozzle, starters, and heat exchangers) and their controls, and engine control actuators. It helps engineers in designing market-specific pressure-feed fluid systems in aircraft engines and in building reliable thermal–hydraulic systems.

Aerospace Engine Equipment solution is based on the Simcenter Amesim multidomain system simulation approach and comes with dedicated thermal and hydraulics libraries, including configurable components which are once connected together, representative for the equipment hydraulic behavior. It provides equipment designers with the ability to specify and control the external design of aerospace engine-related components and systems.

The detailed models are based on experimental results of well-known geometries and are suitable to any new geometry or functional requirement. Straightforward coupling between libraries and components make it possible to easily run behavior analysis and testing on simple components and on wholly integrated fuel systems at any project level.

The aerospace fluids database of standard and customizable components is based on experimental results and ensures the necessary accuracy and validity of models required in aerospace engine design. LMS Imagine.Lab Aerospace Engine Equipment finally helps to improve design quality, while reducing design time and the number of experimental tests needed and associated risks.

2.3.5.4 *LMS Imagine.Lab Environmental Control Systems*

Environmental Control Systems solution helps engineers to design the optimal environmental control system that makes air breathable and comfortable—in terms of pressure, temperature, flow, and humidity. Application areas include the comfort optimization of passengers and crew in vehicles such as aircrafts but also in ships, submarines, trains, and battle tanks. Environmental Control Systems is based on the Simcenter Amesim multidomain system simulation approach and handles multidisciplinary systems for advanced design: gas dynamic, thermal pneumatics, vapor cycle (2 phase flow), AC, controls, life, and environmental sciences.

Environmental Control Systems solution helps to design various systems involved in environmental control systems: bleed system control, global energy management, AC, ventilation circuit, CO_2 bottle discharge, O_2 circuit, and cabin. The solution easily handles high system complexity and takes into account multiple parameters (temperature, humidity, pressure, and change of pressure rate) in dynamics conditions (temperature and pressure variations). It helps aircraft engineers to design systems with higher efficiency, lower weight and lower volume, and significantly optimizes energy consumption.

2.3.5.5 *LMS Imagine.Lab Aircraft Engine*

Aircraft Engine solution defines the exact boundary condition for aircraft systems such as electrical power generation, bleed system, fuel system, and the flight dynamics. Using the Aircraft Engine solution, the aeronautic engineer will be able to assess the dynamic behavior of any kind of jet engine but also to get a first evaluation of the performance of the engine itself. Having access to more accurate boundary conditions for every onboard power network, he will be in a position where it will be easy for him to assess the global efficiency of the Integrated Aircraft in various conditions.

2.3.5.6 *LMS Imagine.Lab Aircraft Fuel Systems*

Aircraft Fuel Systems solution helps engineers to design the complete aircraft fuel system—in terms of pressure, temperature, flow transfer, and ullage composition tracking. Application areas include the thermal behavior of the fuel system, which is of major interest.

Aircraft Fuel Systems solution is based on the Simcenter Amesim multidomain system simulation approach and handles multidisciplinary systems for advanced fuel system design: fuel pressurization, fueling, refueling and defueling of reservoir with complex shapes, and accounting for the aircraft attitude and acceleration. The Aircraft Fuel Systems solution helps to design various systems involved in fuel systems: venting, onboard inert gas generation system (OBIGGS), fuel distribution network, and global energy management.

The solution easily handles high system complexity, accounts for multiple phenomena (wing bending and twisting) in dynamics conditions (temperature and pressure variations). It helps aircraft engineers to design systems with higher efficiency, lower weight and lower volume, and significantly optimizes energy consumption while ensuring the fulfillment of certification authorities' rules.

2.3.5.7 LMS Imagine.Lab Electrical Aircraft

Electrical Aircraft solution helps engineers to design the electrical network of any aircraft. Relying on the Electrical libraries (Electric Motors and Drives, Electromechanical, Electric Basics and Converters, and Aircraft Electrics), the Electrical Aircraft solution is the perfect solution for the aeronautic engineer to succeed in facing the new challenges of the more electrical aircraft.

The full energy balance modeling approach ensures the electrical engineer involved in a complex project for Electrical Aircraft not only to analyze the electrical networks and their components but also to start from the very beginning of the design project by accounting for the thermal integration within the complete aircraft.

The Electrical Aircraft Solution helps the aeronautic engineer in designing safer, more reliable actuators such as the steering or braking system and enables them to better analyze the impact of the network reconfiguration in case of failure. Thanks to its SIMULINK co-simulation capability, it is very suitable to also integrate the control for the generator in order to ensure proper and efficient energy feeding to the network. Postprocessing tools (such as the Fast Fourier transform or the unique linear analysis feature) are bringing the capability for an electrical engineer to a new level to really validate long before the first flight that certification test will be a success [25–30].

2.3.6 LMS Imagine.Lab Thermofluid Systems and Components

The LMS Imagine.Lab Thermofluid Systems and Components solutions come as generic packages that complement the application-oriented offer. The solutions are dedicated not only to the design of hydraulic and pneumatic components but also to complex systems and networks that can be submitted to thermal exchanges, to phase changes, or that can contain gas mixtures.

This enables to address a larger set of applications outside the traditional automotive and aerospace industries, and offer a very flexible, innovative way to design robust thermofluids systems and components.

2.3.6.1 LMS Imagine.Lab Hydraulics

The Hydraulics solution is dedicated to the design of hydraulic systems, networks, and components. This solution helps you to develop new design concepts efficiently.

Thanks to a very large collection of hydraulic components (such as pumps, valves, and actuators) and piping parts (pipes and fittings), it allows the user to find the right answers for the optimal development of complex hydraulic systems.

The solution also offers a powerful and unique tool containing the basic building blocks of any hydromechanical system. This library can be viewed as an engineering language that is able to model any hydraulic components included in automotive, aerospace, and hydraulic industries. Since the models are component based, the interpretation of the model layout is straightforward and intuitive.

The Hydraulics solution is also dedicated to the analysis of the evolution of pressure drops and flow rates in hydraulic networks. Evaluating pressure drops and the distribution of flow rates through a circuit and modifying the design of the system if needed is very easy and intuitive.

Models of hydraulic systems are numerically stiff, extremely nonlinear, and are often modeled with discontinuities. The unrivaled Simcenter Amesim solver automatically and dynamically selects the most adapted integration method based on the particular system dynamics better than any other software on the market.

2.3.6.2 *LMS Imagine.Lab Pneumatics*

The Pneumatics solution is dedicated to the design of pneumatic systems, networks, and components, including thermal exchanges. This solution helps you to develop new design concepts efficiently. Thanks to a very large collection of pneumatic components (such as compressors, valves, and actuators) but also CFD1D piping, it allows the user to find the right answers for the optimal development of complex systems.

The solution also offers a powerful and unique tool containing the basic building blocks of any pneumatic component and system. This library can be viewed as an engineering language that is able to model pneumatic components such as cylinders, compressors, oxygen regulators, dampers, LPG injectors, or any kind of pneumatic valve. Since the models are component based, interpretation of the model layout is straightforward and intuitive.

The Pneumatics solution also includes a set of components that allow you to easily model and analyze the evolution of temperatures, pressures, and mass flow rates in pneumatic networks. Based on a transient heat-transfer approach, the solution is used to model thermal phenomena in gases, and to study thermal evolution in these gases when exposed to different kinds of heat sources.

Models of pneumatic systems are numerically stiff, extremely nonlinear, and are often modeled with discontinuities. The unrivalled Simcenter Amesim solver automatically and dynamically selects the most suited integration method based on the particular system dynamics better than any other software on the market.

2.3.6.3 *LMS Imagine.Lab Gas Mixtures*

The Gas Mixtures solution offers a large set of pneumatic components and moist air models, the combination of which allows the user to model various systems using gas mixtures with up to 20 species. In pneumatic systems where the gas mixture's composition along a circuit is an issue, the use of the Gas Mixtures components becomes necessary. This solution allows the user to model phenomena/components such as Complete pneumatic circuits (pipes, pipe fitting, valves, etc.), Heat exchanges (heat exchangers, convection), and Diffusion phenomena.

The moist air models complete the solution: The condensation phenomenon can be accounted for with the associated interactions between the condensate and the gas mixture.

2.3.6.4 LMS Imagine.Lab Thermal–Hydraulics

The Thermal–Hydraulic solution is dedicated to the design of hydraulic systems and networks. This solution helps you to develop new design concepts efficiently. Thanks to a very large collection of hydraulic components (such as pumps, valves, actuators, and heat exchangers); it allows the user to find out the right answers for the optimal development of complex hydraulic systems.

The solution is also dedicated to the analysis of the evolution of pressure drops, flow rates, temperatures, and heat exchanges in hydraulic networks.

Evaluating performance in terms of heat exchanges, temperature levels, pressure drops, and distribution of flow rates through a circuit and modifying the design of the system if needed is very easy and intuitive.

Composed of models of varying complexity levels, this solution includes both an isothermal and a transient heat-transfer approach used to model thermal phenomena in liquids (energy transport, convection) and to study the thermal evolution of these liquids in hydraulic systems.

Models of hydraulic systems are numerically stiff, extremely nonlinear and are often modeled with discontinuities. The unrivaled Simcenter Amesim solver automatically and dynamically selects the most adapted integration method based on the particular system dynamics better than any other software on the market.

2.3.6.5 LMS Imagine.Lab Two-Phase Flow Systems

The Two-Phase Flow Solution enables users to model complex system networks where the fluid at some point is submitted to phase changes such as AC system, heat pump, or Rankine cycle. This solution is based on a lumped transient heat transfer approach to compute Energy transport through the system; Internal flow convective heat exchanges in single or two-phase conditions; Pressure losses, temperature levels, mass flow rate, and enthalpy flow rate distribution; Gas mass fraction evolutions in the system; Mass transfer between the vapor and liquid phase; and External flow convective heat exchanges (wall/moist air), including the influence of water vapor condensation.

The Two-Phase Flow Solution is based on a basic element approach giving the user the ability to model a maximum of two-phase flow network configurations from a minimum set of components. The Two-Phase Flow Solution allows modeling real transient behavior with compressor stopped conditions with a realistic thermodynamic behavior. A complete modeling methodology helps the user build Stop and Start applications.

2.3.6.6 LMS Imagine.Lab Mobile Hydraulics Actuation Systems

The LMS Imagine.Lab Mobile Hydraulics Actuation Systems solution helps to design fluid power actuation systems for crane, crawler, earthmoving and mining equipment, machine tools, and more. It delivers the required insights into improving product quality, robustness, and reliability, reduce power generation (variable displacement pumps, load-sensing), and to develop new functions (self-leveling, control strategies).

The LMS Imagine.Lab Mobile Hydraulics Actuation Systems provides engineers with a set of cutting-edge features and advanced simulation tools for developing products with components actuated by hydraulic fluid power systems; improving product quality with robustness and reliability; reducing power generation (variable displacement pumps, load-sensing); and developing and optimizing new functions (self-leveling, control strategies) regardless of loads and machines kinematics.

The dynamic behavior of such systems is hard to predict since every subsystem needs to be taken into account. For prototyping early in the development cycle, it is efficient to use a single simulation environment in which users can couple different modules. This results in enhanced compatibility, reduced implementation time, user-friendliness of the interface, and no programming effort.

Amesim helps the visualization of mobile hydraulics systems with AMEAnimation, a 3D visualization tool that seamlessly imports your CAD data, automatically generates a CAD view from your Amesim model, and animates simulation results.

2.3.6.7   LMS Imagine.Lab Thermofluids Systems

The Thermofluids Systems solution is dedicated to the design of thermofluids systems and networks. This solution helps you to develop new design concepts efficiently for a wide range of applications.

LMS Imagine.Lab Thermofluids Systems solution is based on the Simcenter Amesim multidomain system simulation approach and comes with a dedicated set of tools and fluid libraries handling multidisciplinary systems for advanced design: liquid and gas dynamics, gas mixture with variable composition, thermal pneumatics, two-phase flow, AC, and cooling systems.

Engineers can further analyze steady-state and/or transient behavior of complex systems and networks investigating pressures and temperatures distribution, flow balancing, and heat transfer management. The LMS Imagine.Lab Thermofluids Systems solution gives engineers the ability to easily improve existing architectures to evaluate new potential architectures and develop innovative concepts [31–34].

2.3.7   LMS Imagine.Lab Electromechanical

LMS Imagine.Lab Electromechanical helps engineers define straightforward strategies throughout the design process of electrical or electromechanical systems. Mechatronic system engineering has become a key development concern feature due to the increased use of electrical and electronic systems in automotive, aerospace, and other mechanical applications.

LMS Imagine.Lab Electromechanical simulates electromechanical components such as linear actuators and electric motors from the specification to the design and validation of control strategies. Furthermore, it supports different analysis levels of electrical systems, such as power consumption estimation and accurate transient response evaluation or thermal effects.

LMS Imagine.Lab Electromechanical gives engineers the possibility to work on detailed models for electromechanical components such as solenoids to design and validate controls laws of electric machines such as their impact on electric network, and to size automotive 14V board nets.

2.3.7.1   LMS Imagine.Lab Electric Storage Systems

Storing electrical energy is a functional requirement for many multiphysic systems. Batteries or similar electric storage devices provide a continuous electrical energy source for applications, where mobility and reliability is of uttermost importance. Moreover, as electrification is becoming a common engineering challenge to improve energy efficiency, the usage of such reversible devices is steadily increasing.

The Electric Storage Systems solution objective is to accurately evaluate the energy exchanges of such devices in variable environment conditions. Depending on those variations and the analysis to perform, the Electric Storage Systems solution enables to model electric storage systems with various complexities and features. Representing their behavior is especially interesting in multiphysics systems to specify the power and energy needs, size a pack, design a corresponding cooling subsystem, or validate related control systems.

2.3.7.2   LMS Imagine.Lab Electromechanical Components

LMS Imagine.Lab Electromechanical Components helps simulate mechatronic systems for the specification of components, the interaction with their mechanical structure, and the design and validation of control strategies. Designers of electromechanical and electrofluid components get access to an integrated platform to simulate the overall system response and take into account a combination of electrical, magnetic, mechanical, thermal, and fluid phenomena, which are accurately described within the Amesim components.

The Amesim multidisciplinary platform comes with a set of specific libraries of thermoelectric and magnetic models that help model electromechanical actuators such as solenoid valves, torque motors, and piezoelectric actuators. LMS Imagine.Lab Electromechanical Components integrates two complementary approaches: (1) description of the magnetic circuit using simple elements such as reluctances and airgaps (lump parameter) and the (2) use of FEM tables. Strong coupling capabilities between electric/ magnetic/ mechanic models enable component specification and sizing and optimization of dynamic behaviors.

Engineers are able to design multidisciplinary systems and subsystems and evaluate their behavior in function of technological choices. The platform's short computation times lead to faster analysis and refined optimization.

2.3.7.3   LMS Imagine.Lab Electrical Systems

LMS Imagine.Lab Electrical Systems focuses on the electrical power network and its components and on the global electrical power distribution for the automotive and aerospace industries and the power generation and distribution markets. It is able to simulate the interaction of the electrical network with a large diversity of electricity consumers from electrical components to complex subsystems with mechanical, fluid, and thermal aspects.

Based on the Amesim multidomain system simulation approach, LMS Imagine.Lab Electrical Systems brings a comprehensive set of electrical behavioral models for the evaluation of new architectures (sizing of power generation, storage elements, and loads) for the analysis of power consumption and for the design and validation of control laws for electrical machines.

Thanks to its advanced features, LMS Imagine.Lab Electrical Systems significantly helps customers to optimize dynamic performances, check controls, and minimizes energy consumption while studying the impact on the electrical environment (voltage drops, current peaks).

2.3.7.4   LMS Imagine.Lab Automotive Electrics

With LMS Imagine.Lab Automotive electrics, the design and optimization of automotive power networks can be performed. Either for driving comfort (electric power steering, heating systems, multimedia…) or safety (ABS, airbag…), electric and electronic systems are extensively used in vehicle; onboard electric power dedicated to these functions can reach several kilowatts and electric organ variety keeps increasing. At the same time, lowering fuel consumption and carbon dioxide emission is a continuous challenge. Electrical energy consumption is determinant as much as vehicle weight management or engine downsizing.

LMS Imagine.Lab Automotive electrics comes as a set of physics-based libraries such as Electric Basic, Electrical Motors and Thermal, and an applicative library of automotive components such as batteries, alternator, and loads. This solution helps in sizing the power network, checking the wiring sizing, validating control laws for optimization of the electric power management, and estimating the transient behavior of each component and their impact on the whole network.

It is fully integrated into the Simcenter Amesim platform, and therefore takes advantage of its flexibility and its multidomain modeling capabilities.

2.3.7.5   LMS Imagine.Lab Fuel Cells

LMS Imagine.Lab Fuel Cells helps design and optimize fuel cell stacks and systems by offering a suitable modeling environment. With LMS Imagine.Lab Fuel Cells, users can size components, optimize architecture and geometries, and develop and test control strategies. Furthermore, it helps to test different gas mixtures and material solutions, predict produced voltage, and transient evolution of temperatures, pressures, mass flow rates, and gas mass fraction everywhere in the system. LMS Imagine.Lab Fuel Cells uses physical modeling based on an energy-exchange approach between basic elements that can be assembled to represent any configuration of a complete fuel-cell system. It helps create a predictive model of a complete system, and represent both static and dynamic behavior. LMS Imagine.Lab Fuel Cells comes with multidisciplinary libraries to cover each subsystem, complete with all elements to build the complete digital mock-up of a FC system together with adapted tools to analyze and optimize it. It is able to simulate global systems and the stack (coupling between stack, cooling system, feeding system, control system, electrical system, etc.). For example, engineers can easily simulate either a polarization curve or an impedance spectrum, which makes it possible to run sensitivity analyses or parameter optimization to enhance the performance and the efficiency of the complete system or of a specific component.

LMS Imagine.Lab Fuel Cells comes with a set of analysis tools that are able to perform sensitivity analysis and obtain a better understanding of the physical behavior of the system. LMS Imagine also provides consulting services to customize existing solutions to specific needs and support our customers to get the best of our platform. From the model, it is easy to compute a polarization curve or an impedance spectrum of a stack in different operational conditions; test new architecture; test new management strategies that work on the control laws of the different subsystems; and represent break-down configuration to optimize the design; LMS Imagine.Lab Fuel Cells is of particular interest to PEMFC or SOFC system integrators, fuel cell component manufacturers, and electrochemistry research laboratories [35–39].

2.3.8   LMS Imagine.Lab Vehicle Energy Management

Under pressure from legislation and customer buying behavior, automotive OEMs and suppliers are currently devising solutions to improve CO_2 and emission levels and fuel economy. Consumers want it all; however, improved fuel economy, lower CO_2 emission levels, and cars that are fun to drive and safe and ever cheaper. Translated to vehicle engineering targets, this means that seemingly conflicting engineering challenges need to be reconciled to manufacture the cars that will determine the world's future mobility. And time is essential: being early on the market is a major success factor.

LMS proposes LMS Imagine.Lab Vehicle Energy Management solutions that remedy the problem of integrating green engineering into the current development process while continuing to improve the car's drivability and performance. The LMS Vehicle Energy Management solutions are designed to be integrated into the existing vehicle development process and combines mechanical, thermal, electrical, and control engineering in one software environment.

The LMS Vehicle Energy Management solution offers a full implementation and deployment of the Vehicle Energy Management process aimed at the customer; balancing vehicle energy performances with drivability and thermal comfort; across vehicle engineering levels: from concept to detailed engineering and from component to full vehicle; compatible with any vehicle architecture—ICE, hybrid, mild hybrid, and electric; managing the growing complexity of control and electronic systems; reduction of time to market and development costs by frontloading design decisions.

2.3.8.1   LMS Imagine.Lab Vehicle Energy Management and Thermal

The Vehicle Energy Management and Thermal package enables to assess energy management with a focus on thermal aspects. The package enables modeling, sizing, and analyzing thermal management components, subsystems, and subsystem interaction: lubrication, engine cooling, AC, thermal engines, AC and in-cabin systems, energy recovery, and more. It is possible to optimize energy flows under the hood and within the cabin, which are directly or indirectly contributing to polluting emissions, fuel consumption, engine performance, and passenger comfort.

2.3.8.2   LMS Imagine.Lab Vehicle Energy Management and Drivability

The LMS Imagine.Lab Vehicle Energy Management and Drivability package has been created to enable manufacturers and suppliers to design powertrain transmission systems focusing on comfort, performance, losses, and NVH issues. The package enables to evaluate various powertrain architectures in terms of drivability and fuel economy and then to design any kind of transmission systems (manual, automatic, CVT, DCT, etc.) and to optimize their control strategies to guarantee comfort and drivability. It also gives access to detailed analysis that enable to ensure maximum driver comfort through optimal shift transmission or noise quality.

2.3.8.3   LMS Imagine.Lab Engine Integration

The LMS Imagine.Lab Engine Integration package focuses on engine, engine component design, and engine control applications.

The package enables to define the most efficient engine architectures, to design and size engine components and actuators such as fuel injection systems, variable valve actuation systems, starters, catalytic converters, to design the air path and exhaust system, and the combustion chamber, to study the implementation of subsystems and their cross interactions, and to define engine plant models for engine controls development.

Assessing engine performance and analyzing the impact of new technologies on the engine attributes as fuel consumption, emissions, and NVH becomes accessible with the use of Amesim for system modeling and simulation. On the other hand, the package makes it easier for control engineers not only to design, test and (pre)-calibrate engine controls, but also to seamlessly generate HiL simulators though robust real-time simulation [40].

Bibliography

1. qthelp://lmsimagine.lab/ame_dir/doc/pdf/manuals/amesim_ref.pdf
2. http://www.gm.com/index.html
3. http://www.toyota.com/
4. http://corporate.ford.com/innovation/innovation-ideas-submission.html
5. https://group.renault.com/en/
6. http://www.valeo.com/en/the-group/
7. http://www.zf.com/corporate/en_de/press/list/release/release_17344.html
8. http://www.volvotrucks.com/trucks/na/en-us/products/powertrain/Pages/powertrain.aspx
9. http://voith.com/en/group/organization/group-divisions/voith-turbo-164.html
10. http://www.continental-automotive.com/www/automotive_de_en/themes/
11. http://www.delphi.com/manufacturers/auto/powertrain/diesel
12. http://www.nissan.com.japan/
13. http://www.toyota.com/
14. https://group.renault.com/en/
15. http://www.fiat.com/news
16. http://www.ifpenergiesnouvelles.com/
17. http://www.psa-peugeot-citroen.com/en
18. http://delphi.com/
19. https://www.iav.com/us/home?r=de&n=1/
20. http://www.zf.com/corporate/en_de/products/spare_parts/steering_gears_pumps
21. https://www.trwaftermarket.com/de/produkte/lenkung-und-aufhangung-/
22. http://www.doosaninfracore.com/en/main.do
23. http://www.internationaltrucks.com/trucks/
24. www.dta.com.tr/pdf/lms_amesim/LMS_ImagineLab/
25. http://www.fiat.com/
26. http://www.honda.com
27. http://www.calsonickansei.co.jp/english/
28. http://www.safranmbd.com/
29. http://www.dassault-aviation.com/fr/
30. http://www.zodiacaerospace.com/en/zodiac-hydraulics/
31. http://www.kpm-eu.com/
32. http://www.komatsu.com/
33. http://dieselturbo.man.eu/
34. http://www.panasonic.com/global/home.html
35. http://www.schneider-electric.com/ww/en/
36. http://www.borgwarner.com/en/default.aspx/
37. http://www.zodiacaerospace.com/en/zodiac-hydraulics/
38. http://www.boschrexroth.com/en/xc/
39. http://www.hilite.com/products/transmission-products/control-valves.html/
40. http://www.faw.com/product/products.jsp?pros=product_list.jsp&phight=1000&about=
    Commercial/

chapter three

Numerical simulation of the basic hydraulic components

3.1 Flow through orifices

The fluid-power systems can be studied with different tools according to the application domain. The high-level applications need a complete image of the real performances built on realistic mathematical modeling, numerical simulation, experimental identification, and extensive tests performed in real operational conditions. The authors of this book regarded the studied field in close connection with many other scientific and technical domains using multiphysics tools. Some remarkable technical achievements were tuned by simulation and test in the author's laboratory, saving a lot of time for setting different industrial applications.

At the first sight, simulation of a common technical system with Amesim is very accessible. This evident positive feature of the above-mentioned language is pointed out by many commented demos included in different sections of the completely attached documentation; this bonus can help any user to raise his professional level to a competitive one. In order to encourage these people, in this part of the book, we discussed the steady-state and dynamic behavior of the basic hydraulic components involved in flow and pressure control. The simplest fluid-power systems are finally studied.

Some own theoretical contributions are included in the analysis, pointing out the information *compressed* in the language library symbols. This important practical aspect can be revealed by simulating the same dynamics with other common languages as SIMULINK. Following the well-known methodological example of Herbert Merritt, in every analysis we have tried to obtain clear design relations rather than leave a mass of equations to tangle the reader. [1,2].

This first section is devoted to widest spread hydraulic components: *orifices*, also called *restrictions*, can be fixed or variable, and are used in *all fluid systems* for different purposes: measuring, controlling, or limiting the flow, creating a pressure drop used for controlling other devices, and so on. The common mathematical description is usually based on Bernoulli's equation and leads to the form where C_q is the flow coefficient.

$$Q = C_q A \sqrt{\frac{2}{\rho}|\Delta p|} \tag{3.1.1}$$

where:
   Q is the flow rate
   C_q is the flow coefficient
   A is the cross section of the orifice
   Δp is the differential pressure
   ρ is the density at mean pressure

However, if C_q is taken to be constant, the equation of Q against Δp has infinite gradient at the origin, which is numerically dangerous and physically unrealistic. To overcome this problem, a variable C_q is used. It is possible to work with an implicit relationship like this but an explicit formula is used in Amesim. This is provided using another dimensionless number introduced by McCloy [3] as the *flow number* and denoted by λ:

$$\lambda = \frac{D_h}{\nu} \sqrt{\frac{2|\Delta p|}{\rho}} \qquad\qquad (3.1.2)$$

where:
    D_h is the hydraulic diameter
    ν is the kinematic viscosity

The flow coefficient used in any realistic computation is

$$C_q = C_{q\,\text{max}} \; \tanh\left(\frac{2\lambda}{\lambda_{\text{crit}}}\right) \qquad\qquad (3.1.3)$$

For λ greater than λ_{crit} this gives effectively a constant value of C_q. For low λ, the value varies approximately linearly with Δp. A reasonable default value for λ_{crit} is 1000. However, for an orifice with complex geometry, it can be as low as 50. For very smooth geometry, it can be as high as 50,000. This phenomenon is shown in the following example from Figure 3.1.1. A linear pressure source supplies a fixed hydraulic orifice described by the basic Amesim model **flowcontrol01** (OR0000) from Figure 3.1.2.

Pressure source increase from 0 bar to 10 bar in 10 s as shown in Figure 3.1.3. For low flow numbers, the flow remains laminar, and the real pressure drop across the orifice is lower than the hypothetic turbulent one (Figures 3.1.4 through 3.1.7). This flow regime creates the possibility to use a linear relation between the pressure drop and the flow rate in the frame of the linear analysis of a dynamic system that includes a restrictor. The mean flow velocity (Figure 3.1.8) depends on many factors, such as pressure drop, cross-sectional area, edge roughness, access conditions, viscosity, cavitations, and many others.

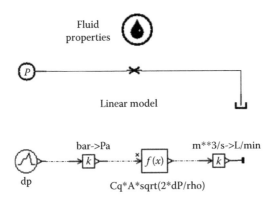

Figure 3.1.1 Flow through a fixed hydraulic orifice with transition at specified flow number.

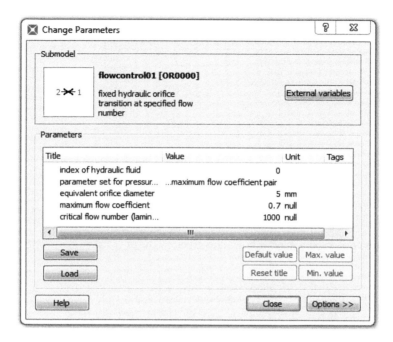

Figure 3.1.2 Parameters window for the **flowcontrol01** fixed hydraulic orifice.

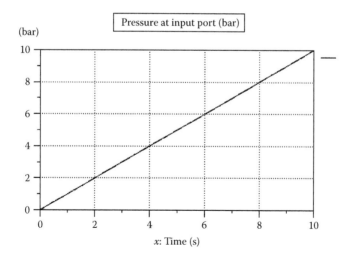

Figure 3.1.3 Pressure at the input port of the fixed hydraulic orifice.

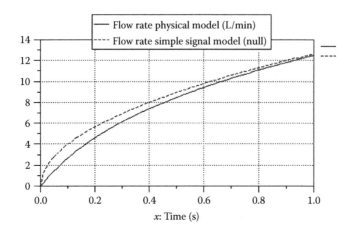

Figure 3.1.4 Comparison between the flow rates given by the physical model and the simplified one at low-pressure drop.

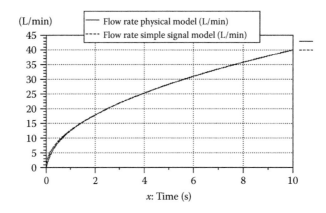

Figure 3.1.5 Flow rate variation over the entire range of simulation.

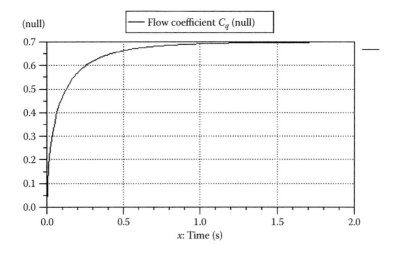

Figure 3.1.6 Flow coefficient variation for low-flow rate.

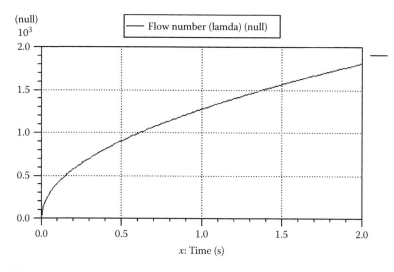

Figure 3.1.7 Flow number variation for low-flow rate.

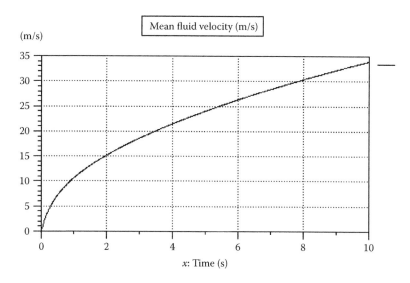

Figure 3.1.8 Mean flow velocity of the fluid in the orifice.

In practice, a lot of variable geometry orifices are used for controlling the flow according to the dynamic needs of constant pressure supply feedback systems. Such a case is studied by simulating the linear opening in 5 s of a small orifice constantly supplied at 10 bar (Figures 3.1.9 through 3.1.16).

A wide category of fluid-power systems are supplied by a constant flow rate source. A typical example is found in the field of the open center automotive-steering systems. A simple simulation network (Figure 3.1.17) reveals interesting details about such a system behavior. A pressure relief valve protects the constant flow rate supply source. The orifice is opened with a constant rate in 5 s (Figure 3.1.18), starting at $t = 1$ s. During the first second of the opening, the pressure relief valve is opened.

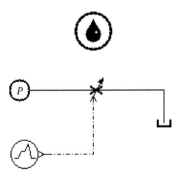

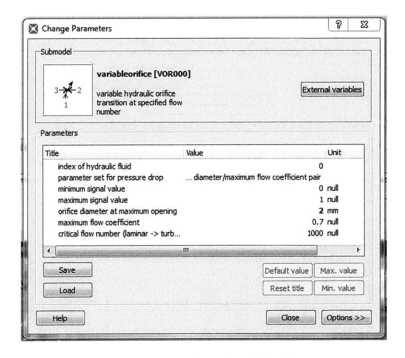

Figure 3.1.9 Simulation network for a variable orifice (VOR000) constant pressure supplied.

Figure 3.1.10 Simulation parameters for a variable orifice (VOR000).

Before the orifice opening, the source flow passes through the pressure relief valve. At the beginning of the orifice opening, the flow still partially passes through the valve. The whole process is clearly shown in Figures 3.1.19 through 3.1.26. A very short time (0.01 s) after the beginning of the opening, the flow coefficient has a linear variation; hence, the flow is a laminar one.

The main conclusion of this chapter is the importance of the assessment of the flow regime in any dynamic behavior study. The use of the components included in the hydraulic libraries of Amesim solves this problem without introducing other difficulties.

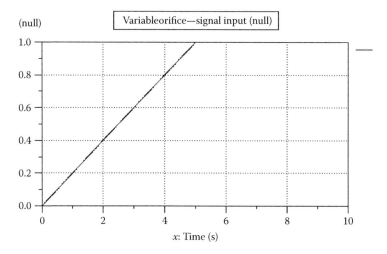

Figure 3.1.11 Signal input for the orifice area.

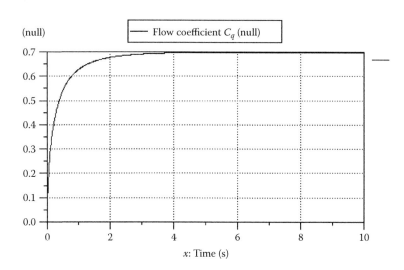

Figure 3.1.12 Flow coefficient variation in time.

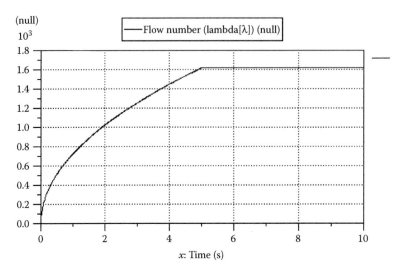

Figure 3.1.13 Flow number variation in time.

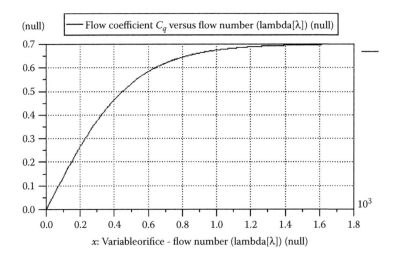

Figure 3.1.14 Relation between flow number and flow coefficient.

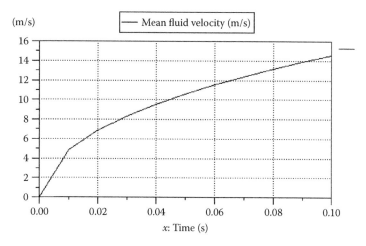

Figure 3.1.15 Mean flow velocity variation at the beginning of the opening.

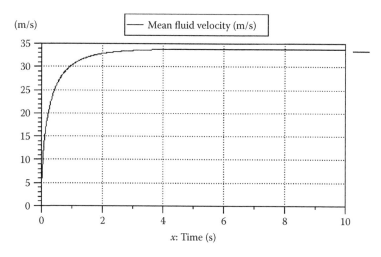

Figure 3.1.16 Mean flow velocity variation during all simulation time.

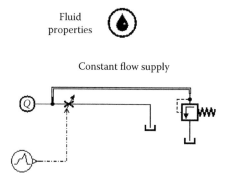

Figure 3.1.17 Simulation network for a variable orifice (VOR000-2) supplied by a constant flow rate source.

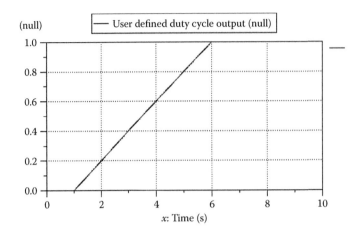

Figure 3.1.18 User defined duty input cycle (relative orifice opening).

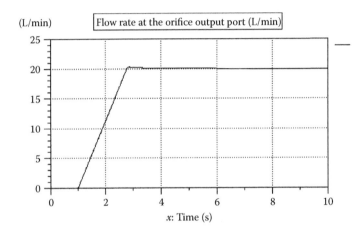

Figure 3.1.19 Flow rate at the output port of the variable orifice.

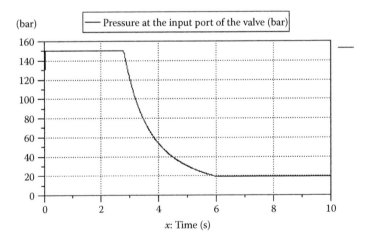

Figure 3.1.20 Pressure at the input port of the variable orifice.

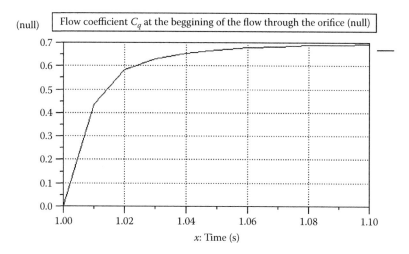

Figure 3.1.21 Flow coefficient at the beginning of the orifice opening.

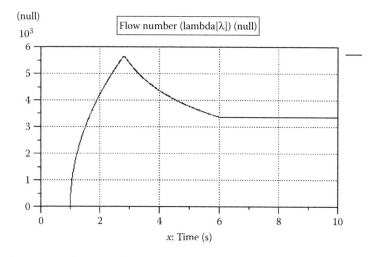

Figure 3.1.22 Variation of flow number (lambda) during the whole transient.

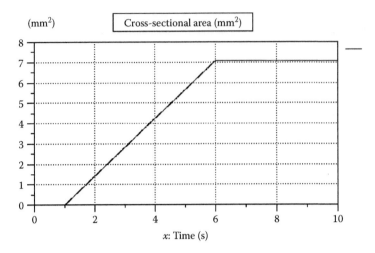

Figure 3.1.23 Cross-sectional area of the variable orifice.

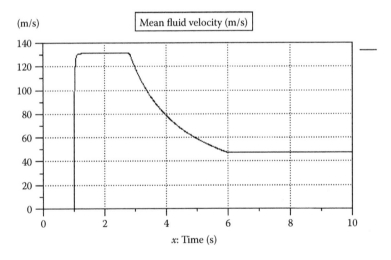

Figure 3.1.24 Mean-fluid velocity in the variable opening orifice.

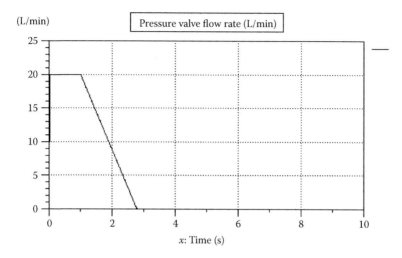

Figure 3.1.25 Flow rate at the output port of the pressure relief valve.

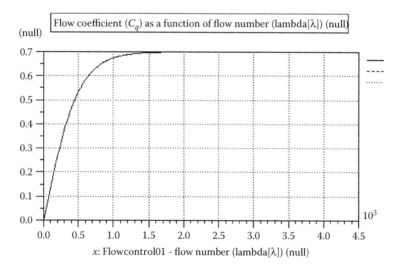

Figure 3.1.26 Flow coefficient as a function of the flow number (usual critical flow number—1000).

3.2   *Three-way flow valves*

All control valves require a supply, a return, and at least one line to the load. Consequently, valves are either three-way or four-way. The two-way valves cannot provide a reversal in the direction of flow. A three-way valve can control a hydraulic actuator with spring assistance, single shaft, and single flow port (Figure 3.2.1). A double side piston motion requires a bias pressure acting on one side of an unequal area piston for direction reversal (Figure 3.2.2). Usually the head-side area is twice the rod-side area, and supply pressure acts on the smaller area to provide the bias force for reversal. This type of control valve is very simple from technological point of view: the central land of the spool needs to match the central chamber of the sleeve within a few microns. This quality is used in small power hydraulic servo to control the servopump or servomotor displacement.

The *generic model* of a three-way flow control valve presented in Figure 3.2.3 allows the study of the flow in both the direction by the aid of a small hydraulic accumulator connecting the two load ports—A and T. The return spring parameters have a small stiffness (10 N/mm) and a symbolic pretension (1.0 N). The constant pressure supply is rated for a common pressure (100 bar).

The input signal is a triangular force of maximum 100 N applied to the spool within a period of 4 s (Figure 3.2.4).

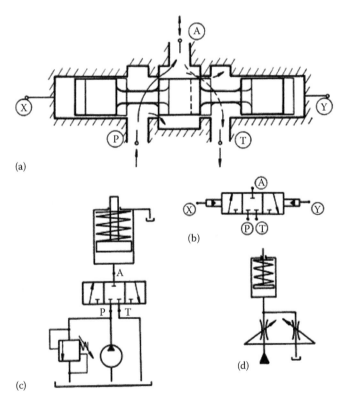

(a)

(b)

(c)

(d)

Figure 3.2.1 Hydraulically operated 3 position 3 ports spool flow valve: (a) operation principle; (b) standard symbol; (c) three-way valve controlling a hydraulic actuator with spring assistance, single shaft, and single-flow port; and (d) equivalent hydraulic diagram of a hydraulic power system, including a three-way flow with a three-port hydraulic valve control valve.

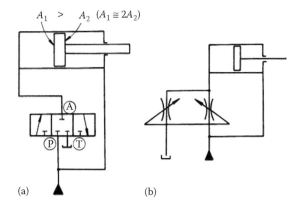

Figure 3.2.2 The control of a hydraulic actuator with single rod and double flow ports motor by a 3/3 flow valve constant pressure supplied: (a) hydraulic circuit and (b) equivalent hydraulic diagram.

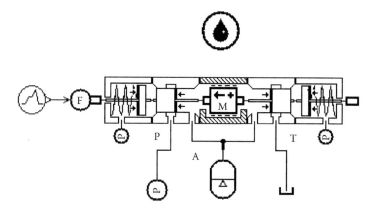

Figure 3.2.3 Simulation network for a three-flow valve.

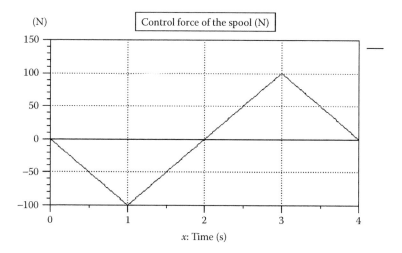

Figure 3.2.4 Control force applied to the spool.

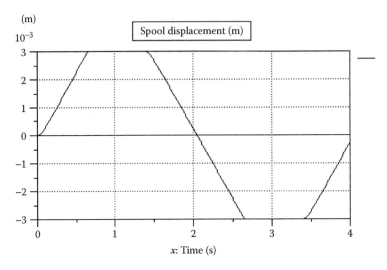

Figure 3.2.5 Spool displacement.

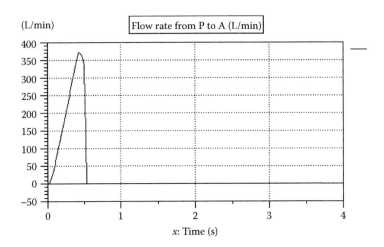

Figure 3.2.6 Flow from P to A.

The result of this input is a quasi-linear trapezoidal displacement with an amplitude of ±3 mm (Figure 3.2.5).

The flow rate sent by the constant pressure supply to the hydraulic motor port is quasi-linear (Figure 3.2.6).

The return flow rate (from A to T) is very different according to the behavior of the accumulator (Figure 3.2.7). The gas pressure is rising quickly (Figure 3.2.8) and is diminishing slowly.

The two opposite flows (Figure 3.2.9) are very different, according to the volume of gas trapped in the accumulator (Figure 3.2.10).

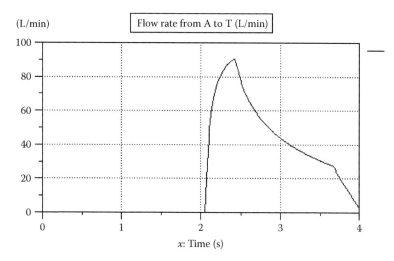

Figure 3.2.7 Flow from A to T.

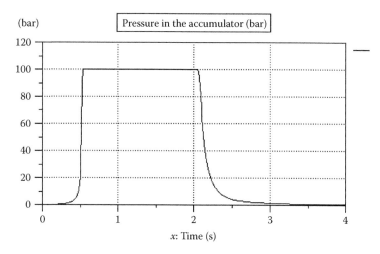

Figure 3.2.8 Pressure in the accumulator.

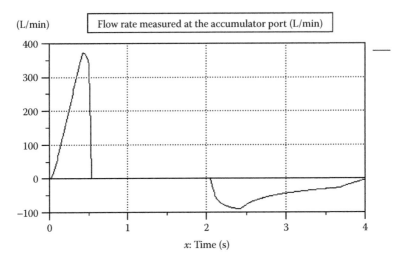

Figure 3.2.9 Flow-rate variation at the accumulator port.

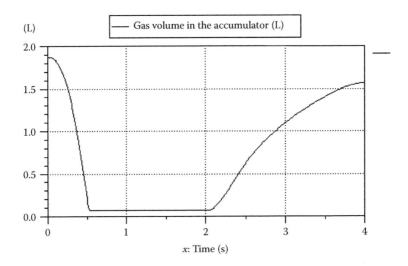

Figure 3.2.10 Gas volume in the accumulator.

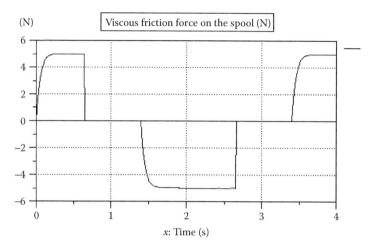

Figure 3.2.11 Viscous force on the spool.

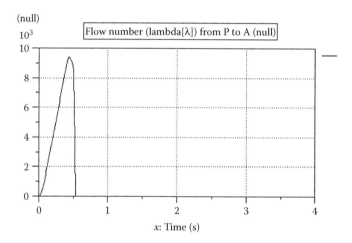

Figure 3.2.12 Flow number (lambda) from P to A.

The viscous friction on the spool cannot be neglected (Figure 3.2.11) because the coefficient of viscous friction (about 1000 Ns/m) corresponds to a normal radial clearance between the spool and the sleeve.

The flow number and the flow-coefficient variations show a turbulent motion during the whole process (Figures 3.2.12 and 3.2.13).

The mean-flow velocity in jets overcomes 100 m/s (Figure 3.2.14) leading to special technologies for the spool and the sleeve that need high geometrical stability, very sharp edges with high hardness.

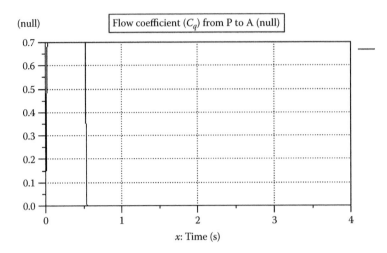

Figure 3.2.13 Flow coefficient for the way from P to A.

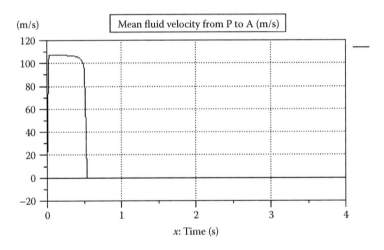

Figure 3.2.14 Mean-flow velocity from P to A.

3.3   *Four-way flow valves*

3.3.1   *The aim of the simulations*

Most of the high-performance hydraulic servosystems are controlled by four-way valves. They can be supplied by constant pressure systems or by constant flow ones. The highest dynamics needs a constant pressure supply of the valve. The constant flow supply offers minimum energy consumption with a good enough accuracy.

This chapter is devoted to the influence of the valve microgeometry on the steady-state behavior of a four-way flow control valve (Figure 3.3.1). All three types of relation between the spool lands and the sleeve grooves (Figure 3.3.2) are investigated according

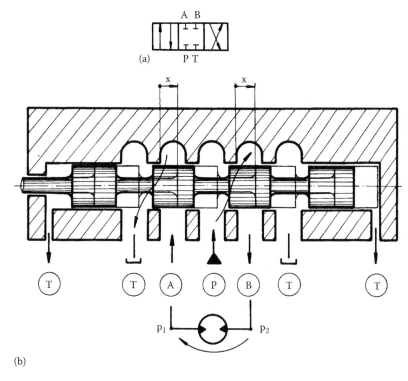

(a)

(b)

Figure 3.3.1 Externally operated 3 position 4 ports spool flow valve: (a) standard symbol and (b) four-way valve controlling a hydraulic motor.

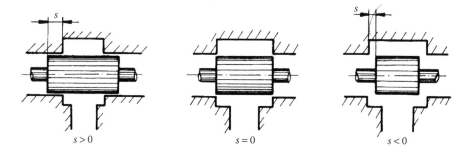

Figure 3.3.2 Geometrical definition of the overlap ($s > 0$), critical lap ($s = 0$), and underlap ($s < 0$) for a spool moving inside a sleeve.

to the hydraulic diagram of the valve presented in Figure 3.3.3. The three main hydraulic characteristic curves are (1) the steady-state flow with no load, (2) the pressure sensitivity (Figure 3.3.4), and (3) and the leakage flow curve around the null operation point (Figure 3.3.5). All the results are obtained by numerical simulation with Amesim using normal practical values for all parameters.

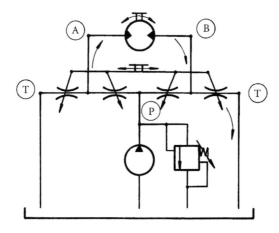

Figure 3.3.3 Hydraulic diagram of a four-way flow valve used for controlling a hydraulic motor.

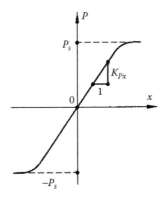

Figure 3.3.4 Flow-valve pressure sensitivity (measured with blocked load lines).

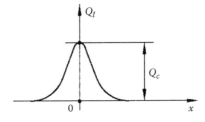

Figure 3.3.5 Leakage flow around the neutral position of the spool.

3.3.2   Critical lap case

The *generic model* of a four-way flow control valve presented in Figure 3.3.6 allows the study of the flow in both directions. The constant pressure supply is rated for a common pressure (210 bar).

The input signal is a sine wave of 1.0 mm amplitude applied to the spool within a period of 10 s (Figure 3.3.7).

The steady-state valve flow characteristic of a valve with critical lap is a linear one with a very small dead band generated by the rounded corners radius (0.005 mm) and the

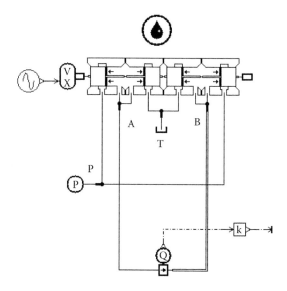

Figure 3.3.6 Simulation network for a four-way flow valve.

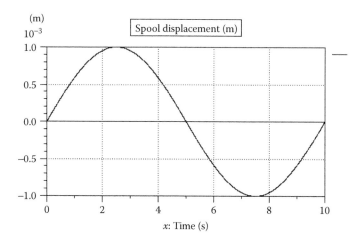

Figure 3.3.7 Spool displacement

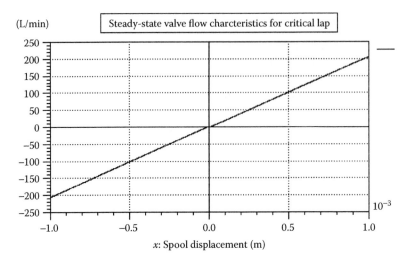

Figure 3.3.8 Steady-state flow characteristics of a four-way flow valve with critical lap.

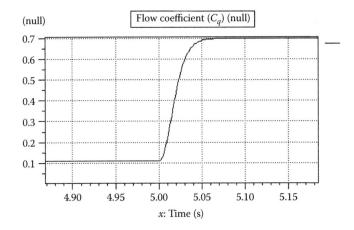

Figure 3.3.9 Flow coefficient variation for small opening.

small clearance on diameter (0.003 mm) for a 10.0 mm spool diameter (Figure 3.3.8). Other relevant parameter evolutions are presented in Figures 3.3.9 through 3.3.13.

The valve pressure sensitivity and the valve leakages when load ports are closed need the simulation model from Figure 3.3.14. The spool stroke is extended on 10% from the nominal one. The results are presented in Figures 3.3.15 through 3.3.19.

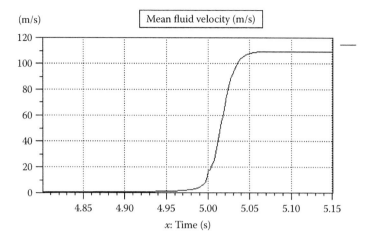

Figure 3.3.10 Mean-fluid velocity for small valve openings.

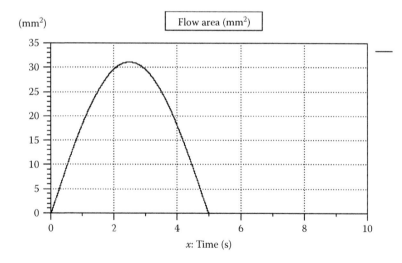

Figure 3.3.11 Flow area for a port (all range of openings).

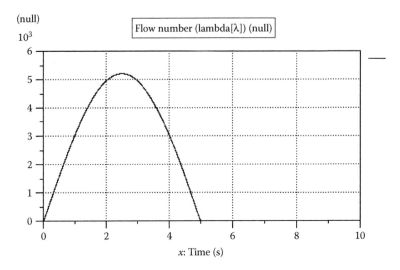

Figure 3.3.12 Flow number variation on half of a spool stroke.

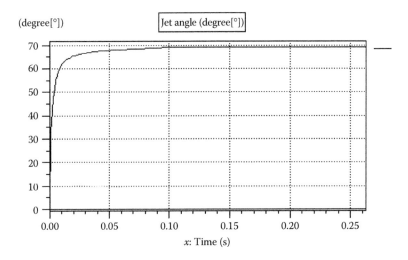

Figure 3.3.13 Jet angle variation at the beginning of the orifice opening.

3.3.3   *Positive lap spool*

A small positive lap of the spool changes the valve behavior around the null point. If we consider a 50 µm positive lap, we obtain, first of all, a larger dead zone of the steady-state flow characteristics, as shown in Figure 3.3.20. All the other quality parameters are strongly influenced by the overlap (Figures 3.3.21 and 3.3.22). For a greater overlap (50 µm), the changes are *dramatic* from the control accuracy point of view (Figures 3.3.23 through 3.3.25).

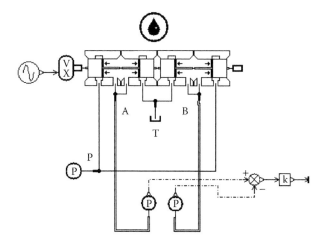

Figure 3.3.14 Simulation network for a four-way flow valve with critical lap and lock load ports.

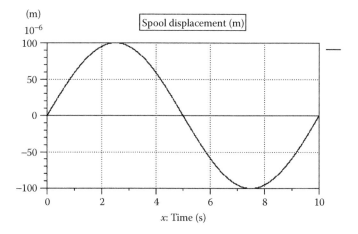

Figure 3.3.15 Spool displacement for the investigation of the lap influence on the pressure sensitivity.

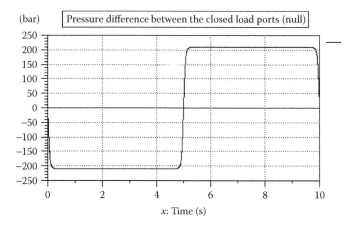

Figure 3.3.16 Pressure difference between the closed load ports when spool is moving slowly in both directions.

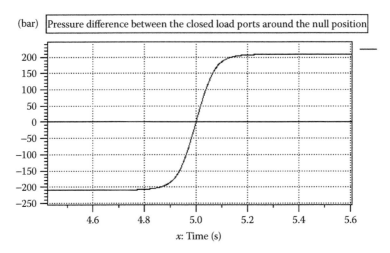

Figure 3.3.17 Pressure sensitivity of a four-way flow valve with critical lap around the null operation point.

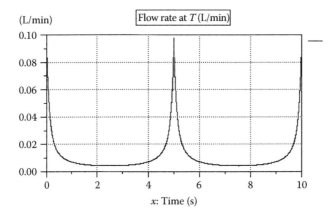

Figure 3.3.18 Overall flow leakages for all the spool stroke.

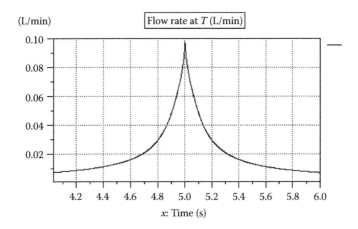

Figure 3.3.19 Overall flow leakages around the null position.

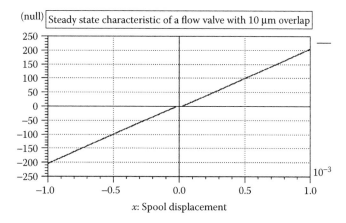

Figure 3.3.20 Steady-state characteristic of a flow valve with 10 μm overlap.

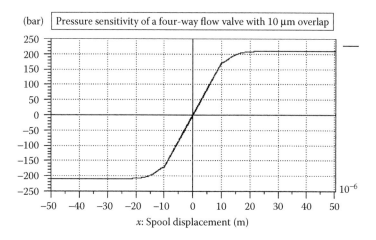

Figure 3.3.21 Pressure sensitivity of a four-way flow valve with 10 μm overlap.

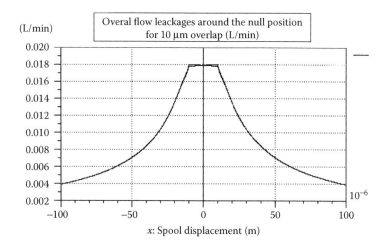

Figure 3.3.22 Overall flow leakages around the null position for an overlap of 10 μm.

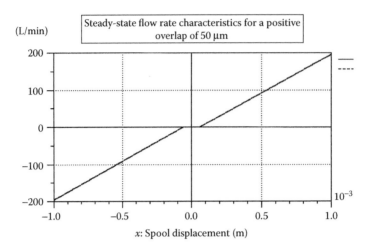

Figure 3.3.23 Steady-state characteristics of a four-way flow valve with 50 μm overlap.

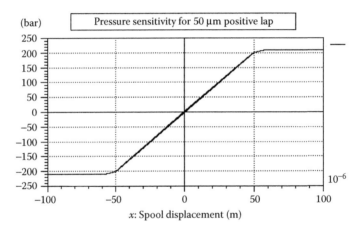

Figure 3.3.24 Pressure sensitivity for a flow valve with a spool overlap of 50 μm.

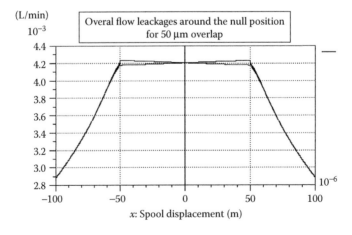

Figure 3.3.25 Overall flow leackages around the null position for a flow valve with 50 μm overlap.

3.3.4   Case of negative lap

The simulations and the experiments show that a small underlap of about 10 μm offers a straight steady-state flow characteristic (Figure 3.3.26), a good pressure sensitivity in the null region (Figure 3.3.27), and a small flow leakage (Figure 3.3.28).

The increase in the negative lap increases the slope of the flow characteristics in the null region (Figure 3.3.29), reducing the stability of any control system. The slope of the pressure sensitivity curve in the null region is reduced (Figure 3.3.30) and the leakages become too high (Figure 3.3.31).

In practice, such a small negative underlap is sometimes accepted from thermal reasons, in connection with the state of the hydraulic fluid use in flight control servomechanisms of the supersonic fighters.

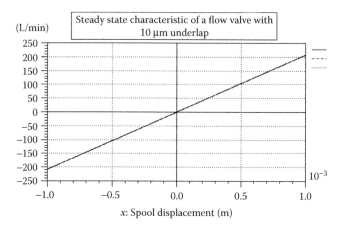

Figure 3.3.26 Steady-state characteristic of a flow valve with 10 μm underlap.

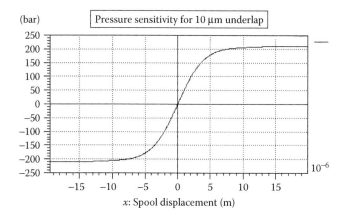

Figure 3.3.27 Pressure sensitivity of a flow valve with 10 microns underlap.

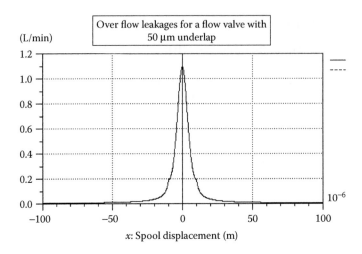

Figure 3.3.28 Overall flow leakages for a flow valve with an underlap of 10 μm.

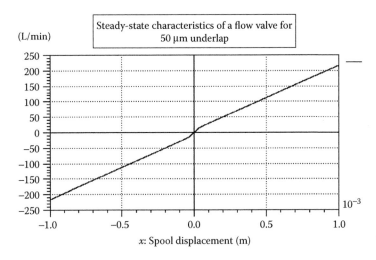

Figure 3.3.29 Steady-state characteristics of a flow valve with 50 μm underlap.

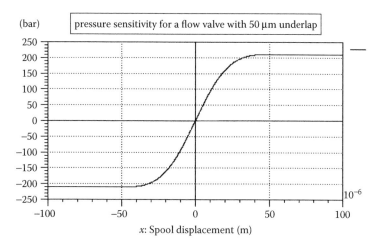

Figure 3.3.30 Pressure sensitivity for a flow valve with 50 μm underlap.

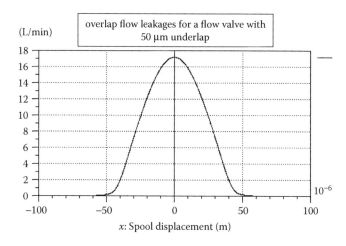

Figure 3.3.31 Overall flow leakages for a flow valve with 50 μm underlap.

3.4   Hydraulic single-stage pressure relief valves dynamics

3.4.1   Problem formulation

All the circuits of the fluid-power systems need protection against the overpressure. The generic name used in practice for the mechanical components accomplishing this function is *pressure relief valves*, although there are some types with different hydraulic diagram and design features, covering different kinds of demands for static or dynamic performances.

This chapter is devoted to the influence of the structure and the microgeometry on the steady-state and dynamic behavior of the pressure relief valves. Both single-stage relief valves and two-stage ones are investigated.

The pressure in the protected circuit is limited by the automatic opening of an orifice created between a case and a mobile obturator, which is actuated by a preloaded spring. The fluid pressure acts on the control surface of the obturator against the spring. The orifice opening and consequently the flow sent out of the circuit depend on many factors, but the most important are the control area and the pretension of the spring. The obturator can be conical, spherical, cylindrical, or plane. A common *industrial* relief valve (Figure 3.4.1) contains a plain seat connected to the input port and a very stiff helical spring pressing a conical obturator. The seat has a very short conical surface in contact with the poppet. For higher pressures, the obturator is a spherical one lying on a very short conical seat (Figure 3.4.2). The ball is slightly guided in order to avoid lateral displacements that generate strong pressure oscillations.

This introductory part presents the behavioral analysis of the single-stage pressure relief valves both in steady-state and in transient regime. The practical result of this analysis is the connection between the structural and dimensional parameters of the modern designs.

One of the main qualities of a pressure relief valve is the tightness for pressures lower than the *cracking pressure*. From this point of view, the conical poppet (Figure 3.4.3) is the best choice.

The close guidance of the cone (a) is useful for preventing the wear of the seat and the cone, but it can lock the poppet. The use of the clearance between the piston and the seat of about 60 μm as valve damper is a good compromise that reduces the manufacturing cost. The free cone can be used as *pilot stage* for small flow rates (1...4) L/min in the frame of the two-stage pressure relief or reducing valves (b). The best combination of components for one-stage relief valves is the conical poppet with a piston as damper. The slope of the steady-state characteristics of such a valve can be reduced by attaching a deflector to the conical poppet (c).

A detailed view of such of relief valve obturator shows the possibility of increasing the valve flow rate by turning the jet in the opposite direction. The hydrodynamic force on the cone is *compensated* by the other flow force, reducing the slope of the steady-state characteristic. However, as it will prove later, this gain is *paid* by reducing the valve stability.

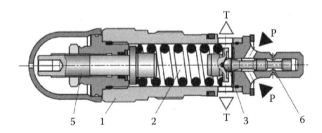

Figure 3.4.1 Single-stage compensated relief valve Type DBDS Version pressure rating 400 bar (poppet seat valve). (Source: Bosch Rexroth AG)

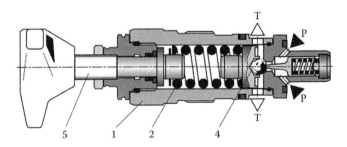

Figure 3.4.2 Single-stage compensated pressure relief valve Type DBDH 10 Version pressure rating 630 bar (ball seat valve). (Source: Bosch Rexroth AG)

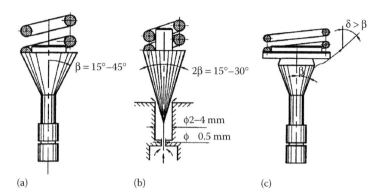

Figure 3.4.3 Common types of conical obturators: (a) with piston guide; (b) with orifice damper; (c) with deflector and piston guide.

The above-mentioned introduction prepares a complete mathematical treatment that points out the facilities of Amesim in solving complex hydromechanical problems. The practical objectives of the following analysis are the following:

1. Optimization of the steady-state behavior of the valve by the microgeometry analysis
2. Optimization of the valve dynamic behavior by the assessment of the valve damper geometry on the sudden input flow variation

3.4.2 Mathematic modeling of the valve dynamic behavior

A realistic study of a pressure relief valve behavior needs the integration of the valve in the simplest hydrostatic transmission protected by this device (Figures 3.4.4 and 3.4.5).

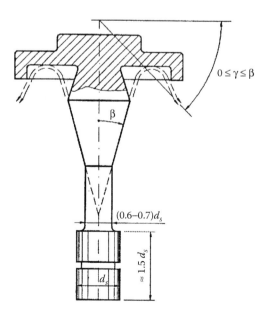

Figure 3.4.4 Poppet with flow deflector and cylindrical damper.

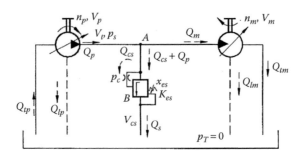

Figure 3.4.5 Hydraulic diagram of the simplest hydrostatic transmission.

A complete system dynamic analysis needs the consideration of all the components dynamics. In order to point out the behavior valve itself, for the moment the other component dynamics is neglected. The evolution of the pressure p_s in the fluid volume V_t included between pump, motor, and valve results from the continuity equation written for the node (A) that connects the three considered components:

$$Q_{tp} - Q_{lp} - Q_{tm} - Q_{lm} - Q_s - Q_{cs} = \frac{V_t}{\varepsilon_e} \cdot \dot{p}_s \qquad (3.4.1)$$

where:
   Q_{tp} is the theoretical volumetric pump flow rate
   Q_{lp} is the pump leakage volumetric flow rate
   Q_{tm} is the theoretical volumetric motor flow rate
   Q_{lm} is the motor leakage volumetric flow rate
   Q_s is the volumetric flow rate exiting through the valve orifice
   Q_{cs} is the volumetric control flow rate of the valve
   ε_e is the equivalent bulk modulus of the liquid included in the volume V_t

The evolution of the pressure from the control chamber of the valve poppet, p_{cs}, results from the continuity equation *attached* to this volume:

$$Q_{cs} - A_{cs} \cdot \dot{x}_s = \frac{V_c}{\varepsilon_e} \cdot \dot{p}_{cs} \qquad (3.4.2)$$

where:
   A_{cs} is the area of the poppet control piston surface
   $\dot{x}_s$ is the poppet speed
   V_c is the average liquid volume from the valve control chamber

The distance between the conical poppet and the seat, x_s, (valve opening) results from the poppet motion equation:

$$m_s \cdot \ddot{\vec{x}}_s = \vec{F}_{cs} + \vec{F}_{hs} + \vec{F}_{ht} + \vec{F}_e \qquad (3.4.3)$$

where:

m_s is the equivalent mass of the spring and the poppet

$\ddot{x}_s$ is the poppet acceleration

F_{cs} is the control pressure force on the damper piston attached to the poppet

F_{hs} is the hydrodynamic force on the poppet generated by the fluid-speed variation inside the valve (steady-state component)

F_{ht} is the pressure force on the poppet generated by the fluid-speed variation inside the valve (transient component)

F_e is the elastic force generated by the helical spring

A way to generate pressure variations in the hydraulic system is to consider the general case of using two-variable displacement volumetric machines. The theoretical volumetric flows of the two machines result from the linear relations:

$$Q_{tp} = n_p \cdot V_p \tag{3.4.4}$$

$$Q_{tm} = n_m \cdot V_m \tag{3.4.5}$$

where:

n_p is the pump shaft speed (variable)

V_p is the pump displacement (variable)

n_m is the motor shaft speed (variable)

V_m is the motor displacement (variable)

The dominant leakage flows of both volumetric machines, Q_{lp} and Q_{lm}, occur in laminar regime:

$$Q_{lp} = K_{lp} \cdot (p_s - p_T) \tag{3.4.6}$$

$$Q_{lm} = K_{lm} \cdot (p_s - p_T) \tag{3.4.7}$$

The constants K_{lp} and K_{lm} depend on the machine type and quality, but for the modern transmissions they are sited in a narrow range. Usually, in preliminary computations, the tank pressure can be neglected.

The flow exiting from the valve can be computed by Bernoulli's equation and the law of conservation of mass:

$$Q_s = K_s \cdot x_s \cdot \sqrt{p_s} \tag{3.4.8}$$

where K_s is the valve constant. For a conical poppet valve and small openings,

$$K_s = \pi \cdot d_s \cdot c_{ds} \cdot \sqrt{\frac{2}{\rho}} \sin\beta \tag{3.4.9}$$

where:

d_s is the valve seat diameter

c_{ds} is the discharge coefficient

β is the half of the conical poppet angle

ρ is the fluid density

The discharge coefficient widely studied by D. McCloy and D.H. Martin depends on the angle β. [3,4].

The valve control flow is generated by the pressure difference $(p_s - p_{cs})$. The modern design uses a cylindrical damper that creates an annular slot with three characteristic dimensions: (1) average diameter d_s, (2) length l_s, and (3) the radial clearance j_s. The usual value of the relative damper clearance is

$$\frac{j_s}{d_s} \cong \frac{0.06 \text{ mm}}{6 \text{ mm}} = \frac{1}{100} \tag{3.4.10}$$

The control flow is very small. This assumption, which will be proved later by an Amesim simulation, allows us to consider this flow as a laminar one (Hagen–Poiseuille plane). Usually, the damper piston is not centered, increasing the flow by 2.5 times:

$$Q_{cs} = 2.5 \cdot \frac{\pi d_s j_s^3}{12 \eta l_s} \cdot (p_s - p_{cs}) \tag{3.4.11}$$

Here, η is the dynamic viscosity of the fluid. The last relationship can be written as

$$Q_{cs} = K_{cs} \cdot (p_s - p_{cs}) \tag{3.4.12}$$

Here, the quantity

$$K_{cs} = 2.5 \cdot \frac{\pi d_s j_s^3}{12 \eta l_s} \tag{3.4.13}$$

is called *damping constant*. From a dynamic point of view, the equivalent mass of the poppet and the helical spring is

$$m_s = m_v + 0.33 m_e \tag{3.4.14}$$

where:

m_v is the poppet mass

m_e is the spring mass

The poppet is accelerated by the force

$$\vec{F}_{cs} = A_{cs} \cdot p_{cs} \cdot \vec{K} = F_{cs} \cdot \vec{K} \tag{3.4.15}$$

Here, $A_{cs} = \pi d_s^2 / 4$ is the control area of the damper and $\vec{K}$ is the versor of the axis Oz.

It is useful to start the computation of the hydrodynamic force, F_{hs} on the lateral surface of the poppet by considering a pure conical closing surface without the deflector mentioned earlier (Figure 3.4.4). In a steady-state valve operation, the momentum theorem applied to the axis-symmetric control surface S_{1-2} (Figure 3.4.6), including the domain of velocity variation takes the form

$$\rho Q_s(\vec{V}_2 - \vec{V}_1) = \vec{F}_{p1} + \vec{F}_{p2} + \vec{F}_{\ell 1-2} + \vec{F}_g \qquad (3.4.16)$$

This vector equation is projected on the poppet axis. Taking into account the axis-symmetry of the flow domain, and neglecting the pressure of the output port ($p_2 \cong p_T \cong 0$), the upper equation gives the force of the solid domain boundaries on the poppet lateral surface,

$$F_{\ell 1-2} = \rho \cdot Q_s \cdot V_2 \cdot \cos \beta \qquad (3.4.17)$$

In this relation, V_2 is the real average velocity in the contracted jet (*vena contracta*) and c_{vs} is the velocity coefficient of the valve orifice:

$$V_2 = c_{vs} \cdot \sqrt{\frac{2p_s}{\rho}} \qquad (3.4.18)$$

The relation (Equation 3.4.17) takes the simple form:

$$F_{\ell 1-2} = K_{hs} \cdot x_s \cdot p_s \qquad (3.4.19)$$

in which K_{hs} is the steady-state hydrodynamic force constant:

$$K_{hs} = 2 \cdot \pi \cdot d_s \cdot c_{ds} \cdot c_{vs} \cdot \sin \beta \cdot \cos \beta \qquad (3.4.20)$$

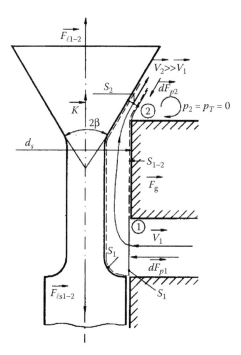

Figure 3.4.6 Control surface for computing the hydrodynamic force on the poppet.

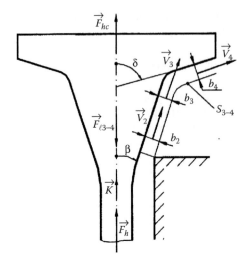

Figure 3.4.7 Compensation of the hydrodynamic force by deflecting the jet of the fluid.

The hydrodynamic force acting on the poppet $\vec{F}_{hs1-2}$ is oriented in the opposite direction as $\vec{F}_{\ell 1-2}$

$$\vec{F}_{hs1-2} = -\vec{F}_{\ell 1-2} = -K_{hs} \cdot x_s \cdot p_s \cdot \vec{K} \tag{3.4.21}$$

The above-mentioned relation points out the valve closure action of the hydrodynamic force that depends on the space—the same as the helical spring. For a constant valve input flow, the force $\vec{F}_{hs1-2}$ has to be added to the elastic one. From a steady-state point of view, this force increases the slope $\partial p_s/\partial Q_s$ of the valve characteristics. This negative effect can be compensated by a deflector attached to the poppet (Figure 3.4.7).

    The axisymmetric jet direction change creates a new hydrodynamic force that *compensates* the effect of the natural one. The magnitude of this force is found by the momentum theorem applied for the surface S_{3-4}. Usual opening pressures (100–400 bar) generate fluid velocities on the deflector of about 140–280 m/s, which increase the poppet stroke. The momentum theory becomes

$$\rho Q_s(\vec{V}_4 - \vec{V}_3) = \vec{F}_{\ell 3-4} = -\vec{F}_{hs3-4} \tag{3.4.22}$$

or

$$\rho Q_s V_2(\cos\delta - \cos\beta) = F_{\ell 3-4} \tag{3.4.23}$$

Here, δ is the deflector angle (greater than the poppet cone one). The positive hydrodynamic force becomes

$$F_{hs3-4} = \rho Q_s V_2(\cos\beta - \cos\delta) \tag{3.4.24}$$

The overall hydrodynamic force on the poppet with deflector depends on the deflector angle δ only:

$$F_{hs} = F_{hs3-4} - F_{hs1-2} = -\rho Q_s V_2 \cos\delta \tag{3.4.25}$$

Without deflector ($\delta = \beta$) relation, Equation 3.4.17 is valid. For a plate deflector, ($\delta = \pi/2$) the hydrodynamic force is null. Usually, the maximum compensation is obtained for $\delta \cong \pi/2 + \beta$, when

$$\vec{F}_{hs} \cong \rho \cdot Q_s \cdot V_2 \cdot \sin\beta \cdot \vec{K} \tag{3.4.26}$$

The transient component of the hydrodynamic force is generated by the acceleration of the fluid in the axisymmetric domain between the poppet and the seat. Usually, this component can be neglected because of the small amount of fluid from this space [5].

The valve helical spring pushes the poppet against the seat by the force:

$$\vec{F}_e = -K_{es} \cdot (x_s + x_e) \cdot \vec{K} \tag{3.4.27}$$

where:
    K_{es} is the static stiffness
    x_e is the pretension

3.4.3 Steady-state characteristics

For an uncompensated poppet, Equation 3.4.3 becomes

$$\vec{F}_{cs} - \vec{F}_{hs} + \vec{F}_e = 0 \tag{3.4.28}$$

or

$$A_{cs} \cdot p_s - K_{hs} \cdot x_s \cdot p_s = K_{es} \cdot (x_s + x_e) \tag{3.4.29}$$

This equation allows computing the valve opening as a function of the fluid pressure in the input port:

$$x_s(p_s, x_e) = \frac{A_{cs} \cdot p_s - K_e \cdot x_e}{K_{es} + K_{hs} \cdot p_s} \tag{3.4.30}$$

The opening of the valve needs the minimum pressure

$$p_{so} = \frac{K_e x_e}{A_{cs}} \tag{3.4.31}$$

Using the above-mentioned result, the relation (Equation 3.4.30) can be written in a simpler form:

$$x_s(p_s, x_e) = A_{cs} \frac{p_s - p_{so}}{K_{es} + K_{hs} \cdot p_s} \tag{3.4.32}$$

The steady-state operation of the pressure relief valve is described by the following relation:

$$Q_s(p_s, p_{so}) = K_s \cdot A_{cs} \cdot \sqrt{p_s} \cdot \frac{p_s - p_{so}}{K_{es} + K_{hs} \cdot p_s} \tag{3.4.33}$$

The graphical image of this relation is a family of curves depending on the opening pressure, which is a function of the spring pretension. Figure 3.4.8 presents the influence of the hydrodynamic force compensation for a typical pressure relief valve with the following parameters: $d_s = 6$ mm, $\beta - 15°$, and $K_e = 100,000$ N/m. Three cases were considered: (1) $\delta = \beta$, (2) $\delta = \pi/2$, and (3) $\delta = \pi/2 + \beta$. The average slope of the uncompensated valve is

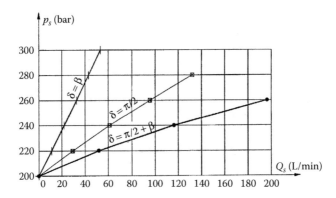

Figure 3.4.8 The steady-state valve characteristics as function of the compensation angle.

very high about 1.8 bar/(L/min). A compensation flat deflector changes the steady-state characteristics as follows:

$$Q_s(p_s, p_{so}) = \frac{K_s \cdot A_s}{K_{es}} \cdot \sqrt{p_s} \cdot (p_s - p_{so}) \tag{3.4.34}$$

and the average slope of the steady-state characteristics reduces about three times:

$$\frac{\Delta p_s}{\Delta Q_s} \cong 0.6 \frac{\text{bar}}{\text{L/min}}$$

A sharp edge deflector as is shown in Figure 3.4.4, turns the jet by 15° only to reduce the slope of the characteristics by six times.

3.4.4 Sizing the valve damper

The optimal value of the radial clearance of the valve damper can be obtained by using the motion equation of the poppet and the continuity equation (mass conservation) written for the control chamber of the valve. We denote $F_e = K_{es} \cdot x_e$ as the reference elastic force developed by the spring, the poppet motion equation becomes

$$m_s \ddot{x}_s = A_{cs} p_s - K_{hs} x_s p_s - K_{es} x_s - F_e \tag{3.4.35}$$

The minimum volume of the fluid from the control chamber is about 0.1 cm³. It can be neglected in relation to the overall fluid volume included between the pump, motor, and valve, which is normally about 1000 cm³. So, we can neglect the compressibility of the fluid from the control chamber. The continuity equation for the control chamber becomes

$$K_{cs}(p_s - p_{cs}) \cong A_{cs} \dot{x}_s \tag{3.4.36}$$

or

$$p_{cs} = p_s - \frac{A_{cs}}{K_{cs}} \cdot \dot{x}_s \tag{3.4.37}$$

Introducing this pressure in the poppet motion equation we obtain

$$m_s \ddot{x}_s = A_{cs} p_s - \frac{A_{cs}^2}{K_{cs}} \dot{x}_s - K_{hs} x_s p_s - K_{es} x_s - F_e \tag{3.4.38}$$

We can now study the relation between the variations of the reference elastic force,

$$\Delta F_e = F_e - F_{eo} \tag{3.4.39}$$

and the variations of the poppet position,

$$\Delta x_s = x_s - x_{so} \tag{3.4.40}$$

considering the pressure from the input valve port as a constant

$$p_s = p_{so} = ct \tag{3.4.41}$$

The poppet motion equation becomes

$$m_s \frac{d^2(\Delta x_s)}{dt^2} = A_{cs} p_{so} - \frac{A_{cs}^2}{K_{cs}} \frac{d(\Delta x_s)}{dt} - K_{hs} p_{so} x_{so} - K_{hs} p_{so} \Delta x_s - $$
$$-K_e x_{so} - K_e \Delta x_s - F_{eo} - \Delta F_e \tag{3.4.42}$$

The static balance of the poppet in any steady-state regime becomes

$$A_{cs} p_{so} - K_{hs} p_{so} x_{so} - K_e x_{so} - F_{eo} = 0 \tag{3.4.43}$$

By eliminating the static regime from Equation 3.4.42, we obtain

$$m_s \frac{d^2(\Delta x_s)}{dt^2} + \frac{A_{cs}^2}{K_{cs}} \frac{d(\Delta x_s)}{dt} + (K_e + K_{hs} p_{so}) \Delta x_s = -\Delta F_e \tag{3.4.44}$$

The Laplace transform applied to this equation gives

$$s^2 m_s \Delta x_s + s \frac{A_{cs}^2}{K_{cs}} \Delta x_s + (K_e + K_{hs} p_{so}) \Delta x_s = -\Delta F_2 \tag{3.4.45}$$

It is useful to introduce the quantity

$$K_{et} = K_e + K_{hs} p_{so} \tag{3.4.46}$$

which can be called *the valve overall elastic constant*. We can now obtain the transfer function of the poppet position loop:

$$\frac{\Delta x_s}{\Delta F_e} = -\frac{1}{s^2 m_s + s \frac{A_{cs}^2}{K_{cs}} + K_{et}} \tag{3.4.47}$$

which can be written in the form

$$\frac{\Delta x_s}{\Delta F_e} = -\frac{1}{K_{et}} \cdot \frac{1}{\dfrac{s^2}{\dfrac{K_{et}}{m_s}} + s \cdot \dfrac{A_{cs}}{K_{cs} \cdot K_{et}}} + 1 \qquad (3.4.48)$$

We now introduce the following fundamentals quantities:

- Natural valve pulsation,

$$\omega_h = \sqrt{\frac{K_{et}}{m_s}} \qquad (3.4.49)$$

- Damping factor of the position loop,

$$\zeta = \frac{A_{cs}^2}{2K_{cs}} \cdot \frac{1}{\sqrt{K_{et} \cdot m_s}} \qquad (3.4.50)$$

Equation 3.4.48 takes the canonic form of a second-order element:

$$\frac{\Delta x_s}{\Delta F_e} = -\frac{1}{K_{et}} \cdot \frac{1}{\dfrac{s^2}{\omega_h^2} + 2\dfrac{\zeta}{\omega_h}s + 1} \qquad (3.4.51)$$

This function describes only the dynamics of the valve position control loop. The optimization of this loop is not enough to obtain the best valve dynamics, but it can be used to size the damper in order to obtain an optimal value for the damping factor ($\zeta_{opt} = 0.7$). From the relation (Equation 3.4.50), we can obtain the necessary value for the damper constant,

$$(K_{cs})_{necesar} = \frac{\pi^2 d_s^4}{32\zeta_{opt}} \cdot \frac{1}{\sqrt{m_s \cdot K_{et}}} \qquad (3.4.52)$$

If the piston of the damper is completely eccentric, the annular damper optimal constant becomes

$$(K_{cs})_{necesar} = (K_{cs})_{efectiv} = 2,5\frac{\pi d_s j_s^3}{12\eta l_s} \qquad (3.4.53)$$

From the last two relations, we can obtain the sizing computation formula for the valve damper:

$$j_s = d_s \cdot \sqrt[3]{\frac{0,6731\eta l_s}{\sqrt{m_s K_{et}}}} \qquad (3.4.54)$$

This result is valid for all the modern relief valves. As a typical example, the pressure relief valve Bosch—dBV10 can be mentioned, which has the following design parameters: $d_s = 6$ mm; $l_s = 9$ mm; $m_v \cong 20$ g; $m_e \cong 90$ g; $K_e = 100,000$ N/m; $\beta = 15°$; $c_{ds} = 0.8$; $c_{vs} = 0.98$; and nominal pressure: $p_{sn} = 350$ bar. We consider a mineral oil as a working fluid with $\eta = 22.9$ Ns/m² at 50°C. The compensation gives $K_{hs} = 7{,}389 \cdot 10^{-3}$ m and $K_{et} = 358{,}615$ N/m. In this case, the normal damper clearance is $j_s = 60$ μm. This value corresponds exactly with the real clearance, and excludes the possibility of locking the poppet. Moreover, it is easy to obtain a tolerance of 1 μm by grinding the whole poppet.

The utility of the annular damper can be proved by computing the diameter of a sharp-edge orifice accomplishing the main task. According to Wuest [2],

$$K_{cs} \cong \frac{\pi d_0^3}{50.4\eta} \tag{3.4.55}$$

The orifice diameter needed for the same damping factor is

$$d_0 = d_s \cdot \sqrt[3]{4,94 \cdot \frac{\eta \cdot d_s}{\zeta_{opt}\sqrt{m_s \cdot K_{et}}}} \tag{3.4.56}$$

For $\zeta = 0.7$, we need a very small orifice: $d_0 = 0.116$ mm. This value is not suited for the industrial valves. The same problem occurs if we replace the radial clearance by a short tube having a diameter d and the length l. In this case,

$$K_{cs} \cong 2.5 \frac{\pi d^4}{128\eta l} \tag{3.4.57}$$

and

$$d = d_s \sqrt{\frac{4\pi\eta d_s}{\zeta_{opt}\sqrt{m_s K_{et}}} \cdot \frac{l}{d}} \tag{3.4.58}$$

For a normal value of the ratio $l/d = 10$, we obtain $d = 0.34$ mm. This small short tube needs a clean fluid and a good technology.

3.4.5   Numerical simulation of the valve dynamics by SIMULINK

The numerical simulations were performed in this case by SIMULINK in order to point out the manner of obtaining detailed results about all the internal variable quantities variations. The simulation was performed for a family of conical poppet pressure relief valves with different degrees of compensation of the hydrodynamic forces and different degrees of damping.

The numerical computations were performed for the following parameters: $d_s = 6$ mm; $K_s = 100,000$ N/m; $\beta = 15°$; $\rho = 900$ kg/m³; $p_{so} = 300$ bar; $j = 6 \cdot 10^{-5}$ m; $l = 9 \cdot 10^{-3}$ m; $\eta = 22.9$ Ns/m².

Three cases of compensation of the hydrodynamic forces were studied: δ = β (uncompensated); δ = π/2 (compensated); and δ = π/2 + β (overcompensated). Figure 3.4.9 presents the steady-state characteristics of the three types studied.

The overall simulation network of the valve together with the *internal* networks is presented in Figures 3.4.10 through 3.4.13.

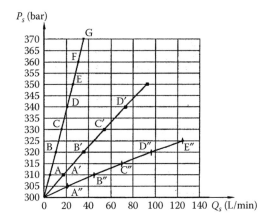

Figure 3.4.9 The steady-state characteristics for uncompensated (A…G), compensated (A′…E′), and overcompensated (A″…E″) pressure relief valves.

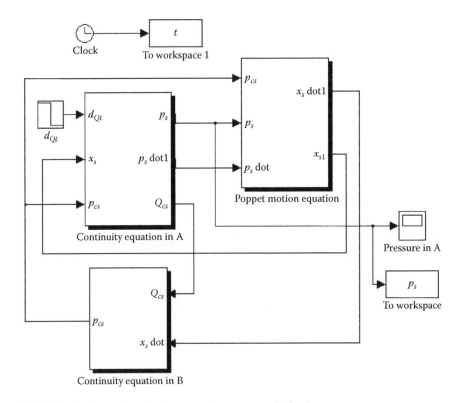

Figure 3.4.10 Simulation network of a poppet pressure relief valve.

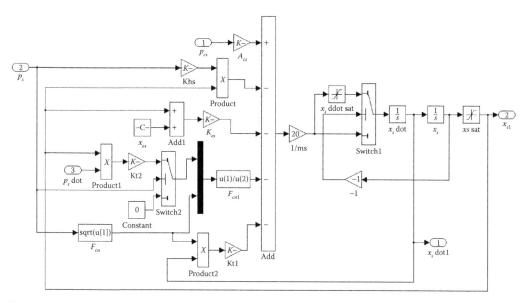

Figure 3.4.11 Simulation network of the poppet motion equation.

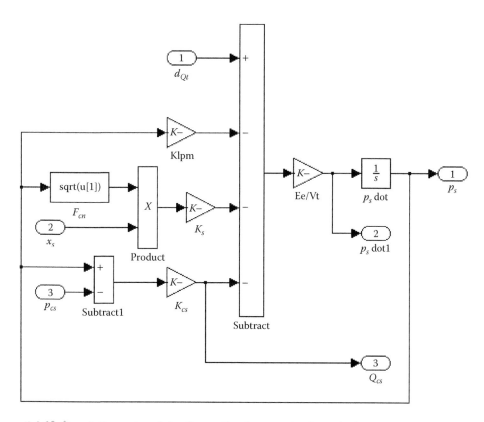

Figure 3.4.12 Simulation network for the continuity equation in node A.

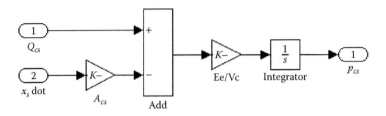

Figure 3.4.13 Simulation network of the continuity equation for the valve control chamber.

The deep understanding of the influence of different geometrical, mechanical, or hydraulic parameters involved in the problem needs a big amount of simulation. Only some of them are presented in Figures 3.4.14 through 3.4.18 in order to prove the following assumptions: Three parameters are dominating the valve behavior: (1) damping degree, depending on the piston radial clearance, (2) liquid volume protected by the valve against overpressures, and (3) the degree of hydrodynamic force compensation according to the poppet deflector shape.

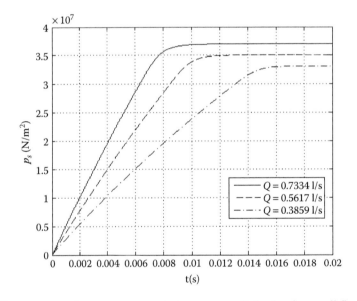

Figure 3.4.14 The response of an uncompensated pressure relief valve for small flow step inputs.

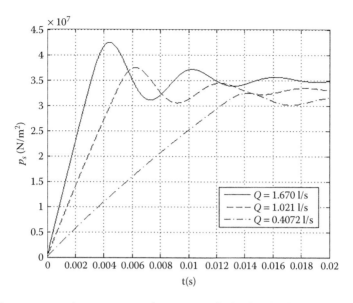

Figure 3.4.15 The response of a compensated pressure relief valve for different flow step inputs.

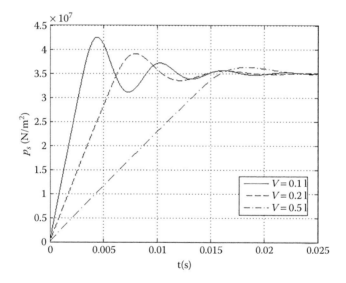

Figure 3.4.16 The response of a compensated pressure relief valve for different fluid volume between pump and motor.

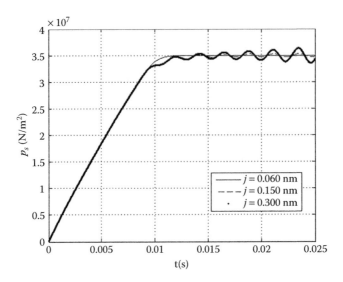

Figure 3.4.17 The influence of the damper clearance on the flow step input response of an uncompensated pressure relief valve.

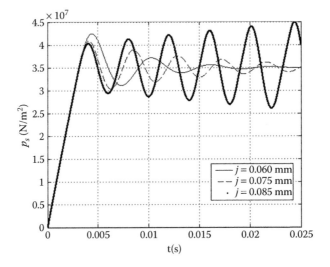

Figure 3.4.18 The influence of the damper clearance on the flow input response of an overcompensated pressure relief valve.

3.4.6 Conclusion

The following conclusions are useful for any kind of activity in fluid power systems:

1. For small flow inputs, the uncompensated pressure relief valve can be described by a first-order element with a well-predictable time constant.
2. All the compensated pressure relief valves are described by second-order transfer functions, even for small flow signals.

3. The damper radial clearance has the most important influence on the overshot and response time of any pressure relief valve. The clearance increase generates the valve instability for any kind of compensation.
4. The volume of the liquid protected by the valve against the overpressure and the fluid bulk modulus of compressibility have a great influence on the valve dynamics. A flexible hose can essentially cut the pressure shocks.

The above-mentioned analysis point out the utility of SIMULINK [12] or other equivalent languages [13] for the simulation of the nonlinear fluid-power components. At the same time, the synthesis of large control systems needs *a big amount of programming work*, leading to the need of a higher level of description of the components of different systems (electric, mechanic, hydraulic, hybrid, etc.). This way, the analysis and the synthesis time can be strongly reduced. The next chapter presents such a vision by the same study for the pressure relief valves by the aid of Amesim.

3.5   *Simulation of a pressure relief valve by Amesim*

3.5.1   *Building the simulation model*

Amesim software includes a complete library devoted to *Hydraulic Component Design*. This software component is a powerful design tool, including the basic building blocks of any hydromechanical system (Figure 3.5.1). This library can be viewed as an engineering language that is able to model any hydraulic components included in automotive, aerospace mobile hydraulic industries. All the models are validated by experiments and are updated according to the last researches developed in the field. This chapter shows how easy it is to optimize the structure and the design parameters of a pressure relief valve without performing a hydromechanical deep study by integrating the set of differential and algebraic equations that describes the valve dynamics.

The hydromechanical structure of the valve is the same as in previous chapter (Figure 3.5.2). It contains a pretension spring, a conical poppet connected by a rod with a cylindrical damper based on a small radial clearance that is precisely controlled. The valve is supplied with oil by a long rigid pipe, which contains a medium amount of fluid in order to reduce the natural overshot generated by a flow input step. The main numerical parameters used to find the dynamic behavior of the valve are the following:

- Diameter of poppet: 15 mm
- Diameter of hole: 6 mm
- Diameter of rod: 3 mm
- Poppet half angle: 15°
- Maximum flow coefficient: 0.7
- Critical flow number: 100
- Spring stiffness: 100 N/mm
- Spring force at zero compression: 400 N
- External piston diameter: 6 mm
- Clearance on diameter: 0.06 mm
- Eccentricity between piston and sleeve: 0
- Contact length: 9 mm, constant
- Equivalent mass of the poppet, rod, damper piston, and 30% of the spring: 0.05 kg
- Coefficient of viscous friction: 5 N/(m/s)

- Spring piston diameter: 15 mm
- Length of the spring chamber: 30 mm
- Volume of oil between the flow rate source and the valve input port: 1000 cm^3
- Nominal flow of the oil supply system: 60 L/min
- Specified oil bulk modulus: 17,000 bar
- Fluid density: 850 kg/m^3
- Fluid viscosity: 50 kg/(m/s)

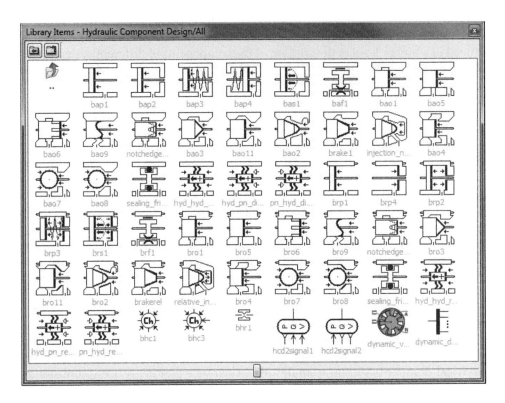

Figure 3.5.1 Amesim Hydraulic Component Design content.

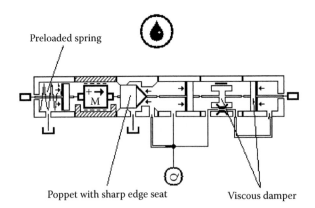

Figure 3.5.2 Simulation network for a conical poppet valve with cylinder damper.

3.5.2   Running the simulation

The low flow rate at the output port of the valve generated by a step input flow rate of 60 L/min is presented in Figure 3.5.3. The small flow rate overshot is generated by the elasticity of the fluid that is included in the valve supplied pipe.

The pressure evolution during the transient has a normal aspect for a second-order hydromechanical system, including mass, spring, and pure viscous damper (Figure 3.5.4).

The poppet lift has a small overshot; the overall aspect corresponds to a three-order system (Figure 3.5.5).

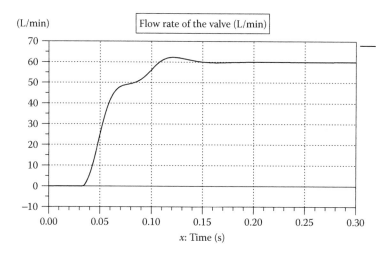

Figure 3.5.3 Flow rate at the output port of the valve generated by a step input flow rate.

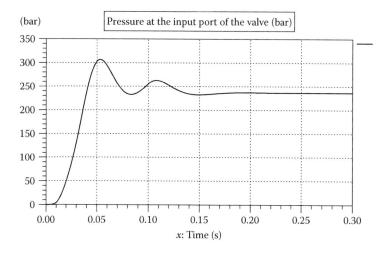

Figure 3.5.4 Pressure evolution in the input port of the valve generated by a flow rate step input.

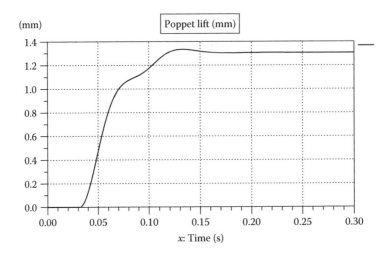

Figure 3.5.5 Poppet lift evolution generated by a flow rate step input.

The pressure force on the control piston (the end surface of the cylindrical damper) has two stages of variation (Figures 3.5.6 and 3.5.7): (1) at the beginning of the flow rate increase, the pressure force increases to the spring pretension force (400N) and (2) in the second stage, the combination of the spring force with that of the hydrodynamic one leads to a third-order type behavior. The overshot has small value due to the fluid volume compressed between the flow source and the valve (1000 cm^3).

The spring force on the poppet starts from the pretension force and trends to the steady-state value in a similar manner with the force on the control piston of the damper.

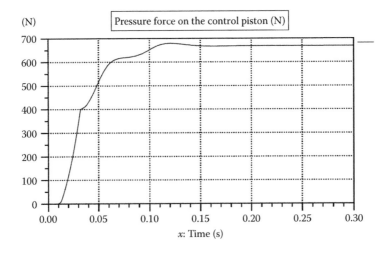

Figure 3.5.6 Pressure force on the control piston of the damper.

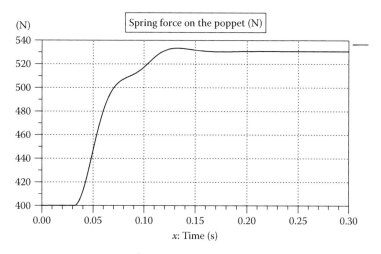

Figure 3.5.7 Spring force on the poppet after a flow rate step input.

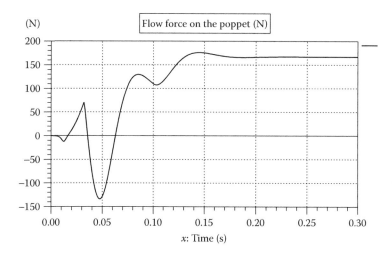

Figure 3.5.8 Flow force on the poppet (excluding the force on the control surface).

The flow force on the active surface of the poppet has a very special variation: a strong negative value at the beginning of the poppet motion (Figure 3.5.8) followed by rapid increase to the normal value corresponding to the low pressure on the cone in the acceleration region of the stream tube going out to the valve.

The fluid speed in the *vena contracta* reaches a normal value for the pressure drop on the valve: 165 m/s (Figure 3.5.9). This value corresponds to a normal flow coefficient of 0.7 and a normal contraction coefficient of 0.98. According to the numerical simulations with CFD methods, the value of the flow coefficient is lower than the one found by experiments using a cone with free exit of the fluid. The exit of the jet is guided by complex solid boundaries lowering the flow rate and the flow coefficient.

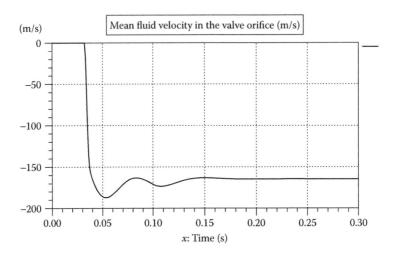

Figure 3.5.9 Mean-fluid velocity in *vena contracta* of the valve annular orifice.

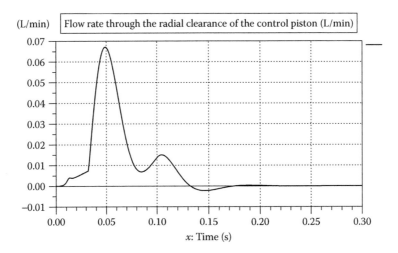

Figure 3.5.10 Flow rate through the radial clearance of the control piston during the transient.

The flow rate through the radial clearance of the control piston (Figure 3.5.10) is a laminar one because of a small maximum fluid velocity in the gap (about maximum 2.5 m/s). Reynolds number in the annular gap of the damping piston remains under 10.

The poppet velocity relative to seat remains small enough to preserve the good contact between the poppet and the seat for a long time (Figure 3.5.11).

The friction force on the control piston reaches small values, but great enough to keep the valve stability (Figure 3.5.12).

The nominal value of the flow coefficient (0.7) is reached quickly after the start of the flow through the valve (Figure 3.5.13).

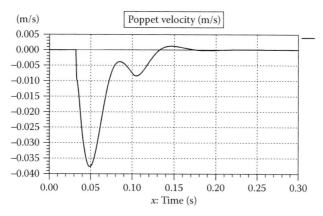

Figure 3.5.11 Poppet velocity relative to seat.

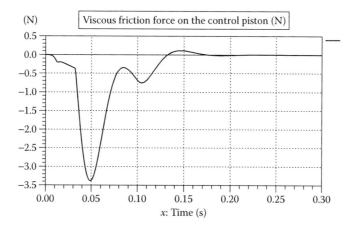

Figure 3.5.12 Viscous friction force on the control piston.

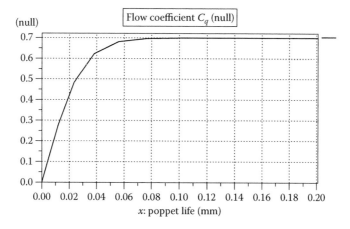

Figure 3.5.13 Flow coefficient variation.

3.5.3   Conclusion

The above-mentioned study shows the facility offered by the Hydraulic Component Design Library in the field of the hydromechanical components and systems. The time saving is achieved without reducing the result's accuracy.

For common studies of the complex hydraulic systems, the best solution is to use the component *Hydraulic relief valve* included in the Hydraulic Library under the name RV000.

3.6   Simulation of the two-stages pressure relief valves

3.6.1   The structure of the two-stages pressure relief valves

This section is devoted to a wide category of pressure relief valves used for limiting the pressure in the hydraulic system with nominal flow higher than 120 L/min. This limit comes from the size of the reference spring. The nominal flow of a single-stage valve can be improved by increasing the size of the main orifice closed by the poppet. This option leads to oversized spring, which cannot be accepted in the common fluid-power technology. However, this measure is widely used in many other fields such as steam generators and water supply systems. The practical solution used for the safety of the fluid-power systems is the combination of a small pressure relief valve (called *pilot stage*) having a high natural frequency with a much bigger valve called *power stage* and having a small natural frequency (Figures 3.6.1 and 3.6.2). If the pressure from the input of the valve opens the pilot, its small flow passes through a control sharp edge orifice sited in the spool of the power stage and pushes the spool against a small stiffness spring generating a big flow area to the tank. The steady-state characteristics of the valve have two slopes: (1) one—very high, corresponding to the pilot and (2) the second—very

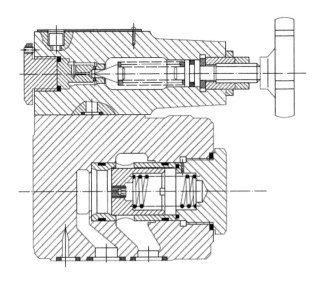

Figure 3.6.1 Two-stages pressure relief valve with conical poppet pilot stage. (Source: Bosch Rexroth AG)

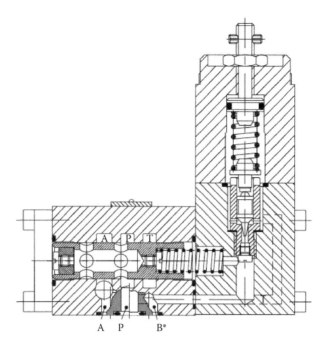

Figure 3.6.2 Two stages pressure relief valve with conical poppet pilot stage and spool power stage. (Source: Parker Hannifin.)

small, corresponding to the power stage large diameter. The main problem of this kind of valves is the stability. Usually, a damping orifice is inserted between the two stages, but the combination of all the calibrated sharp orifices is a delicate problem. In this case, an extended numerical simulation of all the operating regimes generated by different system components is very useful. The strangest thing connected to this kind of problem is the real probability to have some undamped frequency during the operation outside the basic regime.

Most of the manufacturers of the industrial valves prefer the conical unguided small pilot valves due to the small flow gain corresponding to the very small cone angle (Figure 3.6.2). However, a small tilting of the cone opens the orifice at the level needed to leave a small flow rate but great enough to open the power stage (0.5–2.5 L/min). In this case, the power loss can be significant, affecting the overall system efficiency. The problem of the cone stability can be solved as in the case of proportional two-stage pressure relief valve (Figure 3.6.3) by guiding the cone with small round surfaces. Other valve designers are using a spherical ball valve pressed with a shaped piston in a small cone by a very stiff spring (Figures 3.6.4 and 3.6.5). The small price of this solution is compensated by the well-known lateral instability of the ball versus the seat even for small opening of about 20–60 μm. The fine tuning of the damper of such a valve is delicate demanding a high manufacturing precision of all the moving components.

The cartridge configuration of a two-stage pressure relief valve is widely used in mobile applications, where they protect all the hydrostatic transmissions, the steering systems, and so on, and in complex hydraulic control blocks of the industrial equipment such as press, rollers, mills, and so on.

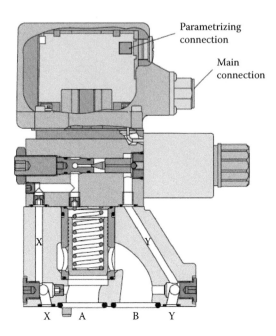

Figure 3.6.3 Two-stage pressure relief valve with proportional conical pilot and integrated amplifier. (Source: Parker Hannifin.)

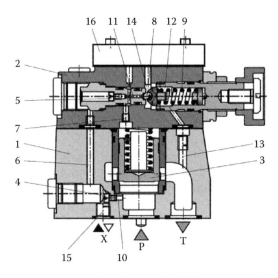

Figure 3.6.4 Two-stage pressure relief valve with ball poppet pilot stage. (Source: Bosch Rexroth AG.)

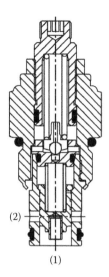

Figure 3.6.5 Two-stage cartridge pressure relief valve with conical ball pilot stage. (Source: Parker Hannifin.)

3.6.2   Simulation of a typical piloted pressure relief valve with conical seat and conical pilot

A deep knowledge of the static and dynamic behavior of a two-stage pressure relief valve needs the complete numerical simulation of the valve itself, and of the valve in the future typical systems. This study begins with the simulation of the dynamics of the typical valve from Figure 3.6.1. The valve is supplied with oil by a rigid pipe, which contains a medium amount of fluid in order to reduce the natural overshot generated by a flow input step. The corresponding Amesim model is presented in Figure 3.6.6.

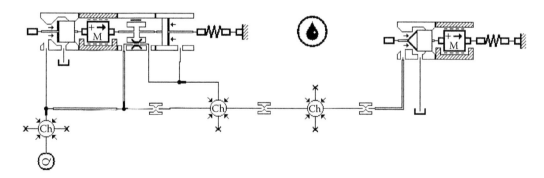

Figure 3.6.6 Simulation network for a two-stage pressure relief valve with conical seat.

The main typical numerical parameters considered to find the dynamic behavior of the valves are the following:

- Diameter of pilot poppet: 10 mm
- Diameter of hole: 5 mm
- Poppet half angle: 15°
- Maximum flow coefficient: 0.7
- Critical flow number: 100
- Spring stiffness: 50–100 N/mm
- Spring force at zero compression: 200 N
- Equivalent mass of the poppet, rod, damper piston, and 30% of the spring: 0.05 kg
- Overall coefficient of viscous friction: 250 N/(m/s)
- Main stage piston diameter: 20 mm
- Diameter of hole: 10 mm
- Clearance on diameter: 0.01 mm
- Eccentricity between piston and sleeve: 0
- Contact length of the spool and the sleeve: 30 mm, constant
- Equivalent mass of the poppet, rod, damper piston, and 30% of the spring: 0.1 kg
- Coefficient of viscous friction: 1000 N/(m/s)
- Higher displacement limit: 3 mm
- Spring stiffness of the spool: 10 N/mm
- Spool spring force with both displacements zero: 400 N
- Maximum flow coefficient: 0.7
- Critical flow number: 100
- Volume of oil between the flow rate source and the valve input port: 1000 cm^3
- Nominal flow of the oil supply system: 300 L/min
- Specified oil bulk modulus: 17,000 bar
- Air–gas content: 0.1%
- Fluid density: 850 kg/m^3
- Absolute viscosity: 51 cP
- Temperature: 40°C
- Control sharp orifice diameter: 1.0 mm
- Critical flow number: 1000
- Main stage damper orifice diameter: 1.0 mm
- Pilot damper equivalent orifice diameter: 1.4 mm
- Maximum flow coefficient: 0.7
- Critical flow number: 1000
- Volume of the oil in the input pipe: 1000 cm^3
- Dead volume of oil between the control orifice and the damper orifices of the spool: 6 cm^3
- Dead volume of oil between the damper orifice of the spool and the pilot damper orifices: 5 cm^3
- Flow rate step input flow: 300 L/min
- Simulation time: 0.1 s
- Print interval: 0.001 s
- Type of integration algorithm: standard
- Number of variables involved in the integration: 137

The pressure variation in the input port of the valve generated by the step input flow is presented in Figure 3.6.7. The pressure evolution during the transient has a normal aspect

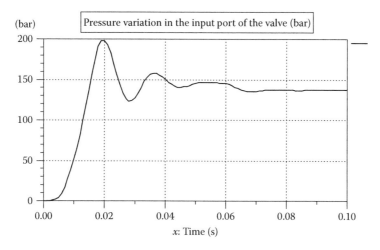

Figure 3.6.7 Pressure variation in the input port of the valve generated by a flow rate step input.

for a second-order hydromechanical system, including mass, spring, and pure viscous damper.

The flow rate at the output port of the valve generated by a step input flow rate of 300 L/min is presented in Figure 3.6.8. The small flow rate overshot is generated by the elasticity of the fluid included in the valve supplier pipe.

The overall force on the spool of the main valve has the same type of variation as the pressure variation in the input port (Figure 3.6.9).

The velocity, displacement, and the flow area of the main valve spool are presented in the following three figures: 3.6.10 through 3.6.12. The displacement of the main spool remains in the physical limit of the conical seat: 3 mm. The four big diameter holes (8 mm) connecting the seat and the output port of the valve are introducing a small pressure drop before the output port of the valve.

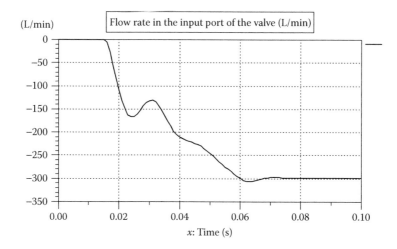

Figure 3.6.8 Flow rate at the input port of the valve generated by a step input flow rate.

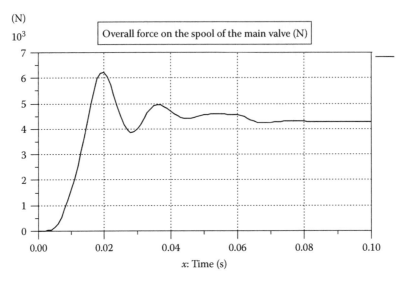

Figure 3.6.9 The overall force on the spool of the main valve.

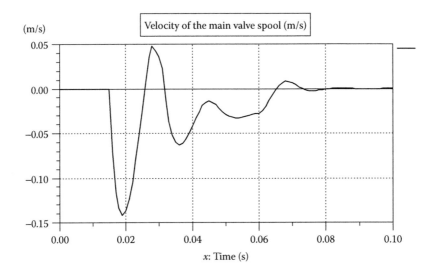

Figure 3.6.10 Velocity of the main valve spool.

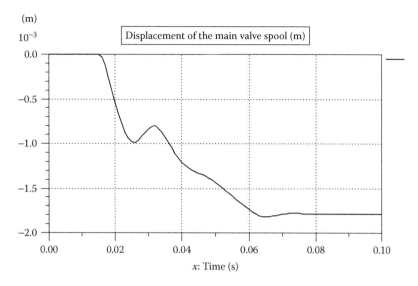

Figure 3.6.11 Displacement of the main valve spool.

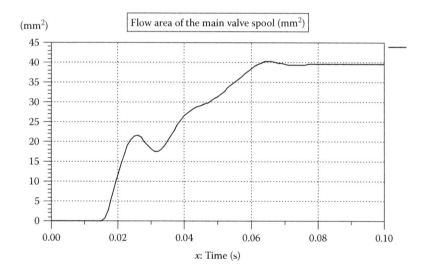

Figure 3.6.12 The variation of the flow area of the main spool.

The high-mean fluid velocity in the main valve orifice (Figure 3.6.13) corresponds to the highest value of the flow coefficient (Figure 3.6.14) and high values of the flow number (Figure 3.6.15).

The spring force on the spool of the main valve shows a second-order dynamic behavior (Figure 3.6.16). The small clearance between the spool of the main valve and the sleeve introduce a high viscous friction force, which increases the stability of the valve (Figure 3.6.17). The flow rate that controls the motion of the spool has a normal variation (Figure 3.6.18) and the final value (1 L/min) is accepted by all the industrial pressure relief valves manufacturers.

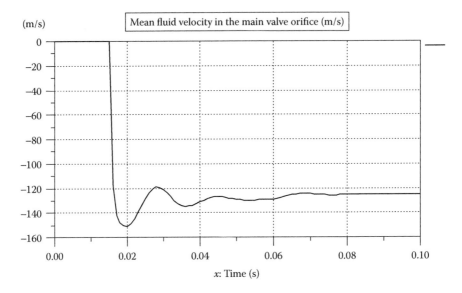

Figure 3.6.13 Variation of the mean-fluid velocity in the main valve orifice.

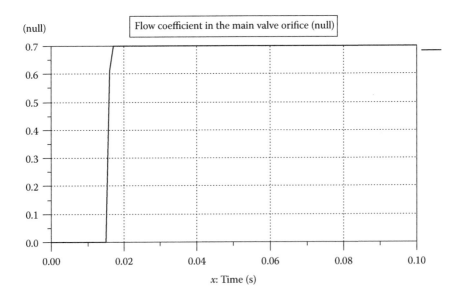

Figure 3.6.14 Variation of the flow coefficient in the main valve orifice.

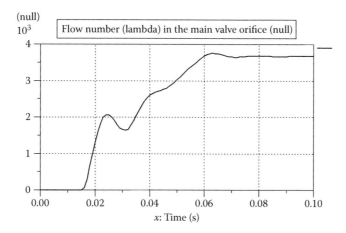

Figure 3.6.15 Variation of the flow number in the main valve orifice.

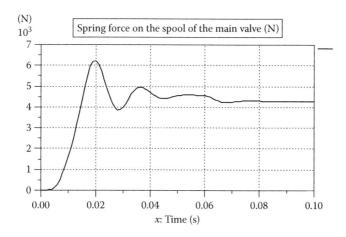

Figure 3.6.16 The variation of the spring force on the spool of the main valve.

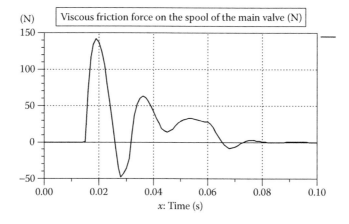

Figure 3.6.17 The variation of the viscous friction force on the spool of the main valve.

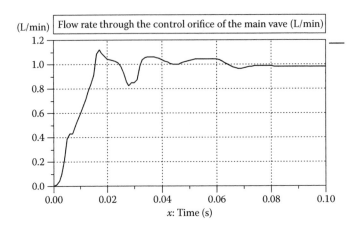

Figure 3.6.18 The variation of the flow rate through the control orifice of the main valve.

The mean fluid velocity through the control orifice (Figure 3.6.19) is limited by the downstream components. The flow coefficient (Figure 3.6.20) remains under 0.6, and the flow number (lambda) remains under 600 (Figure 3.6.21) after the transient generated by the flow rate input step.

The flow rates of the damper of the main valve spool, and the flow rate of the pilot damper have near the same variations (Figures 3.6.22 and 3.6.23), because the volumes of fluid sited between the orifices are very small, and the air content of the fluid has a normal value (0.1%).

The pressure drop on the pilot damper orifice resulting from Figures 3.6.24 and 3.6.25 is important at the beginning of the opening process.

The mean-fluid velocity through the pilot damper orifice has a spike of 38 m/s at the beginning of the valve opening, and then remains at a small value (about 11 m/s), keeping the flow number before the transition limit (Figures 3.6.26 and 3.6.27).

The mean-fluid velocity through the pilot valve orifice reaches in a few milliseconds, that is, 120 m/s and finally remains at a normal value of about 100 m/s (Figure 3.6.28).

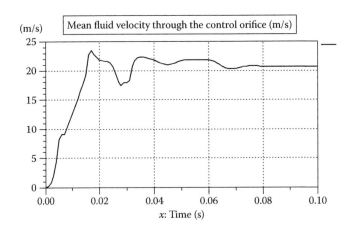

Figure 3.6.19 Variation of the mean-fluid velocity through the control orifice.

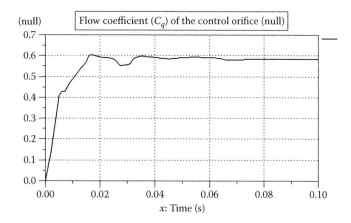

Figure 3.6.20 The variation of the flow coefficient through the control orifice.

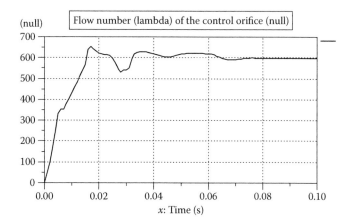

Figure 3.6.21 The variation of the flow number for the control orifice.

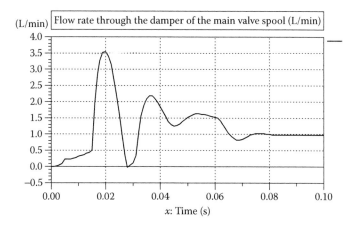

Figure 3.6.22 The variation of the flow rate through the damper of the main valve spool.

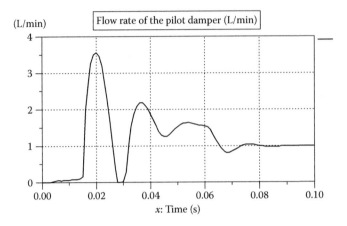

Figure 3.6.23 The variation of the flow rate of the pilot damper orifice.

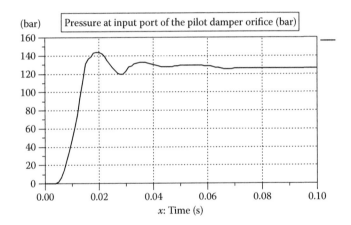

Figure 3.6.24 Pressure variation at the input port of the pilot damper orifice.

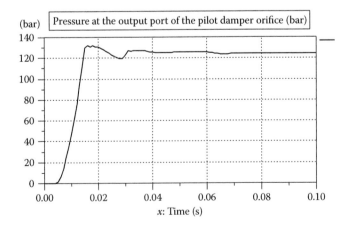

Figure 3.6.25 Pressure variation at the output port of the pilot damper orifice.

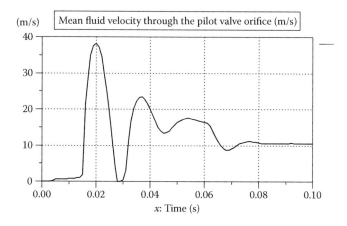

Figure 3.6.26 Variation of the mean-fluid velocity through the pilot damper orifice.

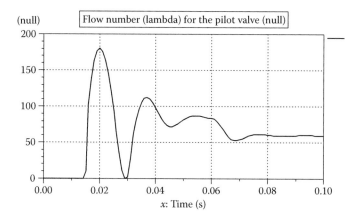

Figure 3.6.27 Variation of the flow number for the pilot valve.

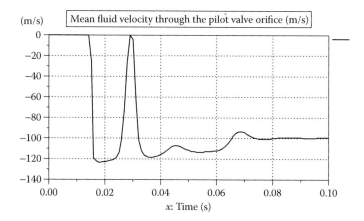

Figure 3.6.28 Variation of the mean-fluid velocity through the pilot valve orifice.

3.6.3   The influence of the geometry of the main valve and the pilot valve

The conical seat of the main valve needs a special manufacturing process. Many new types of two-stage pressure relief valve use a cylindrical spool lying on a flat seat with a very small conical sealing surface. The simulation network for such a valve is presented in Figure 3.6.29.

Such a valve opens quickly (Figure 3.6.30) and the stroke of the spool is very short (Figure 3.6.31).

The last generation of two-stage pressure relief valves uses a spherical pilot valve (Figure 3.6.32) and a flat seat with a very short conical chamfer for guiding the cylindrical spool of the main valve. All the calibrated orifices are independent components that can be tuned for different operational conditions of the valve. Small variations of the orifices diameters are generating important changes of the natural valve frequency. The time response is very small (Figure 3.6.33) and the final opening of the ball pilot valve remains very small (Figure 3.6.34).

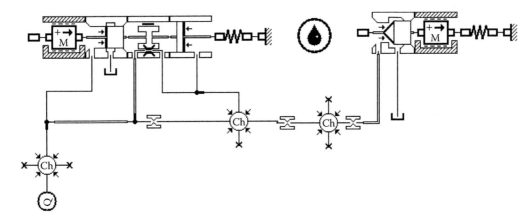

Figure 3.6.29 Simulation network for a two-stage pressure relief valve with cylindrical spool.

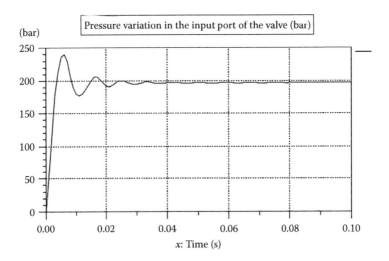

Figure 3.6.30 Pressure variation during the opening of a pressure relief valve with spool.

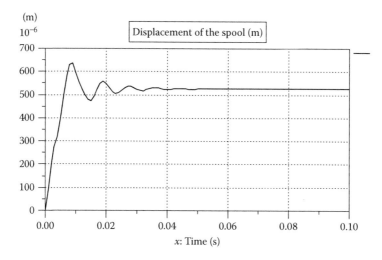

Figure 3.6.31 Displacement of a cylindrical spool during the opening of a pressure relief valve.

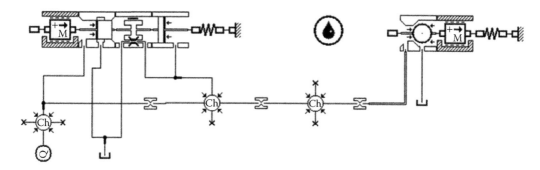

Figure 3.6.32 Simulation network for a two-stage pressure relief valve with ball pilot and flat seat main valve.

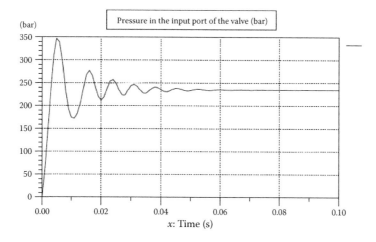

Figure 3.6.33 Pressure variation in the input port of a two-stage pressure relief valve with ball pilot and flat seat main valve.

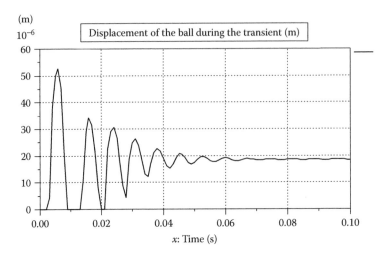

Figure 3.6.34 Pilot valve ball displacement during the opening of the two-stage valve.

It is useful to study the integration of a two-stage pressure relief valve in a simple power system (Figure 3.6.35). A hydraulic cylinder is controlled by a three-way electrohydraulic servovalve by a simple full opening (Figure 3.6.36). The sudden stop of the piston at the end of the stroke opens the two-stage pressure valve.

The displacement of the piston without load is a linear one (Figure 3.6.37).

The reaching of the end of the piston stroke increases the pressure in the input port of the pressure valve (Figure 3.6.38).

The flow rate measured at the output port of the main valve output orifice shows sudden opening both at the opening of the servovalve and at the end of the piston stroke (Figure 3.6.39).

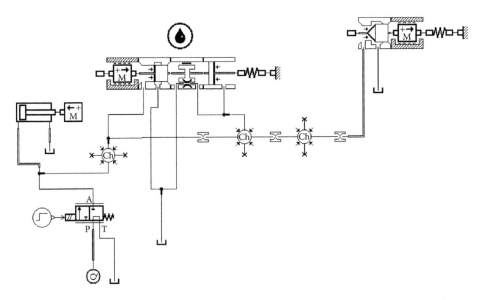

Figure 3.6.35 Simulation network for a simple hydraulic power system with pressure valve.

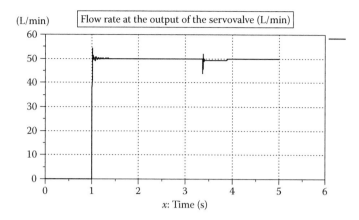

Figure 3.6.36 Flow rate at the output of the servovalve during the operation cycle.

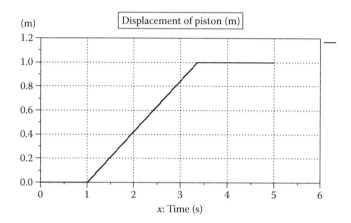

Figure 3.6.37 Displacement of the piston with no load.

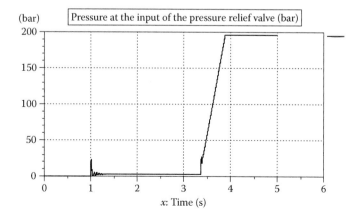

Figure 3.6.38 Pressure variation in the input port of the two-stage pressure relief valve.

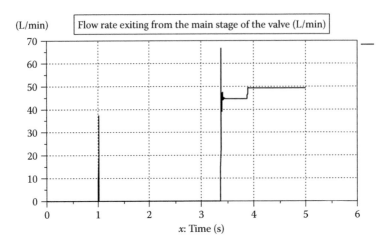

Figure 3.6.39 Flow rate measured at the output port of the pressure valve.

Section 1: Bibliography

1. Blackburn, J.F., Reethof, G., Shearer, J.L. *Fluid Power Control*. Technology Press of M.I.T. and Wiley, Cambridge, MA, 1960.
2. Merritt, H.E. *Hydraulic Control Systems*. John Wiley & Sons Inc., New York, London, Sydney, 1967.
3. McCloy, D., Martin, D.H. *The Control of Fluid Power*. Longman, London, 1973.

Section 2: Bibliography

1. Blackburn, J.F., Reethof, G., Shearer, J.L. *Fluid Power Control*. Technology Press of M.I.T. and Wiley, Cambridge, MA, 1960.
2. Merritt, H.E. *Hydraulic Control Systems*. John Wiley & Sons Inc., New York, 1967.
3. McCloy, D., Martin, D.H. *The Control of Fluid Power*. Longman, London, UK, 1973.
4. McCloy, D., Martin, D.H. *Control of Fluid Power: Analysis and Design*, 2nd ed. Ellis Harwood Limited, Chichester, UK, 1980.
5. Lebrun, M., Richard, Cl. How to create good models without writing a single line of code. *The Fifth Scandinavian International Conference on Fluid Power at Linköping*, Sweden, 1997.
6. Pressure relief valve, direct operated, Type DBD, Technical Catalogue RE 25402, Edition 2016, Rexroth, Bosch Group.
7. Pressure relief valve, pilot operated, Type DB…5X/…XC, Technical Catalogue RE 25802-XC-B2, Edition 2007, Rexroth, Bosch Group.
8. Pressure reducing valve, pilot operated, Type DR, Technical Catalogue RE 26892, Edition 2011, Rexroth, Bosch Group.
9. Pressure reducing valve, pilot operated, Type DR 10 K, Technical Catalogue RE 26850, Edition 2010.
10. Pressure valves, Catalogue HY11-3500/UK, Parker Hannifin GmbH, Hydraulic Controls Division, Kaarst, Germany, 2009.
11. Pressure Control Valves, Catalog HY15-3502/US, Parker Hannifin Corporation, Hydraulic Cartridge Systems, 2014.
12. Vasiliu, N., Vasiliu, D. *Fluid Power Systems*, Technical Press House, Bucharest, 2005. (In Romanian).
13. Guillon, M. *L'asservissement hydraulique et électrohydraulique*, Dunod, Paris, 1972.

Section 3: Bibliography

1. Blackburn, J.F., Reethof, G., Shearer, J.L. *Fluid Power Control.* Technology Press of M.I.T. and Wiley, Cambridge, MA, 1960.
2. Merritt, H.E. *Hydraulic Control Systems.* John Wiley & Sons Inc., New York, 1967.
3. McCloy, D., Martin, D.H. *The Control of Fluid Power.* Longman, London, 1973.
4. McCloy, D., Martin, D.H. *Control of Fluid Power: Analysis and Design*, 2nd ed. Ellis Harwood Limited, Chichester, UK, 1980.
5. Lebrun, M., Richard, C.L. How to create good models without writing a single line of code. *The Fifth Scandinavian International Conference on Fluid Power at Linköping*, Sweden, 1997.
6. Pressure relief valve, direct operated, Type DBD, Technical Catalogue RE 25402, Edition 2016, Rexroth, Bosch Group.
7. Pressure relief valve, pilot operated, Type DB...5X/...XC, Technical Catalogue. RE 25802-XC-B2, Edition 2007, Rexroth, Bosch Group.
8. Pressure reducing valve, pilot operated, Type DR, Technical Catalogue RE 26892, Edition 2011, Rexroth, Bosch Group.
9. Pressure reducing valve, pilot operated, Type DR 10 K, Technical Catalogue RE 26850, Edition 2010.
10. Pressure valves, Catalogue HY11-3500/UK, Parker Hannifin GmbH, Hydraulic Controls Division, Kaarst, Germany, 2009.
11. Pressure Control Valves, Catalog HY15-3502/US, Parker Hannifin Corporation, Hydraulic Cartridge Systems, 2014.
12. Vasiliu, N., Vasiliu, D. *Fluid Power Systems*, Technical Press House, Bucharest, 2005. (In Romanian).
13. Guillon, M. *L'asservissement hydraulique et électrohydraulique*, Dunod, Paris, 1972.

Section 4: Bibliography

1. Guillon, M. *L'asservissement hydraulique et électrohydraulique*, Dunod, Paris, 1972.
2. Merritt, H.E. *Hydraulic Control Systems.* John Wiley & Sons Inc., New York, 1967.
3. McCloy, D., Martin, D.H. *The Control of Fluid Power.* Longman, London, 1973.
4. McCloy, D., Martin, D.H. *Control of Fluid Power: Analysis and Design*, 2nd ed. Ellis Harwood Limited, Chichester, UK, 1980.
5. Vasiliu, N., Vasiliu, D. *Fluid Power Systems*, Technical Press House, Bucharest, 2005. (In Romanian).
6. Rexroth (Bosch Group), *Pressure Relief Valve, Direct Operated, Type DBD*, Technical Catalogue RE 25402, 2016.

Section 5: Bibliography

1. Blackburn, J.F., Reethof, G., Shearer, J.L. *Fluid Power Control.* Technology Press of M.I.T. and Wiley, Cambridge, MA, 1960.
2. Merritt, H.E. *Hydraulic Control Systems.* John Wiley & Sons Inc., New York, 1967.
3. McCloy, D., Martin, D.H. *The Control of Fluid Power.* Longman, London, 1973.
4. McCloy, D., Martin, D.H. *Control of Fluid Power: Analysis and Design*, 2nd ed. Ellis Harwood Limited, Chichester, UK, 1980.
5. Lebrun, M., Richard, C.W. How to create good models without writing a single line of code. *The Fifth Scandinavian International Conference on Fluid Power at Linköping*, Sweden, 1997.
6. Rexroth (Bosch Group), Pressure relief valve, direct operated, Type DBD, Technical Catalogue RE 25402, 2016.
7. Rexroth (Bosch Group), Pressure relief valve, pilot operated, Type DB...5X/...XC, Technical Catalogue RE 25802-XC-B2, 2007.

8. Rexroth (Bosch Group), *Pressure reducing valve, pilot operated*, Type DR, Technical Catalogue RE 26892, 2011.
9. Rexroth (Bosch Group), *Pressure reducing valve, pilot operated*, Type DR 10 K, Technical Catalogue RE 26850, 2010.
10. Parker Hannifin GmbH, *Pressure Valves*, Catalogue HY11-3500/UK, Hydraulic Controls Division, Kaarst, Germany, 2009.
11. Parker Hannifin Corporation, *Pressure Control Valves*, Catalog HY15-3502/US, Hydraulic Cartridge Systems, 2014.
12. Vasiliu, N., Vasiliu, D. *Fluid Power Systems*, Technical Press House, Bucharest, 2005. (In Romanian).
13. Guillon, M. *L'asservissement hydraulique et électrohydraulique*, Dunod, Paris, 1972.

Section 6: Bibliography

1. Blackburn, J.F., Reethof, G., Shearer, J.L. *Fluid Power Control.* Technology Press of M.I.T. and Wiley, Cambridge, MA, 1960.
2. Merritt, H.E. *Hydraulic Control Systems.* John Wiley & Sons Inc., New York, 1967.
3. McCloy, D., Martin, D.H. *The Control of Fluid Power.* Longman, London, 1973.
4. McCloy, D., Martin, D.H. *Control of Fluid Power: Analysis and Design*, 2nd ed. Ellis Harwood Limited, Chichester, UK, 1980.
5. Lebrun, M., Richard, C.W. How to create good models without writing a single line of code. *The Fifth Scandinavian International Conference on Fluid Power at Linköping*, Sweden, 1997.
6. Rexroth (Bosch Group), Pressure relief valve, direct operated, Type DBD, Technical Catalogue RE 25402, 2016.
7. Rexroth (Bosch Group) Pressure relief valve, pilot operated, Type DB...5X/...XC, Technical Catalogue RE 25802-XC-B2, 2007.
8. Rexroth (Bosch Group), Pressure reducing valve, pilot operated, Type DR, Technical Catalogue RE 26892, 2011.
9. Rexroth (Bosch Group), Pressure reducing valve, pilot operated, Type DR 10 K, Technical Catalogue RE 26850, 2010.
10. Parker Hannifin GmbH, Pressure valves, Catalogue HY11-3500/UK, Hydraulic Controls Division, Kaarst, Germany, 2009.
11. Parker Hannifin Corporation, Pressure Control Valves, Catalog HY15-3502/US, Hydraulic Cartridge Systems, 2014.
12. Vasiliu, N., Vasiliu, D. *Fluid Power Systems*, Technical Press House, Bucharest, 2005. (In Romanian).
13. Guillon, M. *L'asservissement hydraulique et electrohydraulique*, Dunod, Paris, 1972.

chapter four

Simulation and identification of the electrohydraulic servovalves

4.1 Simulating the behavior of the electrohydraulic servovalves with additional electric feedback

4.1.1 Problem formulation

This chapter presents the mathematical modeling, numerical simulation, and experimental identification of the last generation of both mechanical and electrical feedback electrohydraulic servovalves. The synthesis of a three-stage servovalve by Amesim, the simulation of the power-stage overlap influence on the servovalves performance, the design of the controller for a servovalve by simulation, and the dynamic identification of the electrohydraulic servovalves by simulation with Amesim are other problems studied in this chapter. The experimental researches concerning these problems, performed by the authors, proved that numerical simulations with Amesim can speed the design process of these important boundary components [1,2].

The developing of any new industrial product needs a long iterative design and test processes, which must be carried out as quick as possible, in order to successfully launching it onto the market with minimum overall cost. The classical procedure of designing and building a prototype, and then carrying out tests is not competitive enough. It allows for various design stages to be included unintentionally in the final product. This occurs because in the design stage not all the possible factors are taken into account, no feedback from the client is possible and little test can be done on the emerging product. This can be corrected by an iterative approach where design, prototype building, and testing stages are being repeated until the customers' demands (both technical and economic) are met. In order to lower the costs of developing a new product, this iterative approach uses inputs not only from design engineers but also from manufacturing specialists, maintenance workers, and the customers to which the product is ultimately addressed. The integration of all these concurrent aspects is possible if the design and prototyping stages take place in a virtual environment such as Simcenter Amesim or LMS Virtual.Lab. This shortens the development phase of a project and minimizes the costs.

In this section, the authors discuss various aspects of the design of new servovalves, and the way of obtaining the best static and dynamic performances. In order to improve the steady-state behavior around the null point, the spool position is measured by a high-resolution LVDT and fed to an internal servocontroller. In certain conditions, this additional feedback can improve the overall performances. The effect of the position feedback was successfully studied for the first time by J-Ch. Mare [3] in 1994 using a General Prediction Control Algorithm implemented on a 386 PC with a Schaevitz data acquisition card.

The main target of this section is to evaluate the real improvements that can be gained by an extra feedback loop using the modeling and simulation environment Amesim. The theoretical conclusions were validated by static and dynamic tests performed with a modern test bench at the Politehnica University of Bucharest, Bucharest, Romania. The laboratory is certified by the National Laboratories Accreditation Body (RENAR). The systematic experiments have shown that the additional feedback is reducing the overall hysteresis of the steady-state characteristics and the time constant.

For a hydraulic servosystem, delivering a good steady-state and dynamic performance is of the utmost importance, because these systems are being employed in domains such as aeronautics, automotive, and heavy industry, where the performance demands are very strict. The performances of any servovalve have a very strong influence over the general performances of the hydraulic system. Servovalves can be regarded as the interface between the hydraulic and the electrical part of an electrohydraulic servosystem.

In the case of hydraulic servosystems, the modeling and simulation of the servovalve plays a major role in accurately simulating the system and obtaining meaningful results. Out of the many designs of servovalve developed over the years by various manufacturers, we have chosen the most common one: a nozzle flapper (Figure 4.1.1) provided with a mechanical force feedback [4].

This classical design, although it has its shortcomings, especially regarding the fluid filtering conditions, is well studied and proven to deliver good performances. It is also technologically easier to manufacture than other designs (as jet-pipe servovalve).

4.1.2 Mathematical modeling

Servovalves are complex devices and therefore a mathematical model for them is going to be complex and extremely difficult to integrate in order to conduct a numerical simulation. The modeling process for a servovalve has to be divided among its three main stages.

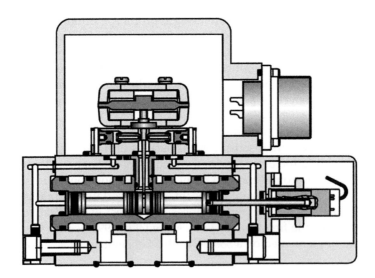

Figure 4.1.1 SE2E type servovalve with additional position feedback from the spool. (Source: Parker Hannifin.)

Each of these subsystems must be modeled by equations that are compatible with each other so that in the end the results can be summed up in one unifying equation and integrated. This task, regarded from a mathematical point of view, is a complicated one, but if individual components can be modeled separately, then a high degree of accuracy can be obtained. So, in a truly concurrent way, we can have electrical engineering models for electrical components such as the torque motor and the electrical drive, mechanical engineering models for the mobile armature, and flexible tube that couple the torque motor with the hydraulic preamplifier, and hydraulic engineering models for the flow rate of fluid through the preamplifier, and the sleeve–spool assembly. This approach is integrated into the Amesim environment, and the model of the servovalve was developed in the Amesim R13 release.

The modeling of the torque motor is based on the electromagnetism theory. The two coils, the mobile and fixed armatures form a magnetic circuit. The equations governing its behavior are given in Equations 4.1.1 and 4.1.2:

$$M_{em}(\theta, i) = K_i \cdot i + K_m \cdot \theta + M_h \tag{4.1.1}$$

where:

M_{em} is the electromagnetic torque of the motor (Nm)
K_i is the motor's torque constant (Nm/A)
i is the command current (A)
θ is the angular displacement of the mobile armature (the angle between the initial and final positions of the mobile armature axis) (rad)
K_m is the motor's electromagnetic constant (Nm/rad)
M_h is the motor's hysteresis couple

The second equation is an electrical one:

$$u(t) = \frac{1}{2} \cdot R \cdot i + L \cdot \frac{di}{dt} + 2 \cdot K_i \cdot \frac{d\theta}{dt} \tag{4.1.2}$$

where:

$u(t)$ is the coil voltage
L is the coil inductance
R is the coil resistance

Converted to the Amesim environment, we have an electromagnetic circuit (Figure 4.1.2), which expects an electrical current as input (ports 6 and 3), and as output it has mechanical displacement at ports 1, 2, 4, and 5.

The motor's torque is transmitted through the flexible tube from the torque motor's mobile armature to the nozzle and flapper system:

$$M_{em}(t) = J \frac{d^2\theta}{dt^2} + \beta \frac{d\theta}{dt} + k_\theta \cdot \theta + M_s(t) \tag{4.1.3}$$

where:

J is the mobile armature's inertial torque
θ is the rotation angle of the mobile armature
k_θ is the elastic tube's rigidity coefficient
β is the viscous friction coefficient
M_s is the resistive torque

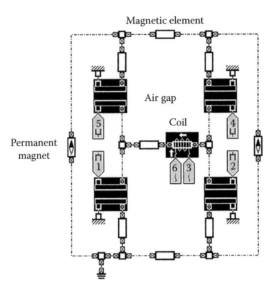

Figure 4.1.2 Torque motor supercomponent in Amesim.

In Amesim, this was modeled as a separate custom component that transforms the torque of the motor into linear movement of the flapper:

$$\begin{cases} f = \dfrac{M \cdot l^2}{2E \cdot I} + \dfrac{F \cdot l^3}{3E \cdot I} \\[2ex] \theta = \dfrac{M \cdot l}{E \cdot I} + \dfrac{F \cdot l^2}{2E \cdot I} \end{cases}$$

(4.1.4)

where:

f is the displacement of the flexible tube's free end
θ is the angle of the tube's medium fiber
E is the elasticity module of the tube's material
I is the inertial moment of the tube's cross section
F is the resultant force acting on the tube
M is the resultant moment action on the tube

This linear movement drives two pistons that transform the mechanical input of the torque motor into hydraulic quantities. The mobile armature movement Equation 4.1.3 and the flexible tubes equilibrium Equation 4.1.4 are used at this stage. In addition at this level, the mechanical feedback of the servovalve has to be taken into account. The mechanical force feedback is implemented through the means of a steel rod connecting the flapper of the hydraulic preamplifier with the spool of the main valve. The value of this force is given by the equation

$$F_{ra} = K_{ra}\left[X_p + X_s + \theta_p \left(l_2 - l_1 \right) \right]$$

(4.1.5)

where:

F_{ra} is the feedback force (N)
K_{ra} is the stiffness of the elastic feedback rod (N/m)

l_2 is the distance between the end of the elastic feedback rod and the axis of the torque motor (m)

X_s is the valve spool displacement

The last factor that needs to be taken into account is the flow of hydraulic oil through the nozzles of the hydraulic preamplifier,

$$Q = C_d \cdot A \cdot \sqrt{\frac{2}{\rho}|\Delta p|} \cdot sign(\Delta p) \tag{4.1.6}$$

where:

Q is the flow through the orifice (m³/s)

C_d is the discharge coefficient

A is the flow area of the orifice (m²)

Δp is the pressure drop on the orifice (N/m²)

The hydromechanical model built in Amesim incorporates the above-mentioned equations and models the mobile ensemble (mobile armature, flexible tube, and feedback rod) as a supercomponent with the input in the form of a torque motor and displacement of the valve spool and the output in the form of displacement of the flapper (Figure 4.1.3). The hydraulic component is modeled using small diameter pistons, which are used to change the volume of some hydraulic volumes that connect simple fixed orifices; thus, a variable pressure drop proportional to the travel of the flapper is obtained.

The whole servovalve model is shown in Figure 4.1.4. The flow from an idealized hydraulic supply represents the input of the system, whereas the output is the pressure differential that is applied on the ends of the pistons used to model the spool of the valve. The valve's spool was modeled using standard Amesim components, a series of six pistons with a concentrated mass in the middle section. The end pistons are driven by the pressure difference created by the hydraulic preamplifier, whereas the middle four pistons with the corresponding hydraulic chamber elements model the spool geometry and the four-way connections of the valve.

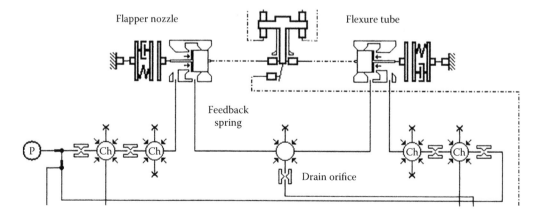

Figure 4.1.3 Hydraulic nozzle-flapper preamplifier modeled in Amesim.

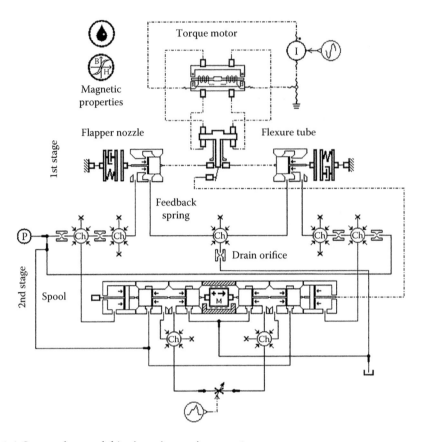

Figure 4.1.4 Servovalve model in Amesim environment.

4.1.3   Simulations results

The theoretical steady-state characteristic was obtained by numerical simulation. The servovalve was excited with a low-frequency sine signal (0.05 Hz) with adequate amplitude (10 V). With this data, the steady-state diagram for this simple servovalve was built (Figure 4.1.5).

As expected from a high-tech flow control valve, the steady-state characteristics of the servovalve are very good. However, a small hysteresis is always present. The saturation and the insensibility around the null are very small. Figure 4.1.6 presents the steady-state characteristics of the valve in the null region obtained for a complete period of sine input. It is clear that even if the modeled valve was considered with a zero overlap, the behavior around the null shows a slight insensitivity that can lead to stability problems if the servovalve is used in high-accuracy positioning system. Therefore, an additional feedback loop was considered: the position of the valve's spool was supplied by a position transducer and the feedback loop was closed electronically through summing blocks.

The model of the servovalve with additional feedback loop is presented in Figure 4.1.7. The simulation was generated by a signal with the same parameters: sine wave function with an amplitude of 10 V and 0.05 Hz frequency. The steady-state characteristic is given in Figure 4.1.8. Figure 4.1.9 presents the servovalve behavior around the null. From the two figures, we can see that the additional feedback loop has three major effects. First, it

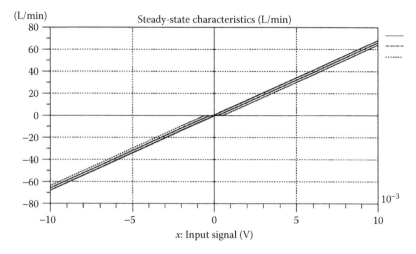

Figure 4.1.5 Steady-state characteristics of the servovalve for different laps.

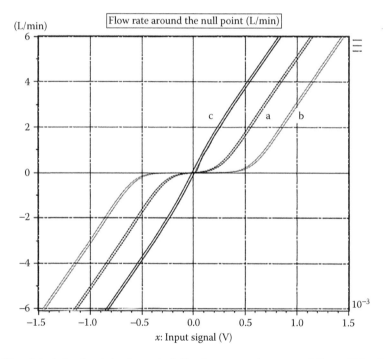

Figure 4.1.6 Simulated steady-state characteristic of a servovalve around the null for different laps: (a) null lap, (b) positive lap (10 μm), and (c) negative lap (10 μm).

dampened the overall gain of the loop; therefore, lowering the maximum flow of the servovalve. Second, the hysteresis of the servovalve was also lowered in comparison with the results from the servovalve without additional feedback. Most important, the insensibility around the null was reduced significantly. This can be regarded as the main advantage of adding the electrical position feedback loop, even outweighing the overall damping of the

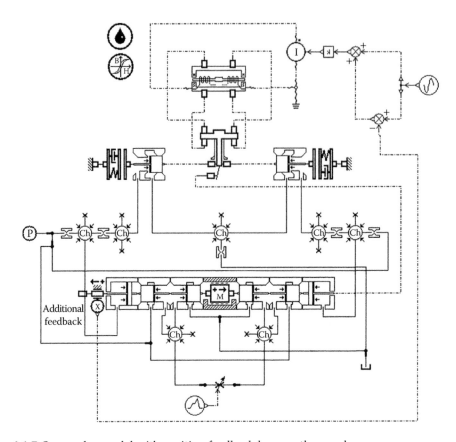

Figure 4.1.7 Servovalve model with position feedback loop on the spool.

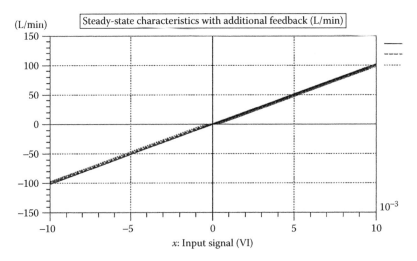

Figure 4.1.8 Steady-state characteristics of the servovalve with additional feedback for different laps: −10, 0, and +10 μm.

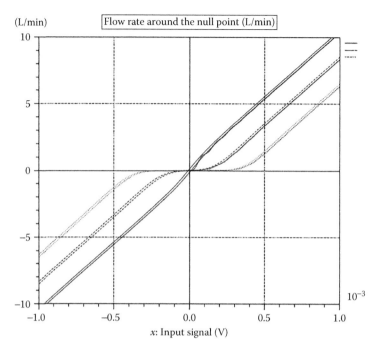

Figure 4.1.9 Steady-state characteristics for the servovalve with physical null lap, overlap (10 μm), and underlap (−10 μm).

system and the necessity for modifying the structural components of the valve in order to compensate the increase in the spool weight.

4.1.4 Experimental results

The experimental researches carried out to validate the Amesim models were performed in the Electrohydraulic Test Laboratory set up by the authors at the Politehnica University of Bucharest, Romania (Figures 4.1.10 and 4.1.11). The laboratory is certified by the RENAR to perform tests on servovalves and other proportional electrohydraulic flow and pressure valves according to the international standards [5,6].

The tests were performed on a test bench that includes a constant pressure supply flow source, a manifold for testing various valves, and a hydraulic cylinder with a very low friction and low inertia piston is used to compute the flow in conjunction with a LabVIEW powered data acquisition system [7–11]. All the experimental results point out that the servovalve with additional feedback loop has lower hysteresis and very little insensibility around the null. According to Figure 4.1.12, the null hysteresis is less than 0.1%!

This servovalve is suited for applications where high-positioning accuracy is necessary and very small levels of input signal are expected for most of the working time. In addition, the dynamic performances of such a servovalve are above the known average. The high cutoff frequency displayed by such valves is due to the improvement in response time obtained by the addition of the electrical position feedback system (Figures 4.1.13 and 4.1.14).

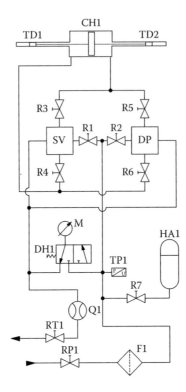

Figure 4.1.10 Test bench scheme for the steady-state characteristics of the electrohydraulic servo-valves and DDVs.

Figure 4.1.11 Servovalves test bench front view.

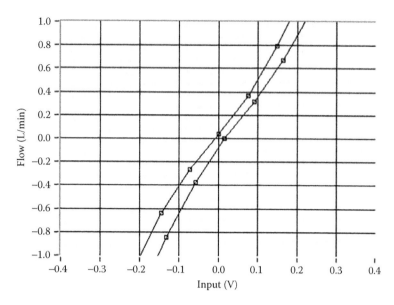

Figure 4.1.12 Experimental steady-state characteristics around the null point for a SE2E servovalve.

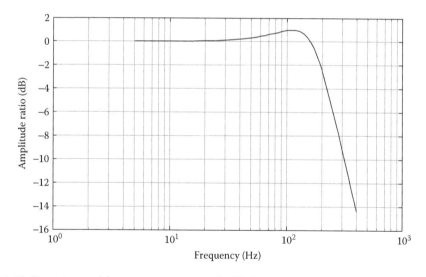

Figure 4.1.13 Experimental frequency response of a SE2E servovalve.

The results supplied by the dynamic tests show that the servovalve has a high cutoff frequency and a very low response time (Figure 4.1.13). The servovalve frequency response is very well described by the following transfer function:

$$H(s) = \frac{88.32s + 1100000}{s^2 + 1100s + 1100000} \qquad (4.1.7)$$

From this simple transfer function, we can compute the natural frequency of the valve: 167.0 Hz, which is a very high one. The damping coefficient has also an optimal value: $\zeta = 0.524$.

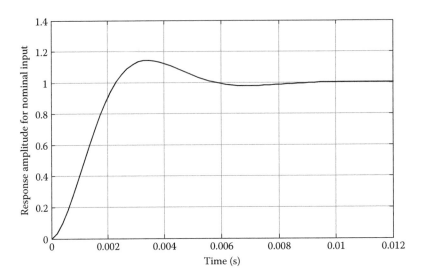

Figure 4.1.14 Step input response of a SE2E servovalve for nominal input.

The modern researches regarding the position additional feedback servovalves are devoted to the use of this type of electrohydraulic amplifier in high-precision redundant servomechanisms [12–15].

4.1.5 Conclusion

The main target of this chapter is the simple implementation of a complex servovalve model in the Amesim simulation environment. So, we have created the possibility of simulating more complex systems, including this kind of servovalve. The use of the position additional feedback in the servovalve can be regarded as a redundant feature, which increases the system performance too.

All the modern studies confirm that the ever closer merging of hydraulics and electronics that we see as a constant trend in proportional hydraulics over the past years is a good way of increasing the control systems performance.

4.2 Simulation with Amesim as a tool for dynamic identification of the electrohydraulic servovalves

4.2.1 Problem formulation

This section presents a new method for experimental identification of the dynamic performance of the industrial electrohydraulic servovalves, which contain a position transducer attached to the spool. The new procedure can replace the traditional method of measuring the input signal and the flow of the valve by considering the displacement of the valve's spool as input for computing the Bode diagram. It combines the analytical model of the servovalve with the process of an experimental identification in order to automate the procedure completely. The method was validated by direct measurement of the spool position for a chirp input signal. The experimental results are found in good agreement with the numerical simulation performed by Amesim software.

The modeling and simulation of any electrohydraulic control systems need the performance specification of all the electric, hydraulic, and hybrid devices. Any manufacturer is obliged to provide those specifications to a certain degree. Although the supplied specifications are of great value in choosing a certain component for a hydraulic control system, they offer a rather poor estimate for the performance level of the entire system.

When a performance estimation of a new system is required, the design engineer must perform a complete numerical simulation. In this case, the specification supplied by the manufacturers will not cover the needs, and a full mathematical model of the hydraulic components is needed. Such mathematical description can be estimated by theoretical means and the coefficient of the equations obtained through a process of system identification. Another path for simulation would be to just identify the transfer function for the modeled object and insert it into the simulation network of the system.

This section describes the implementation of a simplified method for obtaining the transfer function of the electrohydraulic (proportional) servovalves, which contains a position transducer attached to the spool. The methods used to carry the test necessary for the dynamic identification are in compliance with the international standards regarding the testing of the four-way proportional valves. The test bench has been certified by the RENAR.

The software used for controlling the test procedures is written in LabVIEW and enables the full automation of the procedures from the generation of appropriate input signals to recording the signals needed for system identification and computing the Bode diagram [1].

4.2.2   Preliminary simulations

In order to gain a better understanding of the phenomenon involved in testing the electrohydraulic servovalves, we first performed some simulations in Amesim involving not only the valve, but the entire test stand. The aim of the preliminary simulations is to evaluate different methods of finding the dominant hydraulic parameters of the valves and to compute the frequency response diagrams. The reference was set by a test bench configuration built as recommended in ISO 10770:2009 with a low inertia, low friction hydraulic cylinder, and a speed transducer attached to the cylinder's piston (Figure 4.2.1).

The simulation network includes both the hydraulic components: control tools and data analysis tools. Normally, the obtained graphs show that the results of the fine simulations are in good agreement with the manufacturers' specifications. This, in turn, validates the methods for investigating the effects of changing the test bench configuration. Figure 4.2.2 shows the same test bench model and the same servovalve, but the flow metering device was changed from a hydraulic cylinder to a simple metering orifice and two pressure transducers.

The metering orifice was chosen according to the diagram from Figure 4.2.3. The criteria for choosing the orifice diameter were the rated flow of the valve and the desired pressure drop across the orifice. The flow regime is considered turbulent due to the high pressure drop:

$$Q = c_d \cdot A_0 \sqrt{\frac{2}{\rho}(p_1 - p_2)} \tag{4.2.1}$$

4.2.3   Simulation results

The results of the preliminary simulations are given both as time graphs, showing in parallel the input and output variables for both test stand configurations, and as frequency

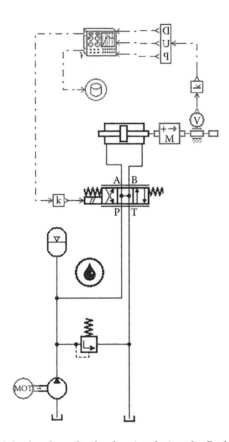

Figure 4.2.1 Test bench with hydraulic cylinder for simulation the Bode diagram.

response graphs, showing the attenuation and phase lag of the output signals for variable frequency. Figures 4.2.4 and 4.2.5 show the practical identity of the results offered by the two methods. Thus, it is possible to say that the simulation proved that as long as the testing conditions are strictly controlled, and the measuring is correctly performed, any significant variable can be considered as the output of the system.

In the next step of the research, we try to establish a more accurate Amesim model for an electrohydraulic proportional servovalve, which includes a position transducer attached to the spool.

Since the flow of a proportional servovalve depends directly by the opening of the spool–sleeve orifices, the position signal for the spool can be considered as an output variable for computing the frequency response:

$$Q = c_d \cdot b \cdot x \sqrt{\frac{2}{\rho} \Delta p} \qquad (4.2.2)$$

where:

    b is the width of the flow orifice

    x is the spool displacement

The proportional servovalve is considered to be set up with zero lap. Figure 4.2.6 shows the simulation network with a complete servovalve hydraulic model, a simplified

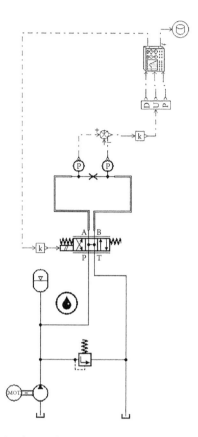

Figure 4.2.2 Simulation network of a test bench with metering orifice for finding the Bode diagram of a servovalve.

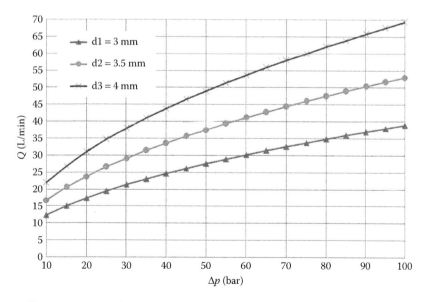

Figure 4.2.3 Flow rate versus the pressure drop for different orifice diameters.

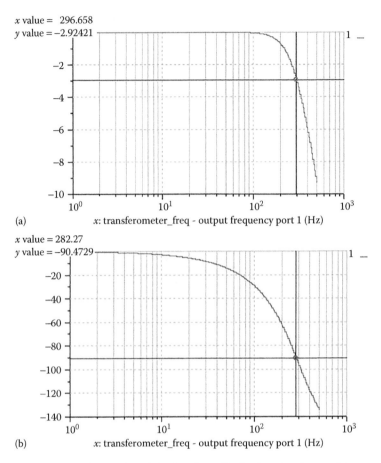

(a)

(b)

Figure 4.2.4 Bode diagram obtained with hydraulic cylinder by simulation: (a) magnitude versus frequency and (b) phase lag versus frequency.

electrical actuator, and a basic hydraulic power source set as specified in the test standard. The frequency response diagram is presented in Figure 4.2.7.

4.2.4   *Experimental results*

The experiments were performed using an original LabVIEW software package developed by the authors with the goal of automating the test procedures for dynamic identification by obtaining the Bode diagram. To this end, it was necessary to design both a data generation program (Figure 4.2.8) and a data acquisition program. Furthermore, the programs needed to be easily changeable and customizable so that they could be used for testing a wide range of electrohydraulic servovalves. The data generation program uses a standard method for producing a voltage signal in LabVIEW. It writes a waveform to the analog output buffer of a data acquisition board. Instead of choosing the waveform from a predefined set, it actually computes the requested signal based on a mathematical formula supplied by the user. In turn, this can induce delays, so it was a necessity to have some measures in order to obtain the desired level of flexibility. The delays were canceled out by carefully

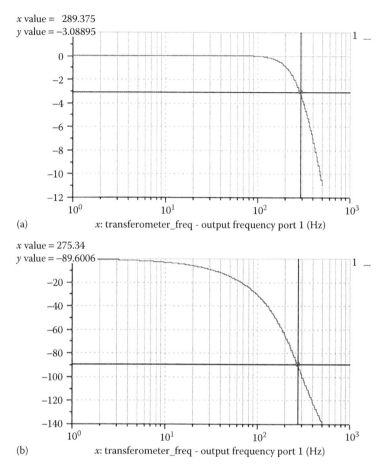

Figure 4.2.5 Bode diagram obtained with metering orifice by simulation: (a) magnitude versus frequency and (b) phase lag versus frequency.

choosing the data to be processed in the main test program. The results from the testing program are presented in Figures 4.2.8 and 4.2.9.

The program for system identification was also written in LabVIEW (Figure 4.2.10). It uses a chirp signal in order to obtain the transfer function for the hydraulic amplifier under test. Validation was done by direct comparison of the identified model and the servovalve responses in the time domain. Figure 4.2.11 presents the step response for both the identified models [10,11].

4.2.5   Conclusion

The important gain of the presented research is the successful implementation of a test method for direct drive valves, which uses a much simpler way to measure the flow of the valve and a high-tech computer system for *on the spot* data analysis.

The implemented method can be readily used for other types of four-way proportional hydraulic valves, and furthermore, the program developed for this work is highly adaptable because new modules can be written and incorporated into it.

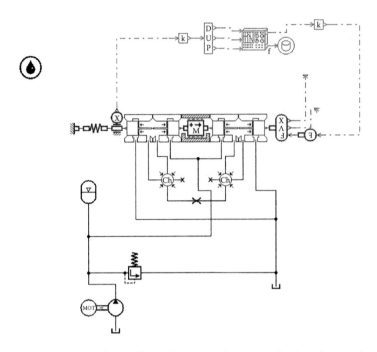

Figure 4.2.6 Model for a servovalve with position transducer attached to the spool.

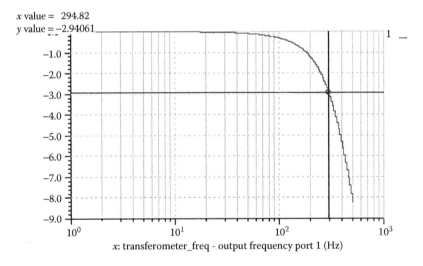

Figure 4.2.7 Frequency response diagram for the model of the proportional servovalve with the position transducer on the spool.

The modules used in the research program can form the basis for further researches into hydraulic system testing and identification.

The identified models are virtually identical as proven by the step-response analysis, and anyone of them can be used for modeling the DFplus industrial servovalve in a hydraulic system simulation.

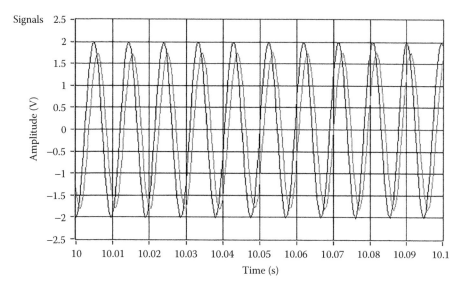

Figure 4.2.8 Time-domain response for PARKER DFplus industrial servovalve given by the test procedure.

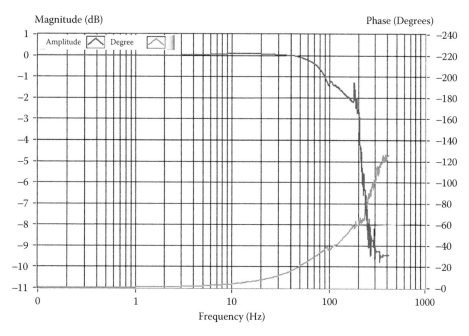

Figure 4.2.9 Amplitude versus frequency and phase lag versus frequency for DFplus servovalve (experimental result for 50% from nominal input).

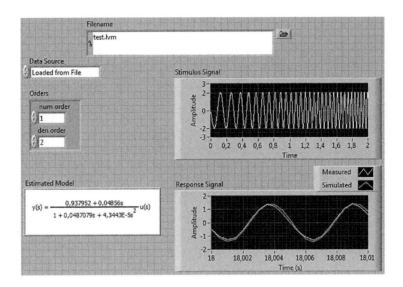

Figure 4.2.10 Dynamic identification of an industrial servovalve showing the estimated transfer function for the second-degree model.

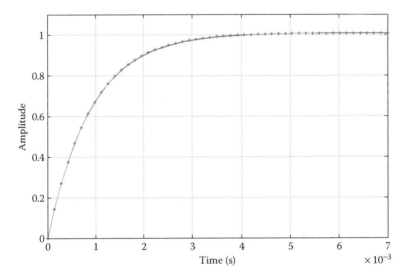

Figure 4.2.11 Step response for the identified models of the DFplus servovalve: first degree model (−) and the second degree one (*).

4.3   Simulation and experimental validation of the overlap influence on the flow servovalves performance

4.3.1   Introduction

This section presents the results of the theoretical and the experimental research carried out on the overlap influence on the steady-state and dynamic performances of the electrohydraulic flow control valves. The valve behavior around the null point has an essential influence on the accuracy and stability of the electrohydraulic servosystems.

A lot of peculiarities of the operational conditions have to be taken into account, such as the environmental temperature variation, the power lost by internal leakages, and so on.

The modern mechatronics systems development needs a large amount of numerical simulations in order to find out the best structure for specified performances. Amesim language [1], officially called Simcenter Amesim™ by LMS Company, now a member of Siemens group, includes a realistic hydraulic components library that can be used to point out the influence of different kind of parameters on the electrohydraulic components and systems performances with high accuracy. The results of the simulation are validated by the author's experiments carried out on different type of servovalves tested in standard conditions. The keywords of this chapter are numerical simulation, hydraulic flow control valves, hydraulic overlap, steady-state, and dynamic performances.

4.3.2   Problem formulation

The modern electrohydraulic servovalves are hybrid equipments used for controlling the flow sent by the pressure supply systems to the hydraulic motors included in position (angle) or force (torque) servosystems.

The industrial requirements of the CIM led to the replacement of the classical two-stage flapper-nozzle servovalves by direct drive servovalves (DDV). This one keeps the flow control valve geometry, but the spool is directly actuated by a linear electromechanical convertor that uses one of the following three principle: (1) moving coil (Figure 4.3.1), (2) solenoids (Figure 4.3.2), and (3) a linear force motor (Figure 4.3.3). The accuracy of the flow control strongly depends on the resolution of the position transducer attached to the spool. Another important feature concerning the dynamic performance is the maximum force supplied by the electromechanical actuator. The mobile equipment, the position transducer, the electromechanical actuator, and the controller are in fact the components of a position loop. The controller of this loop supply the input signal for the internal position loop according to the needs of the governed process.

The overall static and dynamic performances of a servovalve depend on the performances of all the components. The main feature of these electrohydraulic converters is the

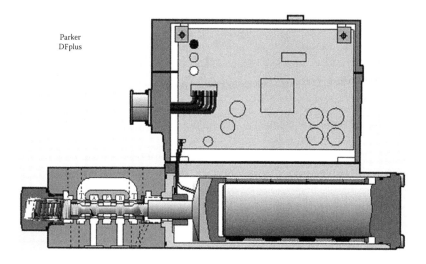

Parker
DFplus

Figure 4.3.1 High-speed industrial moving coil servovalve DFplus. (Source: Parker Hannifin.)

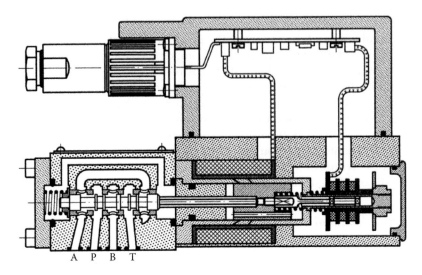

Figure 4.3.2 High-speed industrial solenoid servovalve 4WRPEH 6. (Source: Bosch Rexroth.)

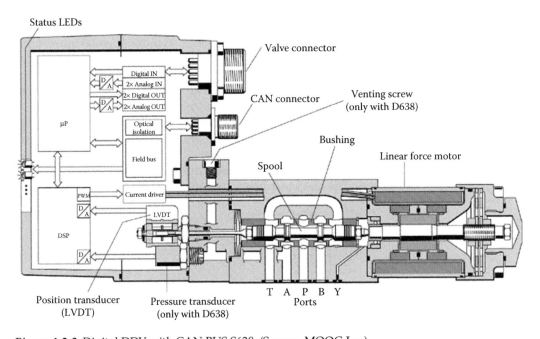

Figure 4.3.3 Digital DDV with CAN BUS S638. (Source: MOOG Inc.)

behavior around the null point. This one strongly depends on the *hydraulic overlap*. The relation between the valve spool displacement, *x*, and the flow sent to the hydraulic motor:

$$Q = f(x, d_s, \Delta p, \delta, j) \qquad (4.3.1)$$

where:

d_s is the spool diameter
δ is the hydraulic overlap
Δp is the pressure drop on the valve

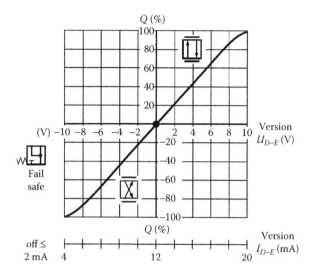

Figure 4.3.4 Typical steady-state flow characteristics of a high-speed industrial servovalve. (Source: Parker Hannifin.)

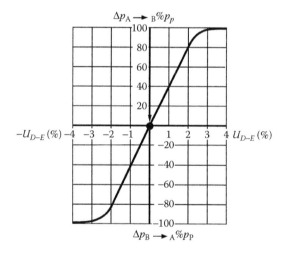

Figure 4.3.5 Typical pressure sensitivity of a high-speed industrial servovalve. (Source: Bosch Rexroth.)

j is the average radial clearance between the spool and the sleeve (bushing)

The static and dynamic performances of the servovalves are supplied by the manufacturers as average complex diagrams as presented in Figures 4.3.4 through 4.3.6. Numerical simulation offers the possibility to refine and explain the aspect of these curves.

4.3.3   Numerical simulation

Amesim software was selected as a current simulation tool. This complex software offers many advantages: rich library of hydraulic symbols and components, which allow the authors to use existing, proven models for well-known components (valves, cylinders);

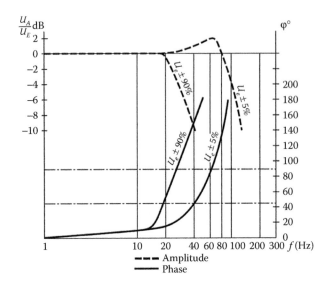

Figure 4.3.6 Typical dynamic performance of a high-speed industrial servovalve. (Source: Bosch Rexroth.)

ability to simulate different part of the system at different levels of complexity, which allows the authors to model different parts of the system at different levels of detail, as required. Amesim models are fully compatible with LabVIEW for Real-Time and Hardware-in-the-Loop (HiL) simulations, can be imported in LabVIEW, and connected to a Real-Time or HiL simulation system. A realistic image of the behavior of a servovalve can be created *before* the prototype manufacturing by numerical simulation, avoiding the *cut-and-try* time- and money-consuming procedure. A realistic Amesim simulation model of a direct-drive servovalve with moving coil is shown in Figure 4.3.7. This model gives the steady-state flow characteristic for different kind of valve overlap.

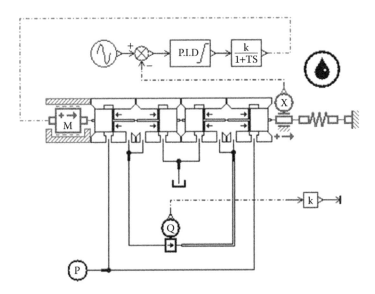

Figure 4.3.7 Amesim model used for finding the steady-state flow characteristics.

A low-frequency sine input (1.0 Hz) is used in order to pass through all the positions of the spool with very small steps. All the three cases encountered in the servovalve practice were investigated (Figures 4.3.8 through 4.3.10).

The results of these simulations are shown in Figures 4.3.11 through 4.3.13. They are in a very good agreement with the experimental ones published by well-known researchers in the field: Michel Lebrun [1], Marcel Guillon [2], Taco Viersma [3], and Jean-Charles Mare [4]. The very small slope of the flow characteristics around the null point shown in Figure 4.3.13 pointed out the small laminar leakages through the radial clearances, which cannot be ignored in the synthesis of the high-accuracy position servosystems.

The high slope of the flow characteristics around the null point of the negative overlap valves (Figure 4.3.12) is used in all the applications where the load random variations have

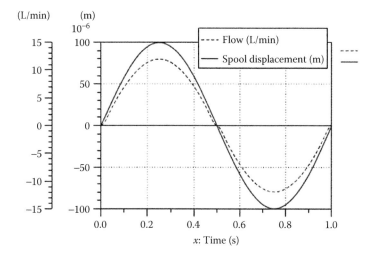

Figure 4.3.8 Low-frequency sine input response of a critical overlap flow control valve.

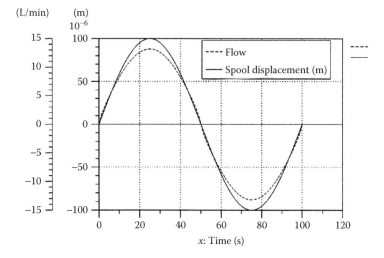

Figure 4.3.9 Low-frequency sine input response of a negative lap flow control valve.

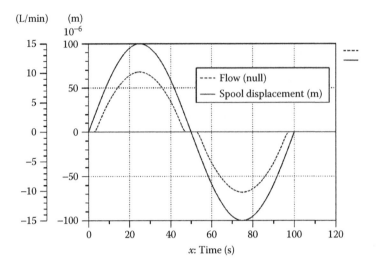

Figure 4.3.10 Low-frequency sine input response of a positive lap flow control valve.

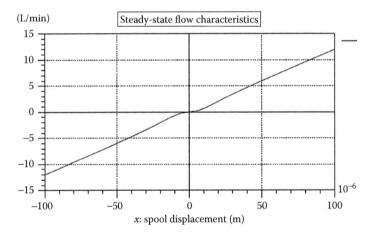

Figure 4.3.11 Steady-state flow characteristics of a critical lap flow control valve.

to be quickly rejected. A typical case is encountered in the automotive steering systems and in other similar applications, which involve the automatic recovery of a variable reference position [1,5]. Usually, these systems are supplied by a constant flow system, in order to keep the response time at the same value for any load variation.

A wide dead band of the flow characteristics (Figure 4.3.13) is useful for increasing the stability of the hydraulic servomechanisms, but it reduces the absolute accuracy. However, it strongly reduces the leakages, which can be important in the overall system power consumption.

All the above-mentioned conclusions have technological consequences leading to specific domains of applications [6–9].

The second series of numerical simulation was devoted to the correlation between the overlap and the valve-pressure sensitivity. The simulation model from Figure 4.3.14 was used in the same cases studied earlier. The results are shown in Figures 4.3.15 through 4.3.17.

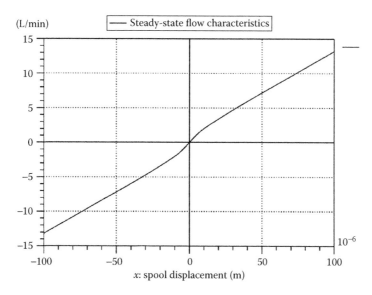

Figure 4.3.12 Steady-state flow characteristics of a negative lap flow control valve (–10 μm).

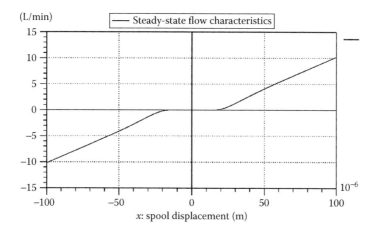

Figure 4.3.13 Steady-state flow characteristics of a positive lap flow control valve (+10 μm).

As expected from the geometrical peculiarities of the valves, the highest pressure sensitivity is supplied by the critical case, which is used in the most performant applications [13].

The negative overlap is always giving a smooth pressure-sensitivity curve used in constant flow supply applications.

The positive overlap is giving a *relay* pressure behavior, which can be useful in slow position-control process. The reduction of the geometrical gain by notches of different shapes leads to a continuous variation of the pressure difference between the hydraulic motor ports. This is the case of the proportional industrial servovalves actuated by proportional force solenoids developed especially by Bosch Rexorth group [15].

The final series of simulations is devoted to the overlap influence on the frequency response of the servovalve. The control library of Amesim includes a Frequency Response

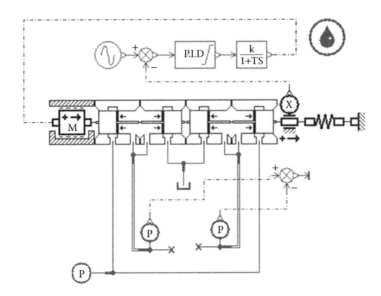

Figure 4.3.14 Amesim model used for finding the pressure sensitivity of a flow control valve.

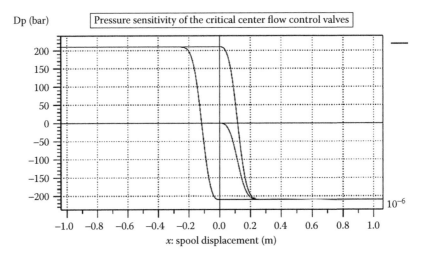

Figure 4.3.15 Pressure sensitivity of a critical lap flow control valve.

Analyzer block called SIGDYNFRA01, which is running as a real device, which generates a constant magnitude input with a variable frequency. For a moving coil actuator of the spool, a first-order transfer function with a few milliseconds constant time is a fair dynamic representation. The simulation model is presented in Figure 4.3.18 and the input signal in Figure 4.3.19.

The force developed by the actuator strongly increases with the input voltage frequency (Figure 4.3.20). The flow passing through the servovalve (Figure 4.3.21) is overpassing 100 Hz for the critical overlap (Figure 4.3.22), increases a little in the case of a negative overlap (Figure 4.3.23), and decreases for a positive overlap (Figure 4.3.24).

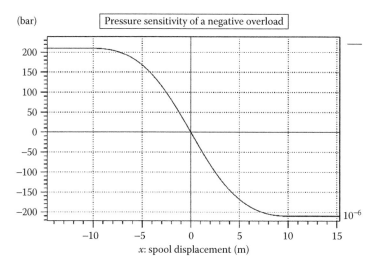

Figure 4.3.16 Pressure sensitivity of a negative lap flow control valve (−10 μm).

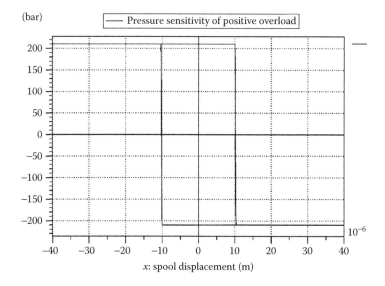

Figure 4.3.17 Pressure sensitivity of a positive lap flow control valve (10 μm).

The realistic choice of the servovalve parameters leads to normal results for this kind of device.

4.3.4   Experimental validation of the simulations

The authors performed the validation of the simulation results in the Fluid Power Laboratory of the Politehnica University of Bucharest, Romania in the section devoted to the electrohydraulic servovalves. The test bench designed for this kind of flow control valves (Figures 4.3.25 and 4.3.26) allows all types of standard tests.

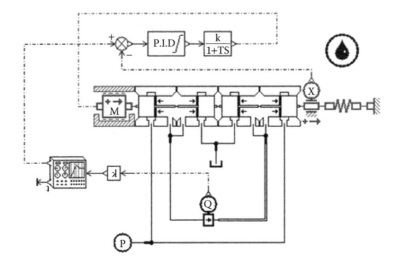

Figure 4.3.18 Frequency response Amesim model for a flow control valve.

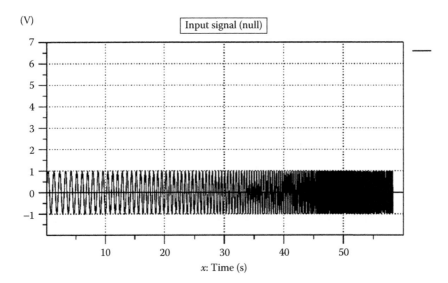

Figure 4.3.19 The sine input applied to the electromechanical actuator for obtaining the frequency response.

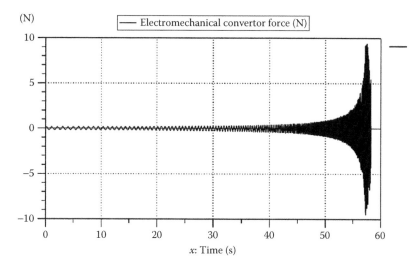

Figure 4.3.20 Force applied to the spool by the electromechanical actuator

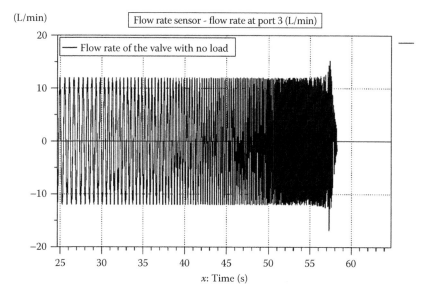

Figure 4.3.21 The flow passing by the valve without load.

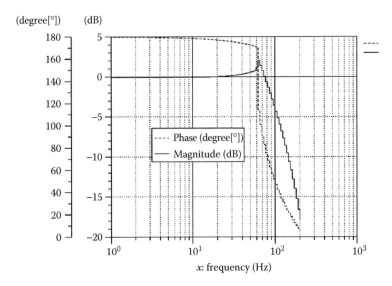

Figure 4.3.22 Frequency response of a critical lap flow valve for a sine input of ±10%.

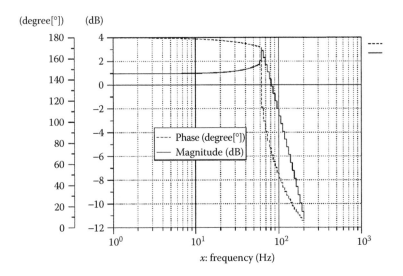

Figure 4.3.23 Frequency response of a negative lap flow valve for a sine input of ±10%.

The steady-state and the frequency response found experimentally in the lab for a direct drive valve with moving coil-type DFplus (Parker) are found very close to the simulated ones with Amesim software (Figures 4.3.27 and 4.3.28). It is important to mention the very small real clearance of the spool (about 2.5 μm), which introduces a good viscous damping.

4.3.5   Conclusion

All the design, test, and identification stages of the commercial servovalves performed by the authors pointed out that Amesim provided a strong solver and numerical core for

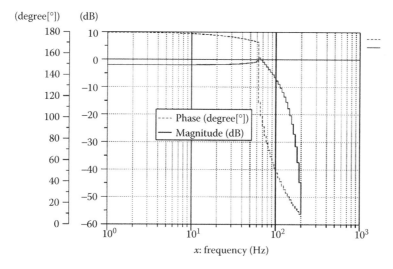

(degree[°])   (dB)

x: frequency (Hz)

Figure 4.3.24 Frequency response of a positive lap flow valve for a sine input of ±10%.

Figure 4.3.25 Overall view of the servovalve static test bench from the performance certification laboratory of the Politehnica University of Bucharest, Romania.

steady-state and transient simulation. As modeling a complex multiphysics system is not the main objective of engineers, it is important to have tools and interfaces, which accelerate and optimize the design. From this point of view, Amesim is a complete software perfectly adapted for model creation and deployment. The users working both in the high-level corporation and in strong research universities continuously extend the wide field of applications [7–10].

Figure 4.3.26 A virtual view of the new servovalve (SOLID WORKS). (Source: Duplomatic.)

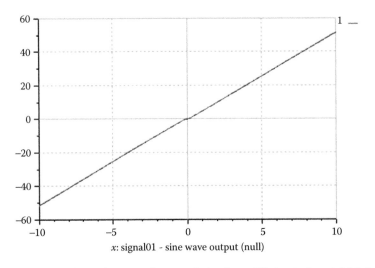

x: signal01 - sine wave output (null)

Figure 4.3.27 Experimental steady-state characteristics for a DFplus industrial high-speed servo-valve ($p_s = 100$ bar, $U_i = \pm 10$ V).

4.4   *Designing the controller of a servovalve by simulation*

4.4.1   *Introduction*

This section presents the research activities aiming to design the controller of a high-flow electrohydraulic servovalve needed for the power stage of the speed governors controlling high-power hydraulic turbines operating under low oil pressure. This is the common case of any Kaplan turbine, but the refurbishing of the old Francis turbines needs the same

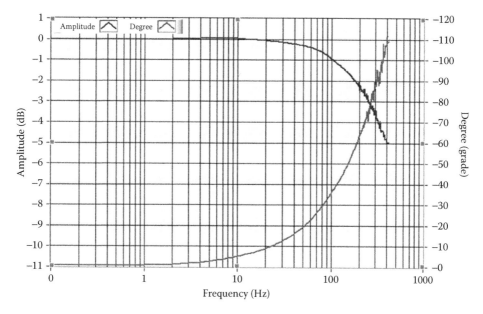

Figure 4.3.28 Experimental frequency response for DFplus industrial high-speed servovalve ($p_s = 100$ bar, $U_i = 10\%$).

valve configuration. Design problems, simulation methods, and experimental researches are briefly presented.

From an industrial point of view, the main idea of the new concept is the use of high-quality industrial electrohydraulic and electronic components only, in order to obtain good performances even under a low-pressure supply. This target generated a new approach to the design by eliminating the pipes between the flow control stages.

A detailed model of a high-power servomechanism containing a three-stage non-linear electrohydraulic proportional servovalve was designed by simulation, taking into account the real geometry of the metering spool windows. The valve dynamic behavior was simulated with SIMULINK and Amesim, and finally the results were compared with some preliminary laboratory measurements. The simulated and the real responses for different inputs were found in good agreement. The keywords of this section are simulation, controller fine-tuning, high-power electrohydraulic servovalves, and speed governors.

4.4.2   Problem formulation

Some important applications of the hydraulic control systems require very large flow and special dynamic performance. A typical practical case can be found in the field of the earthquake simulators of the important buildings, fly simulation tables, automotive dynamics simulators, and so on.

The variation of the speed and the power during the operation of a hydropower unit is achieved by adjusting the water flow that passes through the wicket gates. High-accuracy electrohydraulic servomechanisms actuate these control elements.

The architecture of the power control system directly depends on the hydropower unit size, specific speed, and the required dynamic performance. The last one depends on the quality requirements of the electrical power produced by the unit. The static and the dynamic forces that appear during the hydropower unit operation are rather large. The pistons of the hydraulic cylinders have rather large diameters, whereas the oil pressure is usually in the range 20–160 bar.

The shut down time of a hydropower unit in case of damage is usually lower than 10 s. The current requirements regarding the quality of the electrical power provided by hydropower units are extremely strict, and meeting them requires speed governors to have a better accuracy than 2 mHz. This performance condition requires very precise positioning of the hydraulic cylinders rods keeping a reasonable stability reserve for the speed governor. For medium-sized hydropower units, these contradictory requirements can be satisfied by means of a two-stage proportional flow servovalve such as that presented in Figure 4.2.1. The hydraulic diagram of this servovalve is presented in Figure 4.4.1. The servovalve has critical lap two-slope flow characteristics (Figure 4.4.2). Each metering land of the spool of such a flow control valve has rectangular slots (Figure 4.4.3). The servovalve dynamics (Figure 4.4.4) strongly depends on the supply pressure. The amplitude of the input signals is also an important parameter. A small supply pressure and a high-turbine nominal output lead to a three-stage servovalve. For example, the two servomotors acting

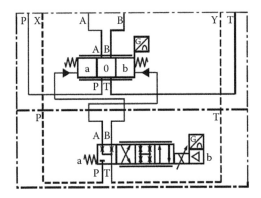

Figure 4.4.1 Hydraulic diagram of the two-stage proportional servovalve.

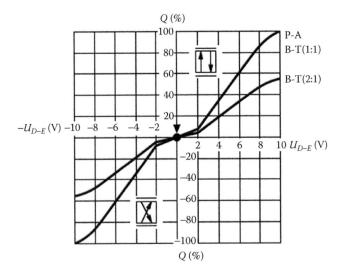

Figure 4.4.2 Flow characteristics of the two-stage proportional servovalve 4WRLE 35.

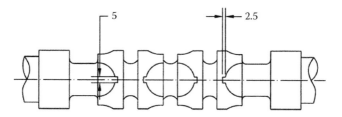

Figure 4.4.3 Typical spool with shaped metering windows.

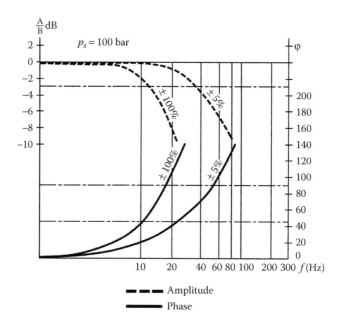

Figure 4.4.4 Typical dynamic performances of 4WRLE 25 proportional servovalves.

as the wicket gates of the Kaplan turbines of the IRON GATES I Hydropower Station from Danube River have a diameter of 600 mm and a stroke of 1200 mm. The servomotor stroking the runner blades has 3120 mm diameter and a stroke of 300 mm.

The pressure supply is about 40 bar and the emergency shutdown occurs in about 6 s. This performance requires a huge oil flow supplied by two control valves with three stages. The final stage is about 200 mm diameter. In such cases, the designer needs always a proper mathematical model and detailed simulations in order to optimize a complex architecture [1–3].

4.4.3   New hardware design

A proportional servovalve type 4WRLE 25 produced by Rexroth [11] was chosen to control a big diameter power stage. For security reasons, this stage was included in the fail-safe control loop of the governor by means of a set of disc springs controlled by an external pressure supplied by the emergency shutdown section of the speed governor.

The power stage has a complex structure, which offers four-flow control ways with a special design of the spool lands shown on Figure 4.4.3. Using a 100 mm spool diameter, the valve typical flow overcomes 7800 L/min under a very small pressure drop (5 bar on a metering edge). The flow of the hydraulic cylinders is normally controlled by the Rexroth servovalve. A sealed position transducer is measuring the spool position continuously, turning the whole assembly into a proportional device included in the position control loop of the wicket gates.

This new combination eliminates the pipes between the control stage and the power stage from the classical design. In such a way, the dynamic performances for the speed control are still very good, in spite of the very low supply pressure of the governor. An emergency signal generated by the speed governor creates a wide connection needed to shut down the turbine in a few seconds in the third stage.

However, an emergency directional flow valve is usually maintained in the hydraulic diagram of the speed governor, in parallel with the three-stage servovalve connected to the hydraulic servomotors. One end of the spool is always connected to the control oil-supply system and the other end can be quickly connected to the tank by a big cartridge valve controlled by a small two-way directional flow valve.

4.4.4   New controller design

The main problem raised by the use of an additional power stage is the stability of the whole valve. The previous research and studies carried out on the same problem by MTS System Corporation in the field of seismic platforms [5] revealed the need of a feed foreword lead term in the basic PI controller (Figure 4.4.5).

The derivative component of the controller with the input for the actuated seismic simulator table offers good dynamic performance.

The only restriction of this servovalve is the very high level of the oil-filtering system because the first stage is a nozzle-flapper one. The recommended filtration level is about 3 microns absolute for the first stage and about 10 microns absolute for the second stage (MTS 100-241-355 Specification).

In the case of a hydropower unit, the oil cleanliness cannot be maintained at the above level. From this reason, the pilot stage of the new servovalve was chosen from the new

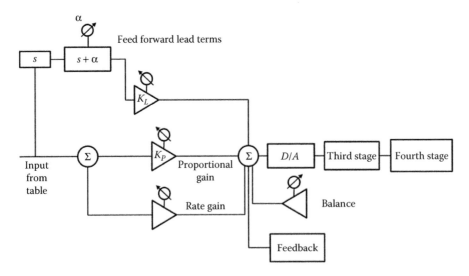

Figure 4.4.5 The MTS controller structure for four-stage electrohydraulic servovalve.

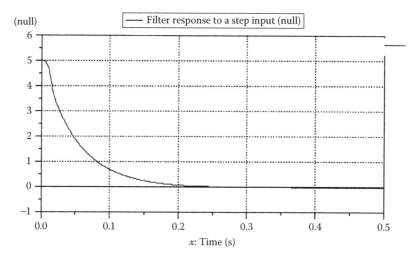

Figure 4.4.6 Step input response of the digital filter used on the error path.

family of the proportional high-speed servovalves, which uses one or a double solenoid for stroking a very light four-way critical lap cylindrical spool.

All the attempts to change the available free design parameters of the third stage of the valve in order to avoid a strong instability pointed out the need of a filter 2/2 on the spool position error. This is the same situation encountered in the synthesis of some fly control system of the big missile such as the Arianne [4]. It comes from the very small dumping factor of the third stage. A typical response to a step input of the filter is shown in Figure 4.4.6.

4.4.5   *Numerical validation of the new design*

Amesim software [6–8] produced by LMS company, a member of Siemens group, was selected as a current simulation tool [9]. This complex software offers numerous advantages: rich library of hydraulic symbols and components, which allows the authors to use existing, proven models for well-known components (valves, cylinders); ability to simulate different parts of the system at different levels of complexity, which allows the authors to model different parts of the system at different levels of detail, as required.

Amesim models are fully compatible with LabVIEW for Real-Time and HiL simulations, can be imported in LabVIEW and connected to a Real-Time or HiL simulation system.

The power stage of the flow valve, which has the greatest influence on the dynamic behavior of the system, was modeled in deep detail [10] at the physical process level, using the Hydraulic Component library of Amesim in order to obtain access to all the internal variables.

The pilot stages, the cylinders, the pressure source, and so on have been modeled at a more concise level using predesigned blocks from the Amesim Hydraulic and Mechanical libraries. Ultimately, the filter needed to improve the valve stability has been modeled by a transfer function because the internal variables are not particularly important for the current simulation. The simplest Amesim model developed by the authors for the three-stage servovalve can be seen in Figure 4.4.7.

Many types of simulation have been performed using this model. First, the authors have simulated a step input signal in order to find the dynamics of each stage of the valve. The evolution of the main parameters of the system for the first simulation run is presented in Figures 4.4.8 through 4.4.13.

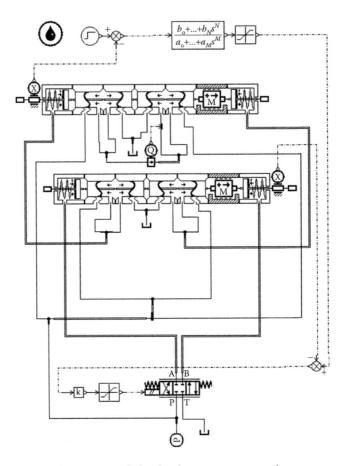

Figure 4.4.7 Amesim simulation network for the three-stage servovalve.

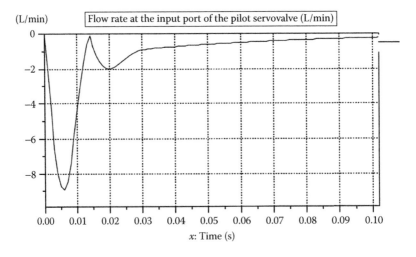

Figure 4.4.8 The flow rate of the pilot stage.

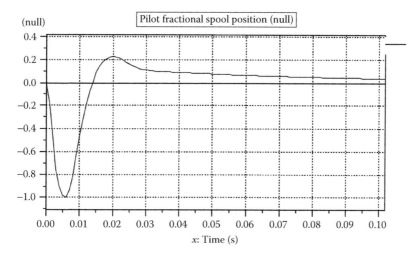

Figure 4.4.9 The first stage spool evolution.

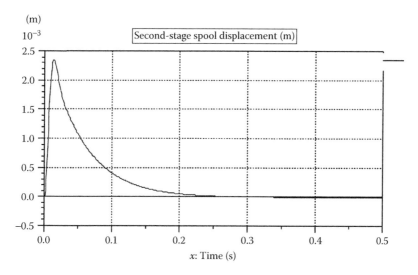

Figure 4.4.10 The second-stage spool stroke.

Second, the authors have simulated a slow linear input signal (Figure 4.4.14) in order to find the steady-state characteristics of the valve. The evolution of the main parameters of the system for a typical simulation is shown in Figures 4.4.15 through 4.4.20. The numerical simulation shows a good dynamic behavior of the servovalve for different kinds of inputs.

The most important quality of this kind of electrohydraulic converter is the specific behavior in the null region and outside this region.

The small slope of the flow curve for small-frequency deviations always occurring during the turbine start is very useful for a very quick start of the hydropower unit.

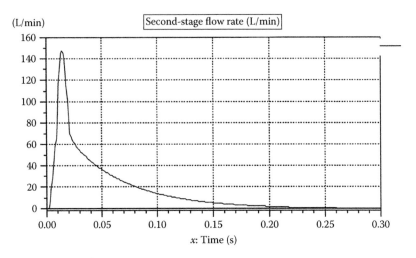

Figure 4.4.11 The second-stage flow evolution.

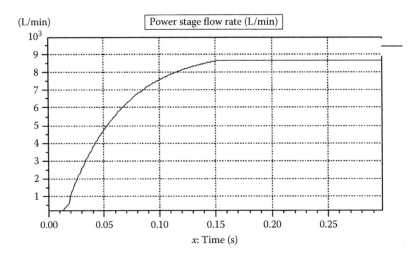

Figure 4.4.12 The servovalve flow evolution.

The big slopes allow a quick shutdown ordered by the speed governor or by the overall power station control unit.

4.4.6   *Experimental validation*

The experimental validation of the new design was performed in the Fluid Power Laboratory of the Politehnica University of Bucharest, Romania. Different types of analog and digital servovalves were tested in the frame of a servomechanism with strong hydraulic load in order to prepare this subsystem for building a fair model for the whole servovalve assembly.

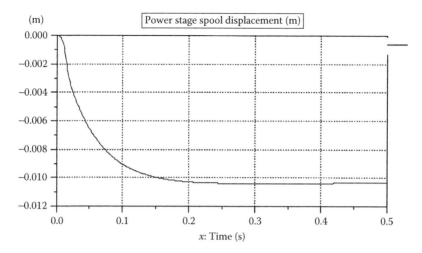

Figure 4.4.13 The power-stage spool evolution.

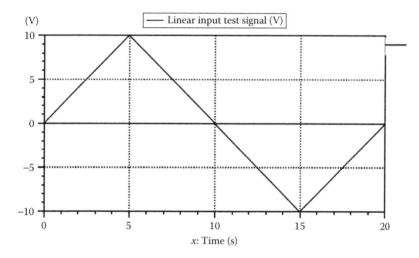

Figure 4.4.14 Linear input voltage applied for finding the steady-state characteristics.

The results obtained in the previous section by numerical simulation have been compared with the experimental data collected by the authors while working on a similar speed governor for the hydropower units of Ramnicu–Valcea hydropower plant [1].

Taking into account, the high-speed response of the Rexroth servovalves [11], used as a preliminary stage in the new three-stage design, the dynamic behavior of the two new governor types is nearly the same.

The only difference comes from the very low-pressure supply used for controlling the hydraulic turbine speed with the new three-stage servovalve. The speed governors dedicated to entirely new Kaplan turbines are sized for 60 bar in order to reduce the size of the hydraulic cylinders, and to improve the control-system performance.

The authors also studied the level of the safety function of the pilot stage of the new servovalve using a new test bench designed for this purpose (Figure 4.4.21). Redundancy

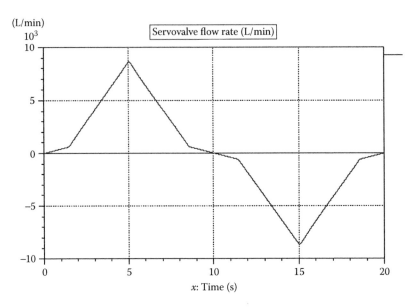

Figure 4.4.15 The power-stage spool flow rate evolution.

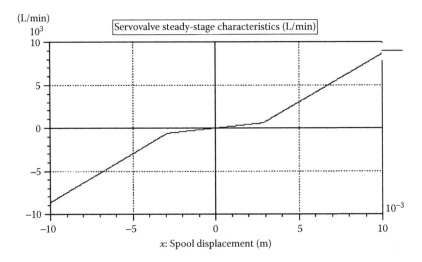

Figure 4.4.16 The steady-state servovalve characteristics.

problems were solved in order to find the best structure of the controllers and the architecture of the supervisor controller. The frequency response of the new type of servovalve was found directly from the Amesim model using a special facility of the language (Figure 4.4.22).

The phase diagram (Figure 4.4.23) shows a poor dynamic of the servovalve due to the very low-supply pressure and the very small slope of the steady-state characteristics

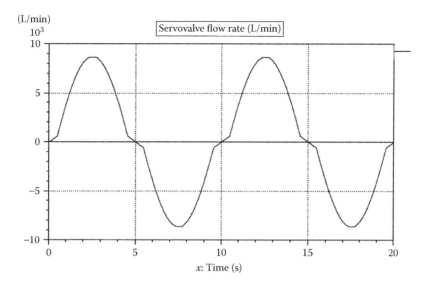

Figure 4.4.17 The servovalve flow for a sine input of 0.1 Hz.

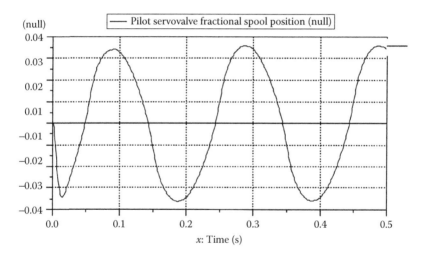

Figure 4.4.18 The pilot spool fractional displacement at the beginning of a sine input of 5 Hz.

around the null region. At the same time, the overall behavior is well suited for the speed control of large hydropower units during start-up.

The latest commercial three-stage servovalve operating under low pressure (26 bar) was designed and manufactured by Duplomatic [10] for the refurbishing of an old hydropower station (Figure 4.4.24).

4.4.7   Conclusion

All the design, test, and identification activities of the new servovalve project pointed out that Amesim provided a strong solver and numerical core for transient's simulation.

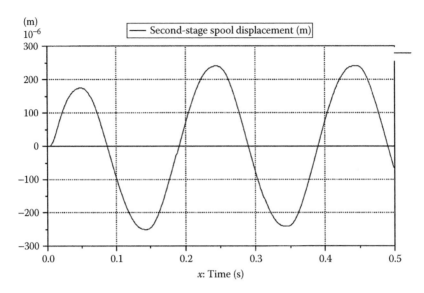

Figure 4.4.19 The second-stage spool displacement at the beginning of a sine input of 5 Hz.

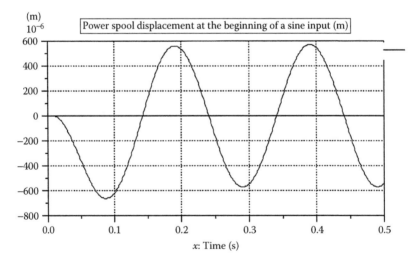

Figure 4.4.20 The third-stage spool displacement at the beginning of a sine input of 5 Hz.

As modeling a complex multiphysics system is not the main objective of engineers, it is important to have tools and interfaces, which accelerate and optimize the design. From this point of view, Amesim is a complete software adapted for model creation and deployment. The wide field of application, including electric powertrain is continuously extended by the users.

Figure 4.4.21 Dynamic test bench for pilot stages (both servovalves and DDVs).

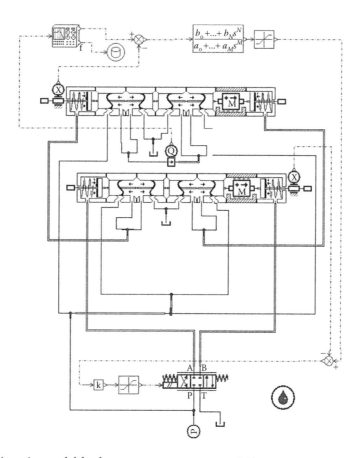

Figure 4.4.22 Amesim model for frequency response computation.

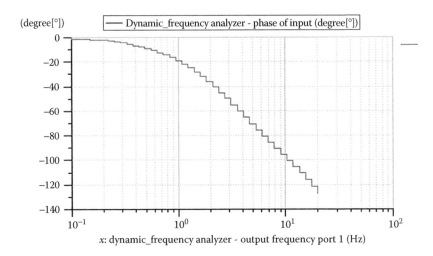

Figure 4.4.23 The phase diagram of the servovalve for a small input signal (10%).

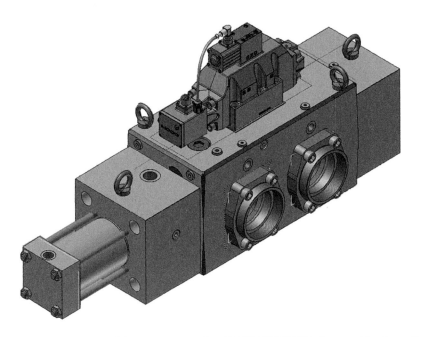

Figure 4.4.24 A virtual view of the new servovalve (SOLID WORKS). (Source: Duplomatic.)

Acknowledgments

The authors are grateful to DUPLOMATIC OLEODINAMICA SpA [12] and to Philippe Geril, the General Secretary of EUROSIS [13], for the permission of publishing this section, first presented in the frame of ESM'2015 as a short release.

Section 1: Bibliography

1. Lebrun M., Claude R. *How to create Good Models without Writing a Single Line of Code.* Fifth Scandinavian International Conference on Fluid Power, Linköping, 1997.
2. LMS INTERNATIONAL. *LMS Imagine.Lab Amesim*, Release 13 User Manual, Leuven, 2013.
3. Mare J.-Ch. *Dynamics of the Electrohydraulic Rotary Servomechanisms.* PhD Thesis, I.N.S.A. Toulouse, 1994.
4. PARKER HANNIFIN. Servovalves series SE. Hydraulic Valve Division, Catalog HY14-1460, Elyria, Ohio, USA, 2004.
5. MOOG. *Specification Standards for Servo Valves.* Internal Standard 117, East Aurora. 1971.
6. MOOG. *Performance Estimation for Electro Hydraulic Control Systems*, Internal Standard 126, East Aurora, 1971.
7. Vasiliu N., Vasiliu D., Călinoiu C., Mihalescu B., Negoiţă, G.C. *E-Learning System for Fluid Power Systems*, Romanian State Grant No.162, 2007–2009.
8. Mihalescu B., Călinoiu C., Vasiliu N. *Modeling, Simulation and Dynamic Identification of the Electro Hydraulic Proportional Valves.* Industrial Simulation Conference ISC'2012, June 4–6, BRNO, Brno University of Technology, Czech Republic, pp. 105–109, EUROSYS-ETI Publication.
9. Mihalescu B. *Contribution to the Experimental Identification of the Electrohydraulic Servovalves*, PhD Thesis, University POLITEHNICA of Bucharest, 2014.
10. Puhalschi R., Vasiliu Daniela, Ion Guţă D.D., Mihalescu B. *Concurrent Engineering by Hardware-in-the-Loop Simulation with R-T Workshop.* The 18th European Concurrent Engineering Conference ECEC 2012, 18–20 April 2012, J. W. Marriott Hotel Bucharest, Romania, pp. 27–32.
11. Vasiliu D., Vasiliu N., Ion Guţă D.D., Bontoş M.D., Mihalescu B., Puhalschi R.C. *Virtual Laboratory for Electrohydraulic Proportional Servovalves.* 6th International Conference on Energy and Environment CIEM 2013, 7–8 November, Bucharest, ISSN 2067-0893.
12. Irimia P.C., Mihalescu B., Feher S., Vasiliu D. *Research on the Electrohydraulic Steering Systems of the Articulated Vehicles*, U.P.B. Sci. Bull., Series D, Vol. 76, Issue 1, 2014, ISSN 1454-2358.
13. Gheamalinga M., Puhalschi R., Mihalescu B., Ganziuc Al. *Modelling and Simulation of Electric Vehicles,* International Middle Eastern Simulation Multiconference, Muscat, Oman, 2014.
14. Feher S., Vasiliu N., Călinoiu C., Mihalescu B., Puhalschi R. *A New Type of Electro Hydraulic Valvetrain System*, 20th European Concurrent Engineering Conference, Novotel, Bruges, Belgium, April 28–30, 2014, pp. 63–67.
15. Vasiliu D., Vasiliu N., Călinoiu C., Mihalescu B., Bontoş M., Puhalschi R. *Using Virtual Laboratories in Concurrent Engineering*, 20th European Concurrent Engineering Conference, Novotel, Bruges, Belgium, April 28–30, 2014, pp. 73–79.

Section 2: Bibliography

1. Vasiliu N., Vasiliu D., Călinoiu, C., Drăgoi C., Mihalescu, B., Negoiţă G.C. *E-Learning System for Fluid Power Systems.* Scientific Report No.162, National Research Authority, 2009.
2. MOOG - Internal Standard 117. *Specification Standards for Servovalves*, East Aurora, 1971.
3. MOOG - Internal Standard 126. *Performance Estimation for Electrohydraulic Control Systems.* East Aurora, 1971.
4. MOOG - *Electrohydraulic Valves—Applications, Selection Guide, Technology, Terminology, Characteristics.* East Aurora, 2010.
5. ISO 10770-1-2009. *Hydraulic fluid power-electrically modulated hydraulic control valves. Part 1. Test methods for four-way directional control valves.*
6. National Instruments Corporation. *LabVIEW Course.* Manu, Texas, 2010.
7. Kehtarnavaz N. *Digital Signal Processing System Design: LabVIEW—Based Hybrid Programming.* Academic Press Elsevier, San Diego, CA, 2008.
8. Landau I.D., Zito G. *Digital Control Systems Design, Identification and Implementation.* Springer-Verlag, 2006.
9. Ogata K. *Modern Control Engineering.* Pearson Education International, 2002.

10. Mihalescu, B., Călinoiu C., Vasiliu N. *Modeling, Simulation and Dynamic Identification of the Electro Hydraulic Proportional Valve,* The European Multidisciplinary Society for Modelling and Simulation Technology Conference ESM'2011, Guimaraes, Portugal, 24–26 October, pp. 320–325, 2011.

11. Mihalescu, B. Researches on the experimental identificetion of the electrohydraulic servo-valves. PhD Thesis, University POLITEHNICA of Bucharest, 2014.

Section 3: Bibliography

1. Lebrun, M., Vasiliu, D., Vasiliu, N. *Numerical simulation of the Fluid Control Systems by AMESim.* Studies in Informatics and Control with Emphasis on Useful Applications of Advanced Technology, Volume 18, Issue 2, pp. 111–118, 2009.

2. Viersma, T.J. *Analysis, Synthesis and Design of Hydraulic Servosystems and Pipelines.* Elsevier Scientific Publishing Company, Amsterdam, 1980.

3. Guillon, M. and Thoraval, B. *Commande et asservissements hydrauliques et électro-hydrauliques.* Technique et documentation – Lavoisier, Paris, 1992.

4. Mare J.-Ch. *Les actioneurs aeronautiques 1.* ISTE Editions, 2016.

5. Rösth, M.R. *Hydraulic Power Steering System Design in Road Vehicles.* PhD Thesis, Linköping University, 2007.

6. Vasiliu, N., Călinoiu, C., Vasiliu, D. *Modeling, Simulation and Identification of the Electrohydraulic Speed Governors for Kaplan Turbines by AMESim.* Symposium on Power Transmission and Motion Control-PTMC, Bath, UK, ISBN 978-0-86197-140-4. 2007.

7. Dardac, L.T., Vasiliu, N., Călinoiu, C. *Electrohydraulic VVT System for High Power Diesel Engines.* Romanian Patent no. RO126878, 2014.

8. Negoiţă, G.C. *Researches on the Dynamics of the Hydrostatic Transmissions.* PhD Thesis, University POLITEHNICA of Bucharest, 2011.

9. Popescu, T.C. et al. *Numerical Simulation—a Design Tool for Electro Hydraulic Servo Systems,* in *"Numerical Simulations, Applications, Examples and Theory",* INTECH PRESS, Zieglergasse 14, 1070 Vienna, Austria, 2011.

10. Vasiliu, C. RTS of the Electric Powertrains. PhD Thesis, University POLITEHNICA of Bucharest, 2011.

11. BOSCH - Automation Technology. *Servo Solenoid Valves. Technical Specification 13/2.* Stuttgart, 1999.

12. Imagine. *Numerical Challanges posed by Modeling Hydraulic Systems.* Technical Bulletin 114, 2001.

13. LMS INTERNATIONAL, *Advanced Modeling and Simulation Environment,* Release 13 User Manual, Leuven, 2013.

14. MOOG. *D680 Series Mini Direct Drive Valve Piloted.* Servo-Proportional Control Valves with Integrated Electronics ISO 4401 Size 05 to 08, 2008.

15. www.boschrexroth.com

16. www.duplomatic.com

17. www.eaton.com

18. www.sauer-danfoss.com

19. www.moog.com

20. www.hydac.com

21. www.parker.com

22. www.siemens.com

23. www.mathworks.com

24. www.dspace.com

25. www.adwin.de

26. www.fluidpower.net

Section 4: Bibliography

1. Vasiliu N. and Călinoiu, C. *Electrohydraulic digital Speed Governor for Hydropower Units*, Romanian Patent no. 120101, 2003.
2. Călinoiu, C., Negoiă, G., Vasiliu, D. and Vasiliu N., *Simulation as a Tool for Tuning Hydropower Speed Governors*, ISC'2011, Venice, Italy, 2011.
3. Vasiliu N., Călinoiu C. and Vasiliu D. Modelling, *Simulation and Identification of the Electrohydraulic Speed Governors for Kaplan Turbines by AMESIM*, Symposium on Power Transmission and Motion Control - PTMC 2007, Bath, U.K., ISBN 978-0-86197-140-4.
4. Mare, J-C., Modelling of Aerospace Actuation Systems, European AMESIM Conference, Strasbourg, 2006.
5. Rood E.O., Chen H.S., Larson R. and Npwak F., *Development of High Flow, High Performance Hydraulic Servo Valves and Control Methodologies in Support of Future Super Large-Scale Skating Table Facilities*, 12 WCEE 2000, New Zeeland.
6. Lebrun M., Vasiliu D. and Vasiliu N. *Numerical Simulation of the Fluid Control Systems by AMESIM*, Studies in Informatics and Control with Emphasis on Useful Applications of Advanced Technology, 2009, Volume 18, Issue 2, pp. 111–118.
7. IMAGINE, *Numerical Challenges Posed by Modelling Hydraulic Systems*, Technical Bulletin 114, 2001.
8. LMS, *Advanced Modelling and Simulation Environment*, Release 13 User Manual, Leuven, 2013.
9. Popescu T.C., Vasiliu D. and Vasiliu N. *Numerical Simulation—a Design Tool for Electro Hydraulic Servo Systems*, in "Numerical Simulations, Applications, Examples and Theory", INTECH PRESS, Zieglergasse 14, 1070 Vienna, Austria, 2011.
10. Costin, I., Vasiliu N., Călinoiu C., Vasiliu D. and Bontoş D.M. *Synthesis of a three Stages Servo Valve by AMESIM*, 29th European Simulation and Modelling Conference ESM'2015, October 26–28, 2015, Holiday Inn, Leicester, United Kingdom.
11. REXROTH, *Servo Solenoid Valves. Technical Specification 13/2*, Automation Technology, Stuttgart, 1999.
12. http://www.duplomatic.com/assets/Sistemi/ENERGY-4.pdf
13. www.eurosis.org

chapter five

Numerical simulation and experimental identification of the hydraulic servomechanisms

5.1 Signal port approach versus multiport approach in simulating hydraulic servomechanisms

5.1.1 Problem formulation

Many different modeling and simulation software packages were created to perform studies in the fields of automotive, aerospace, robotics, offshore, and general hydraulics engineering, but none of them offered the full range of capabilities needed for practice. There were deficiencies in the numerical capabilities of the following: graphical interfaces and general modeling concept.

In the signal port approach, a single value or an array of values are transferred from one component block to another in a single direction. This is fine when the physical engineering system behaves in the same way such as the one with a control system. However, problems arise when power is transmitted. This is because modeling of components that transmit power leads to a requirement to exchange information between components in both directions. In order to use a signal port approach in this situation, two connections must be made between the components where physically there is only one. This leads to a great complexity of connections and this means that even very simple models involving power transmission appear complex and unnatural.

In contrast to the signal port approach, with the multiport approach, a connection between two components allows information to flow in both directions. This makes the system diagram much closer to the physical system. Normally there are two values involved, and the theory of bond graphs provides a good theoretical background into the relationship between these values and the power transmitted. However, there is no limitation in the number of quantities involved. There may be one quantity or three or more quantities. When there is only one quantity, the situation is just like the one with signal ports. Thus, signal ports can be regarded as a special case of multiport case [1–3]. Amesim has always used the multiport approach, and Figure 5.1.1 shows part of a simple electrohydraulic system using multiport block diagrams. Figure 5.1.2 shows the same system with signal ports. The control of the valve is identical in both cases as for this port, the multiport reduces to a signal port. However, for the hydraulic and mechanical ports, the extra connections needed for the signal port approach are apparent.

This *preliminary section* uses the study by SIMULINK® of the important category of the electrohydraulic position control systems to underline, in the next chapters, the utility of the Amesim multiport approach in the analysis and synthesis of the complex technical systems.

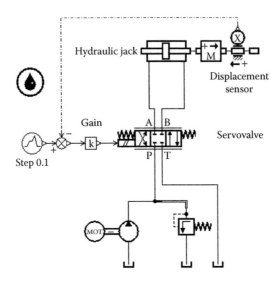

Figure 5.1.1 The multiport approach.

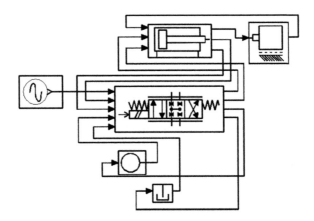

Figure 5.1.2 The signal port approach.

5.1.2 *Mathematical modeling of an electrohydraulic servomechanism controlling the displacement of a servopump*

An industrial bent axis variable displacement pump [4] needs an electrohydraulic servomechanism to tilt the barrel case (Figure 5.1.3).

The servomechanism (Figure 5.1.4) includes a spring-centered hydraulic cylinder, an industrial servovalve, a linear variable differential transformer (LVDT), and a digital or analog servocontroller [4].

The simplest realistic mathematical model of such a system contains the following equations.

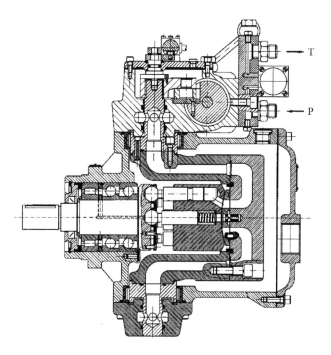

***Figure* 5.1.3** Main section of a bent axis variable displacement servopump for closed circuits (laboratory model).

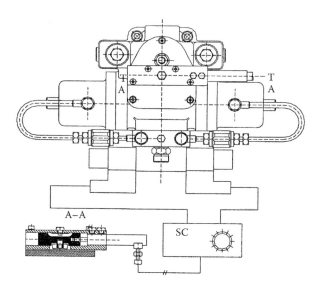

***Figure* 5.1.4** Top view of an electrohydraulic servomechanism set-up on a bent axis pump.

5.1.2.1   *The steady-state characteristics of the servovalve main stage (four way, critical centre, spool valve)*

$$Q_{SV}(x,p) = c_d A(x) \sqrt{\frac{p_S - P}{\rho}}$$ (5.1.1)

where:

    x is the spool displacement from the neutral position
    P is the pressure difference between the ports of the hydraulic cylinder
    $A(x)$ is the metering ports surface
    c_d is the discharge coefficient of the metering ports
    p_s is the supply pressure (constant)

The above-mentioned relation can be written in the following form:

$$Q_{SV}(x,P) = c_d \pi d_s x \sqrt{\frac{p_s}{\rho}} \sqrt{\frac{1-P}{p_s}} = K_{Qx} x \sqrt{\frac{1-P}{p_s}} \tag{5.1.2}$$

Here

$$K_{Qx} = c_d \pi d_s \sqrt{\frac{p_s}{\rho}} \tag{5.1.3}$$

is the *flow valve gain*.

5.1.2.2   The spool motion equation

The servovalve manufacturers specify for each device the transfer functions adequate to slow, normal, and high-speed control process. For slow control process, the servovalve can be regarded as a proportional device, with a single constant—the spool displacement–current gain:

$$K_{xi} = \left. \frac{\partial x}{\partial i} \right|_{x=0} \tag{5.1.4}$$

Hence the spool motion follows the input current, i, without any practical lag:

$$x = K_{xi} i \tag{5.1.5}$$

For normal control process, a servovalve can be regarded as a first-order lag device as follows:

$$\frac{x(s)}{i(s)} = \frac{K_{xi}}{T_{SV} s + 1} \tag{5.1.6}$$

The corresponding differential equation is

$$T_{SV} \frac{dx}{dt} + x = K_{xi} i(t) \tag{5.1.7}$$

Here T_{SV} is the servovalve time constant. For high-speed control process, the designer is obliged to consider the servovalve as a second-order lag device:

$$\frac{x(s)}{i(s)} = \frac{K_{xi}}{(s/\omega_n)^2 + 2s\zeta/\omega_n + 1} \tag{5.1.8}$$

where:

ω_n is the natural frequency

ζ is the damping coefficient

5.1.2.3 The position transducer equation

The modern inductive (LVDT) or magnetostrictive position transducers together with their bridges behave as first-order lag devices; they have a very small time constant, which can be neglected for industrial electrohydraulic control process:

$$U_T = K_T y \tag{5.1.9}$$

where:

K_T is the transducer constant

y is the piston displacement from the null position

5.1.2.4 The error amplifier equation

This stage computes the following error, ε, as a difference between the input signal, U_i, and the position transducer output, U_T, and applies the proportional–integral–derivative (PID) control algorithm to find the solenoid control voltage, U_c. For a PID error compensator,

$$U_c(s) = \varepsilon(s) \, K_P \left[1 + \frac{1}{(sT_i)} + \frac{sT_d}{(\tau s + 1)} \right] \tag{5.1.10}$$

The latest control algorithms are based on the fuzzy and adaptive fuzzy systems theory [7].

5.1.2.5 The servocontroller current generator equation

The current generator of the servocontroller is so fast than it can be regarded as a proportional device:

$$i = K_i U_C \tag{5.1.11}$$

Here $K_i[A/V]$ is the *medium* conversion factor.

5.1.2.6 The continuity equation

This equation is the connection between the servovalve flow and the derivative of the pressure drop across the hydraulic cylinder:

$$Q_{SV} = A_p \dot{y} + K_l P + \frac{A_p^2}{R_h} \dot{P} \tag{5.1.12}$$

where:

A_p is the piston area

K_l is the leakage coefficient between the motor chambers

R_h is the hydraulic stiffness of the cylinder:

$$R_h = 2 \frac{\varepsilon_p}{V_t} A_p^2 \tag{5.1.13}$$

where:

ε_e is the equivalent bulk modulus of the oil

v_t is the total volume of the oil from the cylinder and the connections

5.1.2.7 The piston motion equation

The pressure force F_p has to cover the stroking hydraulic force modeled by a spring force F_e, inertia of the moving parts m_e, and the friction force F_f:

$$m_c \ddot{y} = F_p - F_e - F_f \qquad (5.1.14)$$

Here, m_c is the tilting case of the variable displacement pump, and

$$F_p = A_p P \qquad (5.1.15)$$

$$F_e = 2(K_{e1} + K_{e2})(y + y_{0e1}) = 2K_e(y + y_{0e}) \qquad (5.1.16)$$

The friction force has *mainly* a static component, F_{fs}, and a viscous one, F_{fv}:

$$F_{fs} = F_{fs0} sign\, \dot{y} \qquad (5.1.17)$$

$$F_{fv} = K_{fv} \dot{y} \qquad (5.1.18)$$

5.1.3 Numerical simulation with SIMULINK

The main nonlinearity in the previous mathematical modeling is included in the servovalve main stage. A linear solution can be obtained using a linear form of the steady-state flow characteristics of the servovalve:

$$Q_{SV} = K_{Qx} x - K_{QP} P \qquad (5.1.19)$$

The results supplied by the linear model are useful for estimating the stability only. For high-amplitude input signals, anyone has to use the numerical simulation. Figure 5.1.5 presents the simulation network of the servopump. The computations were performed for the following typical parameters: $K_{Qx} = 0.88$ m²/s; $p_s = 70$ bar; $K_{xi} = 0.02$ m/A; $T_{SV} = 0.01$ s; $U_0 = 1$–10 V; $K_T = 500$ V/m; $A_p = 1.256 \cdot 10^{-3}$ m²; $K_l = 9.1 \cdot 10^{-13}$ m⁵/Ns; $K_{ie} = 0.005$ A/V; $R_h / A_p^2 = 1{,}6 \cdot 10^{13}$ N/m²;

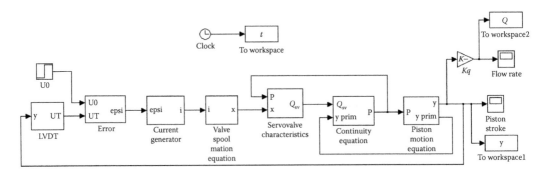

Figure 5.1.5 Simulation network for a linear electrohydraulic servomechanism in SIMULINK.

$m_c = 20$ kg; $F_{fs0} = 125–250$ N; $y_{0e} = 5$ mm; $p = 0....300$ bar; $K_e = 100,000$ N/m; $K_Q = 0.076$ m²/s; $V_{pmax} = 63$ cm³/rev; and $n_p = 25$ s⁻¹.

Figures 5.1.6 through 5.1.11 present the servomechanism response for typical step inputs. The influence of a sudden increase in the load of the hydraulic cylinder is also studied (Figure 5.1.12). The influence of the main parameters of the system components are presented in the last figures of this chapter (Figures 5.1.13 through 5.1.15).

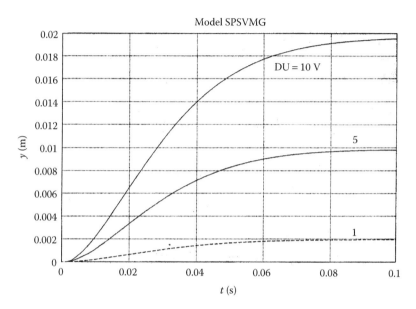

Figure 5.1.6 The response of the servomechanism for different input step signals.

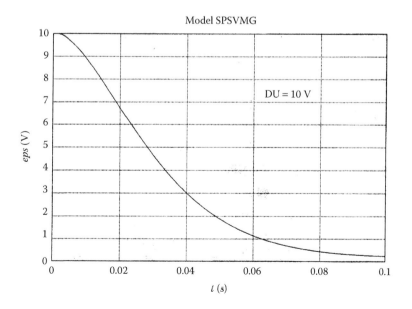

Figure 5.1.7 The error evolution for a step input of 10 V.

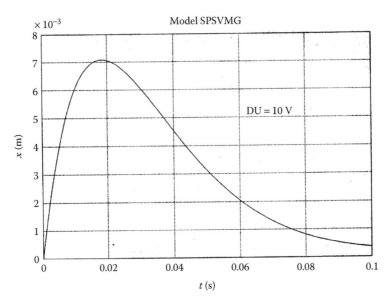

Figure 5.1.8 Evolution of the servovalve spool for the maximum input signal.

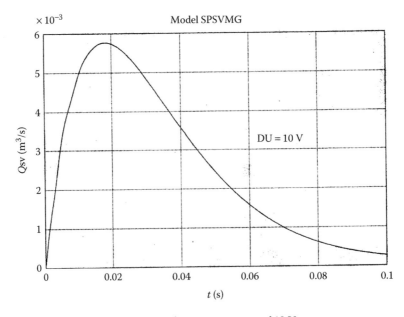

Figure 5.1.9 Servovalve flow rate variation for an input step of 10 V.

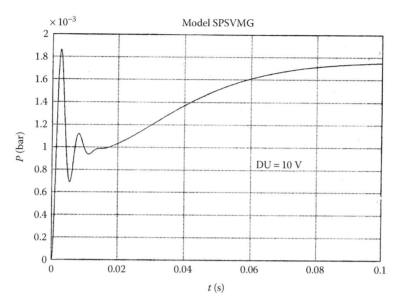

Figure 5.1.10 Pressure drop on the hydraulic cylinder for a step input of 10 V.

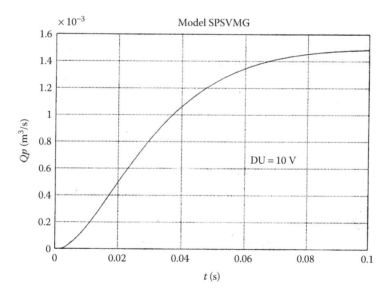

Figure 5.1.11 Servopump flow rate variation for the maximum input voltage.

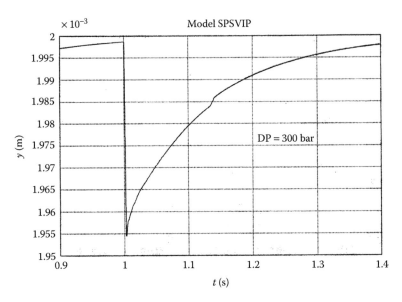

Figure 5.1.12 Servomechanism error compensation after a sudden increase in the pressure at the output port of the servopump from 0 to 300 bar.

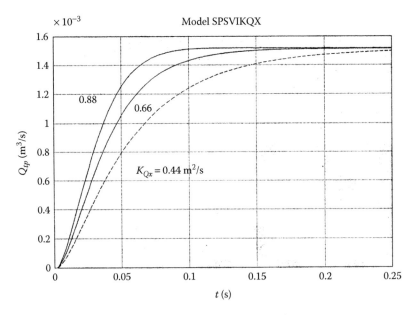

Figure 5.1.13 The influence of the servovalve flow rate gain on the step input response.

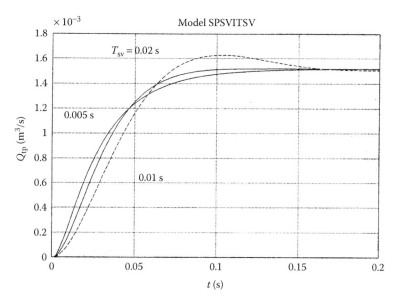

Figure 5.1.14 The influence of the servovalve time constant on the servopump response for a constant flow rate gain.

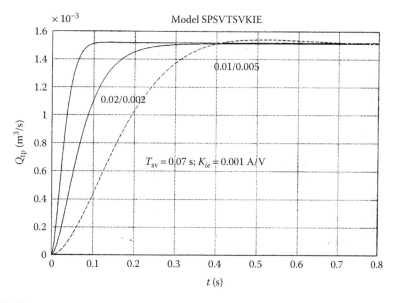

Figure 5.1.15 The servopump response for different servovalve overall performances.

5.1.4   Experimental results

Many experiments were performed to obtain a reasonable combination between the stroking systems of the barrel tilting case and the servovalve. The test started with a Rexroth three-way double-pressure proportional valve, which is controlled by a feed-foreword algorithm; then a BOSCH critical closed center four-way proportional valve was used,

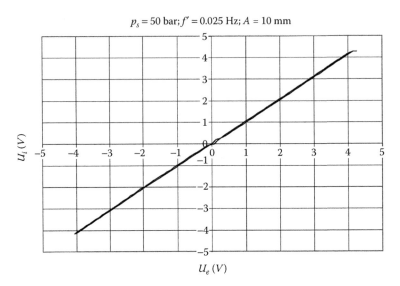

$p_s = 50$ bar; $f' = 0.025$ Hz; $A = 10$ mm

Figure 5.1.16 Typical steady-state characteristics of the servopump [4].

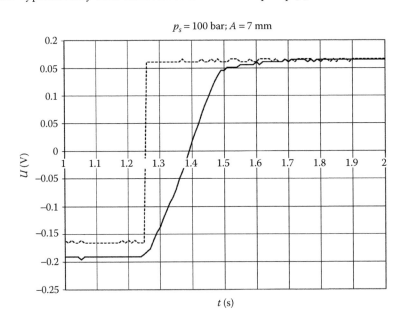

$p_s = 100$ bar; $A = 7$ mm

Figure 5.1.17 Typical step input response of the servopump around the null point generated with a medium-frequency industrial servovalve (DDV) [4].

followed by a Moog mechanical feedback two-stage servovalve. Finally, a Bosch Rexroth DDV type 4WRPEH 6 C B40L-3X/M 24/A1 was used with the best results regarding the ratio of price–performance. A typical steady-state characteristic of the tested servopump is presented in Figure 5.1.16. The small hysteresis obtained with an analog servocontroller can be neglected both in industrial position and in speed control loop. The response time is also small enough around the null flow of the servopump (Figure 5.1.17). These good results explain the use of this type of servopump in heavy environmental conditions instead of the modern swash plate ones.

5.1.5 Conclusion

The simulation performed by SIMULINK shows a simple manner of use in simple cases, where the information path is simple and clear. On the other hand, as will be seen in the next chapters of this book, the realistic models developed by the tools available in Amesim need a simpler form in order to be simulated in real time, as required for building electronic control units (ECU) devoted to mechatronic applications. The high-speed advanced technical process such as aerospace path control or automotive driverless motion requires high-level integrating algorithms, which can be found in the large-scale models developed by the high-tech company with Amesim [5–8]. In more complex processes, there is a possibility to concentrate the equations in a logical sequence, which is very useful in the simulations with hardware in the loop, which is widely used today to develop and test ECU for any mechatronic system [9].

5.2 Dynamics of the electrohydraulic servomechanisms used in variable valve trains of the diesel engines

5.2.1 Problem formulation

This section presents modeling, simulation, design, and experimental identification of a new electrohydraulic valve train drive system for internal combustion engines patented by the authors [1]. Both hydraulic scheme and design features are presented, in close connection with the peculiarities of the classical diesel engines for locomotives, ships, and so on. The numerical simulation was performed by Amesim and SIMULINK. The experimental investigations were carried out in the Fluid Power Laboratory of the University Politehnica of Bucharest, Romania, using original components manufactured by Hidraulica Brasov Company [2], combined with high-performance electrohydraulic industrial servovalves and high-speed Real-Time NI PXI digital controllers [3,4]. The actual system performances are found in good agreement with the requirements of the high-power modern diesel engines. The practical conclusions of this research are very useful in a concurrent engineering approach of any heavy automotive application. The keywords of this section are: EHVS, diesel engines, modeling, simulation, experimental tuning, and concurrent engineering.

5.2.2 EHVS features

For over a century, the internal combustion engine has been the motor of the industry and man's everyday life. Since the first models, a constant evolution is bringing the engines closer to perfection. This progress is mainly the result of global environmental concerns about an increase in the pollution with gases and a decrease in the fossil fuel resources.

The main disadvantage of the classical engine design is the use of the traditional camshaft in the actuation system. To create even more economical and cleaner engines, new solutions of performance improvements are required. A new approach is the use of an electrohydraulic valve train system to eliminate the camshaft and to create a reliable and fully flexible valve actuation system. Industrial successful applications were reported by Robert Bosch since 2004 [10,11] and Sturman Industries since 2005 [12].

The engines requirements for a flexible valve-train drive system led the engineers to various designs of electrohydraulic, electromechanical, and electropneumatic systems used by several engine manufacturers. Unlike conventional camshaft-driven systems and cam-based variable valve-timing techniques, a pure electrohydraulic system offers full management on the valve timing and lift control, which is fully independent of crankshaft position and engine speed.

Some practical implementation of the EHVS developed by Sturman Industries in 2008 for GENERAL MOTORS CORPORATION has shown the following benefits of camless engines, which is confirmed during official assessment [13]: independent valve lift and timing control allow mixed-charge combustion; low parasitic loss, since most of the conventional valve train (including the camshaft), is eliminated; high-speed engine operation (up to 12,000 rpm); homogeneous charge compression ignition combustion is enabled; cylinder deactivation and cylinder balancing are enabled; scalable to any engine size, and can be retrofitted to older engines as all the technology is in the head; compatible with any fuel type; a square lift pattern compared to the lobe profiles that a camshaft requires; can be used to create an internal exhaust gas recirculation (EGR) system; potential up to 60% thermal efficiency; and can be used to help eliminate the need for an exhaust after treatment.

Although flexibility of a camless valve train is limitless, their construction is more complex and expensive and requires deeper studies.

On the other hand, the cam-based systems that are well documented by Berthold Grünwald [6] and many other specialists are more reliable and robust, but they do not offer the required flexibility in operation.

5.2.3   EHVS types and performances

One method of adjusting valve timing and lift, given a fixed cam profile, has been to incorporate a *lost motion* device in the valve train linkage between the valve and the cam as described by Vorih and Egan, in 2001 [5].

Lost motion is the term applied to a class of technical solutions for modifying the valve motion prescribed by a cam profile with a variable length mechanical, hydraulic, or other linkage means. In a lost motion system, a cam lobe may provide the *maximum* (longest dwell and greatest lift) motion needed over a full range of engine operating conditions.

A variable length system (Figure 5.2.1) may then be included in the valve train linkage, intermediate of the valve to be open and the cam providing the maximum motion, to subtract or lose part or all of the motion imparted by the cam to the valve.

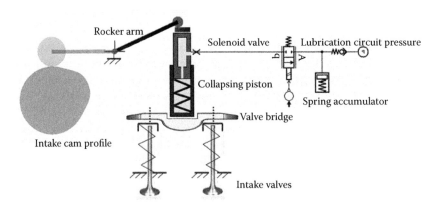

Figure 5.2.1 Electrohydraulic lost-motion system for valve actuation (Amesim demo model for VVA).

This variable length system (or lost motion system) may, when expanded fully, transmit all of the cam motion to the valve, and when contracted fully, may transmit none or a minimum amount of the cam motion to the valve.

An electrohydraulic valve train system with two solenoid valves with two way was developed by Robert Bosch in 2004 (Figures 5.2.2 and 5.2.3).

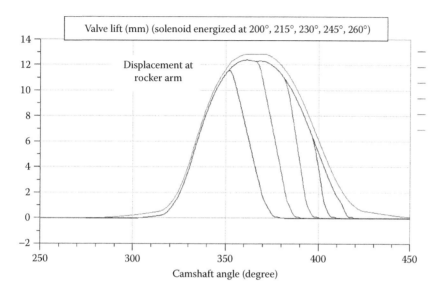

Figure 5.2.2 Valve lift versus cam shaft angle for different angles of solenoid activation (simulations with Amesim model for VVA).

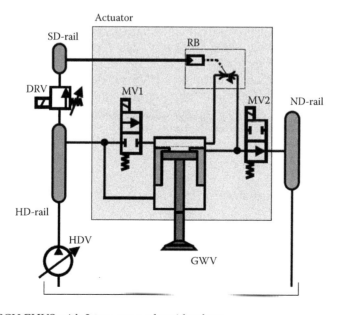

Figure 5.2.3 BOSCH EHVS with 2 two-way solenoid valves.

The new EHVS was tested on all types of internal combustion engines and it elimi-nates the valve closing spring. It has the following major advantages:

1. *In the matter of fuel consumption/efficiency*: Minimization of gas exchange losses (*dethrot-tling*, with manifold injection); valve cutoff, cylinder cutoff, multistroke operation; and reduction of idle speed (more stable due to less residual gas)
2. *In the matter of torque curve/power*: Operating point-dependent optimization of valve timing and stroke exhaust gas; operating point-dependent variation of the charge motion; variable internal EGR; less assembly groups; reduced size of the engine; more efficient engine brake; new starting strategies; more efficient engine brake; and new starting strategies

The simultaneous and uncoupled optimization of the fuel consumption and emissions require the independent control of fresh mixture, residual gas, and charge motion. For this, the opening and closing angles and the stroke of the individual gas exchange valves have to be controlled cycle-to-cycle conform. The system's fail safety, espe-cially the fail-safe position and closed gas exchange valve, has to be considered in the system's design.

In order to achieve a low-cost serial production with a high number of pieces, a use of the control principle and the partial components with all known engine types and com-bustion concepts has to be aimed. All material, partial components, and production pro-cesses necessary for the system's assembly have to be selected in compliance with the criteria availability and process safety.

The different engine types that are all to be covered with the same principle require a broadly applicable range relating to the parameters: force, stroke, and valve veloc-ity. Operational parameter's combination is very wide: high force (>1000 N), velocity (10 mm/2,5 ms), stroke (>10 mm); valve velocity and braking behavior adapted to operat-ing point; valve-free travel is not always to be assumed; and cylinder head with actuators has to be checked as module. After the installation of the whole module on the engine block is possible the exchange of individual actuators installed on the mounted cylinder head at temperatures between −40°C and 140°C. The repeatability of the valve stroke of this system is remarkable (Figure 5.2.4).

Another practical implementation of the EHVS was developed by Sturman Industries from California in 2008 for General Motors engines and for International DT466 6.0L Power Stroke. The *Sturman Camless Engines* uses a new type of three-way high-speed directional valve with a high-timing accuracy, which is specially developed for this type of drive.

5.2.4   New patented design description

The new design was implemented on a standard cylinder head of a 2,400 HP diesel engine from a Sulzer 12LDA28 locomotive (Figure 5.2.5). Two camshafts actuated the four valves against twin helical springs.

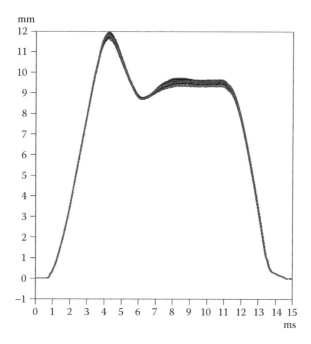

Figure 5.2.4 Valve stroke repeatability for BOSCH EHVS: 50 strokes obtained for the same valve with a constant supply pressure of 250 bar.

Figure 5.2.5 Head of a cylinder used in a diesel engine from Sulzer 12LDA28 locomotive.

The main idea of the new design, presented in Figures 5.2.6 through 5.2.9, is the use of a four-way high-speed industrial servovalve D1FP produced by Parker Corporation, the first prototype used a plunger for valve opening. The two original springs were used for closing the valve. This simple design was not useful because the resonance between the hydraulic cylinder and the springs was strong and unpredictable.

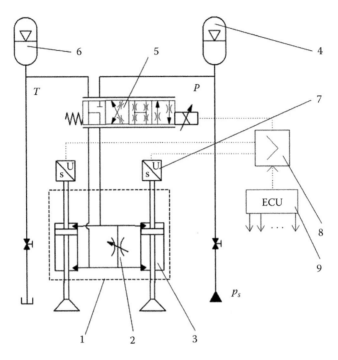

Figure 5.2.6 Hydraulic scheme of the new EHVS: 1—valves actuation block; 2—damping hydraulic resistance; 3—hydraulic cylinder; 4—high-pressure hydraulic accumulator; 5—high-speed indus-trial servovalve; 6—low-pressure hydraulic accumulator; 7—displacement transducer; 8—data acquisition board; and 9—digital controller.

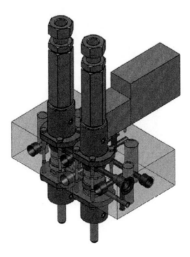

Figure 5.2.7 Partial view of the electrohydraulic servomechanism (SolidWorks).

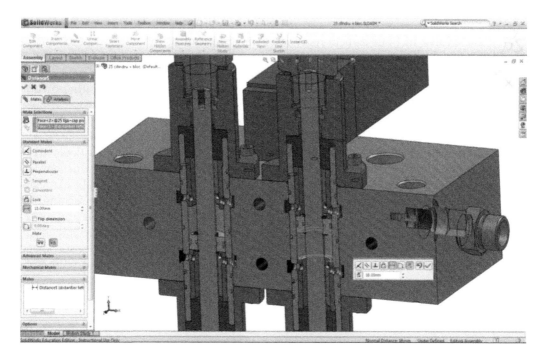

Figure 5.2.8 Cut view of the hydraulic cylinders (SolidWorks).

Figure 5.2.9 Partial view of the EHVS test bench in the Fluid Power Laboratory of U.P.B.

The final design eliminates the springs and uses a symmetric hydraulic cylinder for driving each valve in both directions. The optimal control of the valve position needs a feedback supplied by a special type of eddy current displacement transducer (VRVT) connected with the piston rod. The valve motion is controlled by a PT1 servocontroller connected to a NI PXI controller. The LabVIEW software is used for generating a composed input signal. The servovalve was supplied by an axial pistons swash plate variable-displacement pump from series PV46 (Parker) delivering 45 L/min for 250 bar. Two bladder accumulators were used for avoiding strong pressure surges in the hydraulic lines. The synchronous drive of two valves with the same servovalve needs both higher flows and a high-speed flow divider. This additional unusual device can be avoided by a mechanical linkage between the valves (a bridge). The numerical simulation performed by Amesim and SIMULINK languages (Figures 5.2.10 and 5.2.11) has shown that a simple rectangular input signal cannot be used for obtaining high-frequency valve motion (Figure 5.2.12). A composed input signal (Figure 5.2.13) using both positive and negative voltages can generate the proper valve opening for different frequencies, according to the needed speed range of the engine. Figure 5.2.14 shows different valve lift curves generated with composed signals.

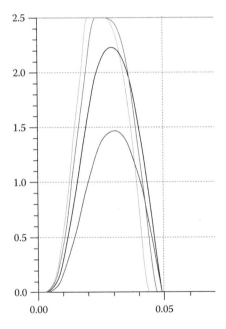

Figure 5.2.10 Amesim simulation of the dynamic behavior of the servosystem: small valve opening [mm] versus time [s].

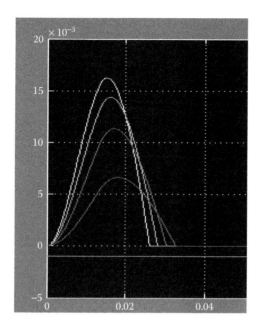

Figure 5.2.11 SIMULINK simulation of the dynamic behavior of the servosystem: large valve opening [mm] versus time [s].

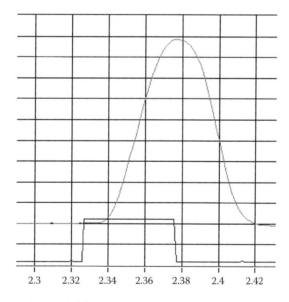

Figure 5.2.12 Experimental valve lift [mm] for a rectangular input signal ($U = 1$ V and $P_s = 100$ bar).

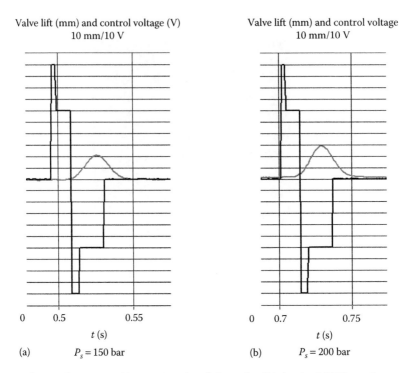

Valve lift (mm) and control voltage (V)
10 mm/10 V

Valve lift (mm) and control voltage
10 mm/10 V

0   0.5   0.55   0   0.7   0.75

t (s) t (s)

(a) P_s = 150 bar   (b) P_s = 200 bar

Figure 5.2.13 Optimal composed input signal and the valve lift for f = 6.25 Hz and two supply pressures. In strong dark: input signal of the servovalve, between –10 V and +10 V; In light dark: valve lift, between 20 mm and 30 mm.

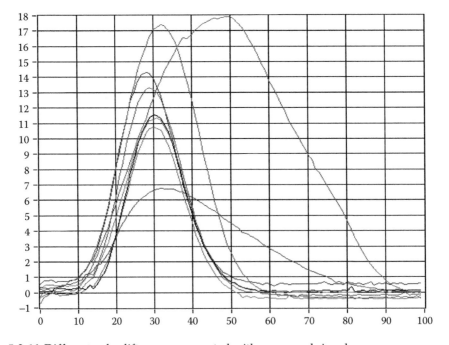

Figure 5.2.14 Different valve lift curves generated with composed signals.

5.2.5   Conclusion

The preliminary researches have shown a real potential for the new design to accomplish all the conditions needed to control a diesel engine. However, a big amount of research is still needed for implementing this new idea on a real high-power engine. At the same time, the high performances obtained by the development of other electrohydraulic automotive applications are encouraging the authors to refine the results by both numerical simulations and practical implementation of the new solution.

The continuous expansion of the piezoceramic actuators in the field of on/off servovalves shows the way to be followed for controlling small- and medium-power automotive engines speed [12]. At the same time, the recent developments in the control field of large-power diesel engines for ships propulsion and power cogeneration lead to the conclusion that the author's choice of the high-speed voice coil servovalves as first stage for the variable valve timing (VVT) systems are an open innovation field. Some preliminary reports of Parker Hannifin fluid power laboratories regarding the overall electrohydraulic control of the four-stroke diesel engines of 100,000 kW [14] show an increased confidence in the use of the electrohydraulic servovalves in the future architecture of all critical systems of this important technical field.

The whole research activity connected to the VVT problems revealed the great utility of the numerical simulation in the frame of any concurrent engineering approach.

5.3   Modeling and simulation of a hybrid electrohydraulic flight control servomechanism

5.3.1   Problem formulation

The *more electric* aircrafts will involve simpler, cheaper, and lighter power generation and distribution architectures: typically, two electric circuits and two hydraulic circuits. To achieve this goal, called "Fly-by-Wire", the modern aircrafts use electrohydraulic servovalves controlled by digital signals. In addition, on top of this, *Power-by-Wire*, which uses electric power to actuate flight control surfaces, has been developed. These developments have been ongoing for years, leading to two types of actuators: electrohydrostatic actuators (EHA) and electromechanical actuators (EMA). This section provides a demonstration showing Amesim capability [1,2,3] to model EHA in a multiscale approach. Functional model of an EHA was divided into two parts: (1) electromechanical and (2) electrohydraulic (Figures 5.3.1 and 5.3.2).

The main idea of this new type of servoactuator is to avoid the troubles produced many times by the electrohydraulic servovalves due to their very small control or dumping orifices, which is sited usually between 0.125 and 0.5 mm. The loop between a small high-pressure pump and a hydraulic cylinder is a direct one that avoids completely the loose of the null point of the servovalves. This idea is not a new one; it has been introduced long time ago in the field of the metal factories with long hydraulic cylinders. The main difficulty of promoting this idea in the aerospace field comes from the volume and the weight of the assembly of electromotor, the pump, the oil cooling system, and the hydraulic accumulator, which prevents the cavitation of the pump. The first step was the use of a DC motor and an internal gear pump for high pressure.

The new generation of the brushless motors increased the security but introduced a very sophisticated power electronics system. Today, the A380 and other giant planes are using this system as a backup one (Figure 5.3.3), and the normal service remains in the classical structure.

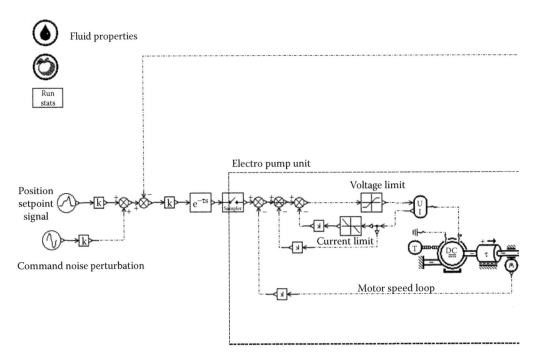

Figure 5.3.1 Electromechanical part of the system. (From Guillon, M. *L'asservissement Hydraulique et électrohydraulique*. Dunod, Paris, France, 1972. With Permission.)

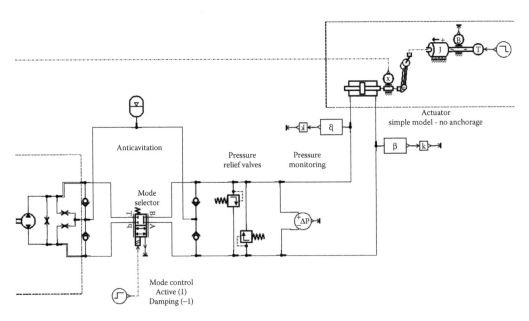

Figure 5.3.2 Electrohydraulic part of the system. (From Guillon, M. *L'asservissement Hydraulique et électrohydraulique*. Dunod, Paris, France, 1972. With Permission.)

Figure 5.3.3 Electrohydrostatic actuator for the ailerons of A380. (From Airbus Industries, Toulouse, France.)

5.3.2   *Preliminary study*

In this section, we study the behavior of a DC version, used for the beginning a core hydraulic system, and many simply hypothesis, as shown in Figure 5.3.4. The hydraulic

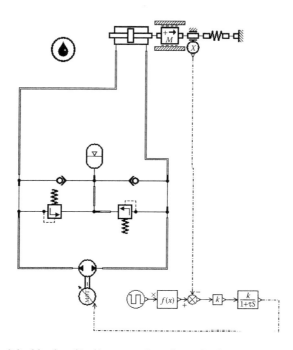

Figure 5.3.4 The simplified hydraulic diagram of an electrohydrostatic servomechanism.

cylinder has a piston of 40 mm, a rod of 20 mm, a stroke of 200 mm, and a mass of 5 kg. The pump is a small displacement one: 5 cm³/rev and a maximum speed of 3000 rev/min.

Figures 5.3.5 through 5.3.19 show a clear image of the whole transient generated by a square input that generates the nominal stroke by a proper choice of the first-order lag introduced on the control path of the DC motor.

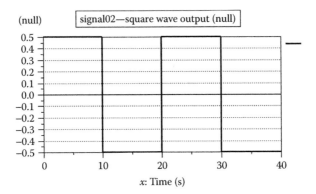

Figure 5.3.5 Input signal of the servosystem.

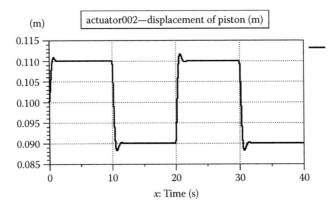

Figure 5.3.6 Cyclic piston displacement.

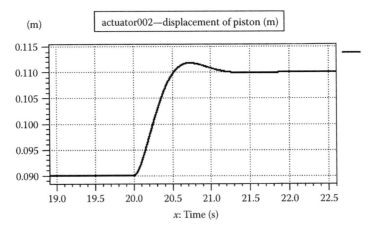

Figure 5.3.7 Piston displacement in a motion cycle.

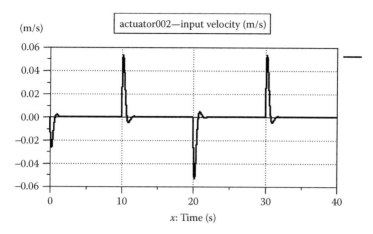

Figure 5.3.8 Piston speed cyclic evolution.

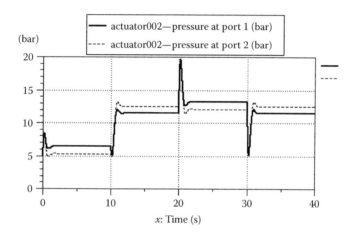

Figure 5.3.9 Pressure in the cylinder chambers.

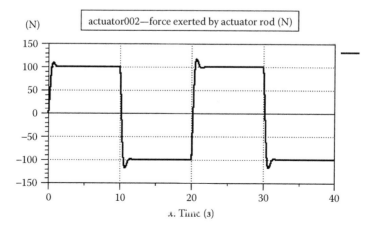

Figure 5.3.10 The actuation force on the cylinder rod.

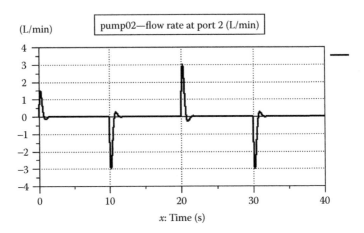

Figure 5.3.11 Pump flow rate.

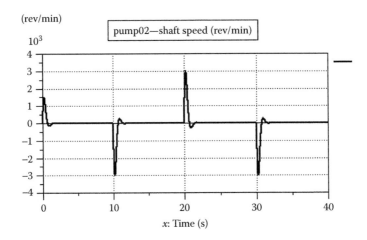

Figure 5.3.12 Pump speed variation.

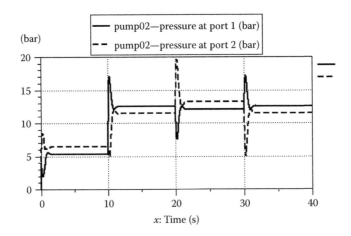

Figure 5.3.13 Pressure variation in the pump ports.

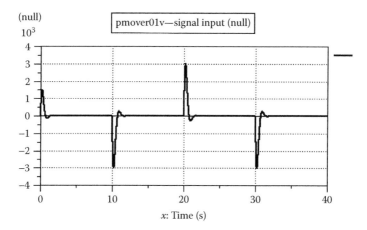

Figure 5.3.14 Input signal for the electrical motor.

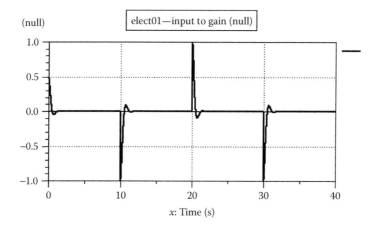

Figure 5.3.15 Positioning relative error evolution.

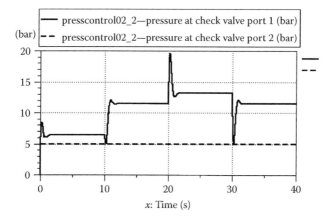

Figure 5.3.16 Pressure in the input port of the check valve no. 1.

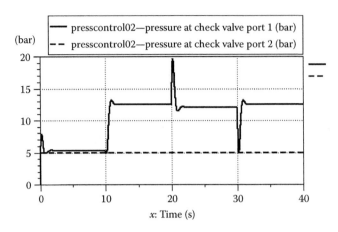

Figure 5.3.17 Pressure in the input port of the check valve no. 2.

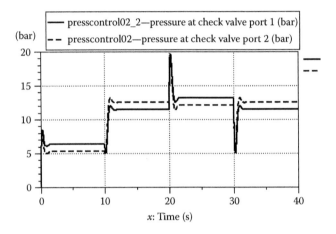

Figure 5.3.18 Pressure in the input ports of the pump check valves.

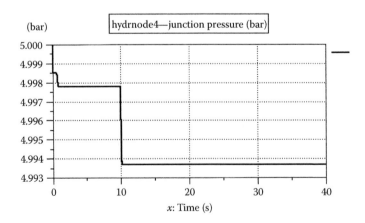

Figure 5.3.19 Pressure in the input ports of the check valves connected to the hydraulic accumulator.

5.3.3   Conclusion

The hybrid electromechanical and hydrostatic integrated servomechanisms offer a good level of security in special conditions due to the lack of the servovalves and the lack of the hydraulic power supply. At the same time, the complexity and the price of such units are very high, thus explaining their application in special cases only. The progress obtained with high speed in the power electronics field will facilitate their application in many other domains. The fast dynamic investigation of their performance with Amesim increases the chance of introducing these apparently complex servosystems in many other high-security levels of operation domains [4–6].

5.4   Increasing the stability of an electrohydraulic flight control servomechanisms by a hydraulic damper

5.4.1   Problem formulation

The modern flight control and guidance systems developed for giant civil or fight aerospace vehicles stated many complex problems for engineers, designers, and manufacturers. First of all, the ratio between the actuation force or torque and the mass of the control device has to be as high as possible. The limited space and weight; the large range of temperature and pressure; and the absolute security requirements are severe conditions for any kind of aerospace vehicle, from civil airplanes and helicopters to missiles or other defense vehicles. On the other hand, the tightening of the environment protection conditions leaded the civil airplane manufacturers to increase continuously the size and capacity of these important transport meanings. Today, the market leader from this point of view is A380, which needs only 3 l of kerosene for each passenger who is transported for more than 100 km.

The main control systems accomplishing the aforementioned conditions are the electro-hydraulic ones. Most of the moving components of a flight control systems are hydraulically driven because of the highest ratio force/mass that can be reached. For example, the rudder of A340 needs 225 kN and is actuated by an electrohydraulic servomechanism weighting 100 kg.

Hydraulic actuators are now electronically controlled by digital optical fiber systems, thereby giving them a good immunity regarding the electromagnetic disturbances. Figure 5.4.1 shows the main components of A380 controlled by electrohydraulic

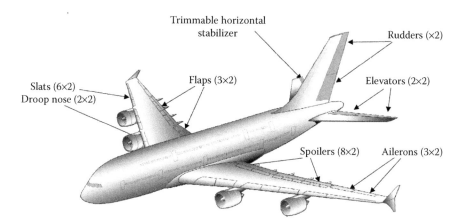

Figure 5.4.1 A380 flight control system. (From Xavier Le tron. A380 flight controls overview. *Presentation at Hamburg University of Applied Sciences*, 2007. With Permission.)

actuators. Moreover, the landing gear, the brakes, the engines' thrust reverser, and the wheel brakes are also hydraulically actuated.

One of the most important problems facing the aerospace engineers who design electrohydraulic servomechanisms is the control stability. The interaction between the aerodynamic forces, the low structures' stiffness, and the hydraulic fluid compressibility generates oscillations that have to be limited to allow a proper control of any mechanical system of the plane [2]. There are many technical solutions for solving these problems but they are useful for moderate aerodynamic loads and mass [3,4]. Mostly, all of them need a lot of tuning generated by the uncertain structural stiffness and natural frequencies of the driven components, which is connected to the hydraulic fluid compressibility. The resonant frequencies have to be as high as possible, and the amplitude of the oscillations has to be limited from fatigue considerations. The previous long experience in the domain generated different technical solutions for solving this problem: use of servovalves with shaped windows in the sleeve, well-controlled fluid leakages between the hydraulic cylinder chambers, hydromechanical transient filters, hydraulic dampers in parallel with the hydraulic cylinder, and so on. The last solution, presented in Figure 5.4.2, was adopted by Airbus [2] for the rudders.

The main advantage offered by a damper working in parallel with the hydraulic cylinder is the possibility of eliminating the steady-state error in any position of the rudder. The calibrated orifice of the damper can be easy tuned by numerical simulation and can be validated on the *Iron Bird* test bench without many corrections after the flight test. The damper design can be a classical one using the same technology as that of the hydraulic actuator. A sharp-edge orifice placed in a mobile sleeve can be changed without dismantling other components of the assembly. The simulation network for the whole servomechanism is presented in Figure 5.4.3.

The main components of the servomechanism have the followings features: hydraulic cylinder bore, 100 mm; rod diameter, 50 mm; maximum stroke, ±100 mm; nominal system pressure, 350 bar; damper cylinder bore, 40 mm; damper rod diameter, 20 mm; servovalve nominal flow, 60 L/min; damper nominal orifice diameter, 1 mm; equivalent rudder mass, 500 kg; and aerodynamic force gradient, relative to the servomotor rod, 1000 N/mm. The hydraulic fluid used in preliminary numerical simulations is a synthetic noninflammable one, with a high bulk modulus (17,000 bar), a common density (850 kg/m³), and a usual absolute viscosity of 51 cP at 40°C. The ratio of air and gas content considered in the next dynamic computation is about 0.1%.

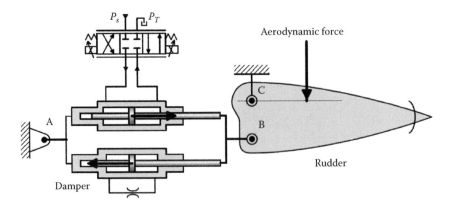

Figure 5.4.2 The electrohydraulic control system of the rudder with hydraulic damper. (From https://www.ac-paris.fr/serail/jcms/s1942790/ee2-lg-clg-paul-valery-portail. With Permission.)

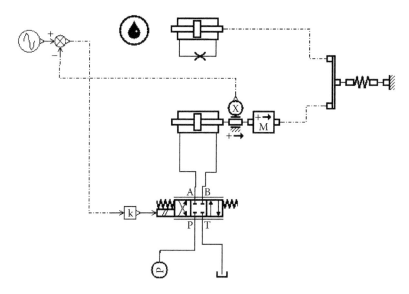

Figure 5.4.3 Amesim simulation network for a sine input signal.

5.4.2   *Sine input response of the servomechanism*

Three types of input signals were used to obtain a complete image of the dynamic behavior of the servomechanism: a constant amplitude, constant frequency sine input, an input small step, a constant amplitude, and increasing frequency sine input. The first category of simulation performed for a sine wave of 0.5 Hz and 50% amplitude from the nominal one is presented in Figures 5.4.4 through 5.4.15.

The piston motion amplitude reaches 92% from the theoretical one (Figure 5.4.5).

The maximum piston velocity is about 0.14 m/s. The pressures in the actuator chambers have periodical variations in opposition between 135 and 215 bar, the average value (175 bar = 350/2 bar) respecting the fundamental law of the critical center constant pressure-supplied servovalves [3].

The force developed by the actuator rod reaches 46,500 N.

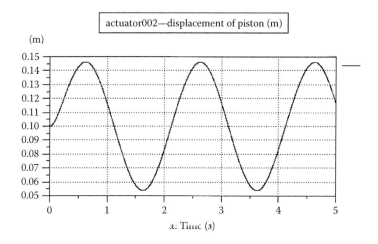

Figure 5.4.4 Displacement of the actuator piston.

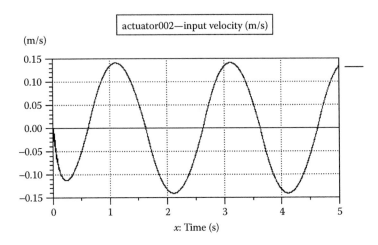

Figure 5.4.5 Piston velocity evolution.

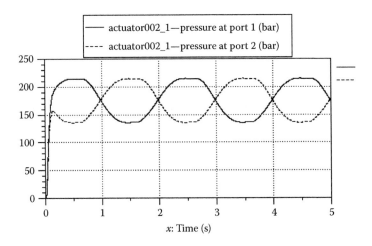

Figure 5.4.6 Pressure evolution at the actuator ports.

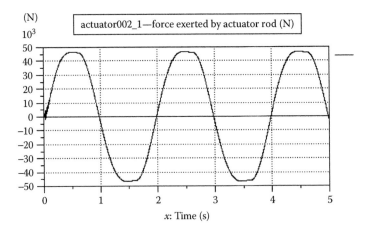

Figure 5.4.7 Force exerted by the actuator rod.

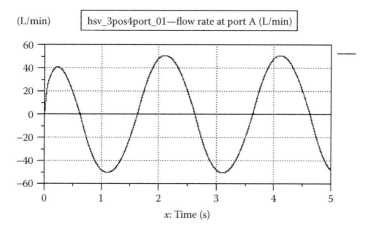

Figure 5.4.8 Servovalve flow rate at port A.

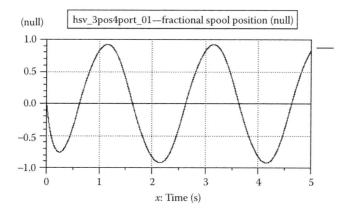

Figure 5.4.9 Fractional servovalve spool position.

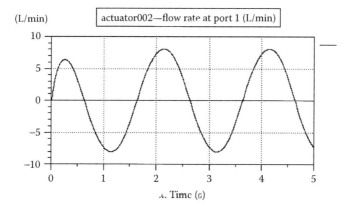

Figure 5.4.10 Damper flow rate.

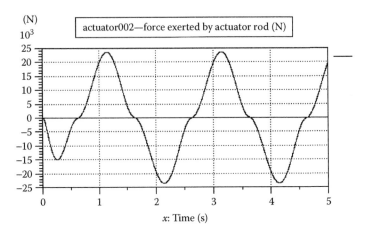

Figure 5.4.11 Damping force variation.

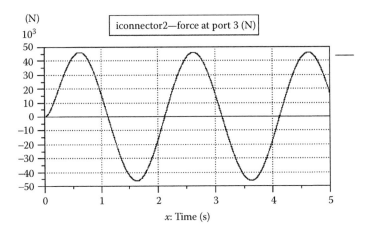

Figure 5.4.12 Variation of the force given by the combination between the actuator and the damper.

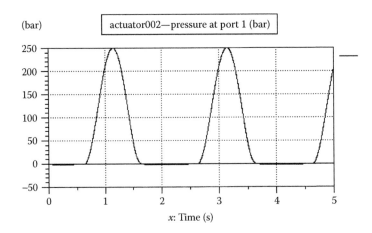

Figure 5.4.13 Pressure at port 1 of the damper.

The servovalve delivers maximum 51 L/min to the actuator—about 85% from the nominal flow (Figure 5.4.8).

The servovalve spool displacement reaches 92% from the maximal stroke (Figure 5.4.9).

The maximum flow rate generated by the damper piston reaches 16% only from the actuator one (Figure 5.4.10).

The maximum force developed by the damper (about 23500 N) reaches 51% from the force developed by the actuator one (Figure 5.4.11).

The overall force applied on the rudder by the combination between the actuator and the damper follows close enough the input position signal (Figure 5.4.12).

The pressures in the chambers of the damper are pulsating from 0 to 250 bar (Figures 5.4.13 and 5.4.14).

The servomechanism follows a low-frequency sine input (Figure 5.4.15).

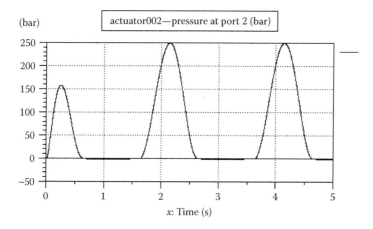

Figure 5.4.14 Pressure at port 2 of the damper.

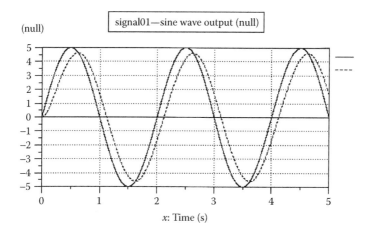

Figure 5.4.15 Sine input response of the servomechanism for $f = 0.5$ Hz.

5.4.3   Step input response of the servomechanism

Another useful way to obtain practical information about the servomechanism behavior is to feed it with a small step input (Figure 5.4.16).

The response is a first-order one (Figure 5.4.17).

The constant time of the actuator is small enough (0.12 s) for the size of the airplane rudder (Figure 5.4.18).

The pressure variations in the actuator's ports are typical for a third-order system (Figure 5.4.19).

The high-speed servovalve flow rate has a normal variation for a well-damped control system (Figure 5.4.20).

According to the small step input, the fractional spool position of the servovalve is about 50% from the nominal one (Figure 5.4.21).

The flow rate at the output port of the damper shows high-frequency small-amplitude variations (Figure 5.4.22). The force exerted by the actuator rod has a strong variation around the final value (Figure 5.4.23). The same phenomenon occurs with the force developed by the damper rod (Figure 5.4.24).

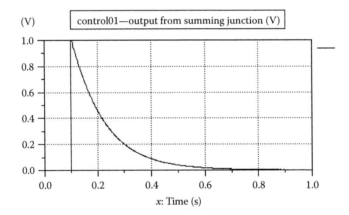

Figure 5.4.16 Output from summing junction.

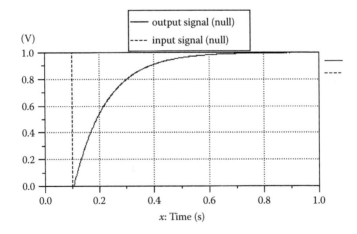

Figure 5.4.17 Input and output signals from the servomechanism.

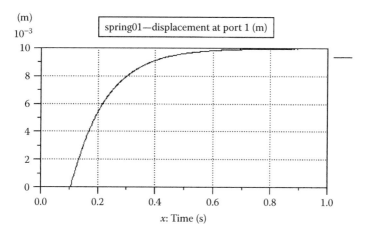

Figure 5.4.18 Piston displacement for a step input.

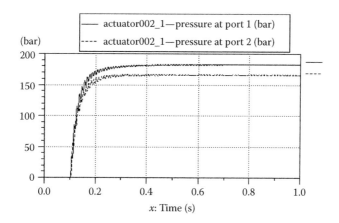

Figure 5.4.19 Pressures in the actuator's ports.

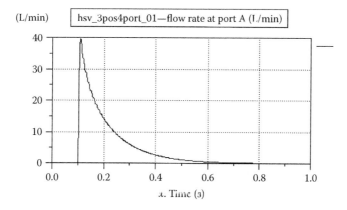

Figure 5.4.20 Flow rate at the port A of the servovalve.

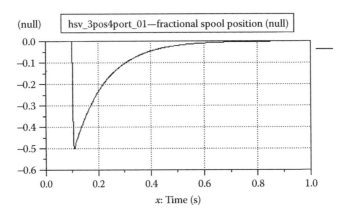

Figure 5.4.21 Fractional servovalve spool position.

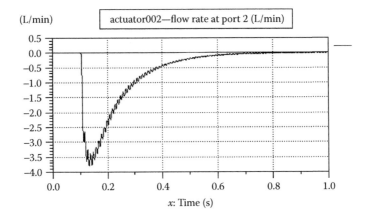

Figure 5.4.22 Flow rate at the output port of the damper.

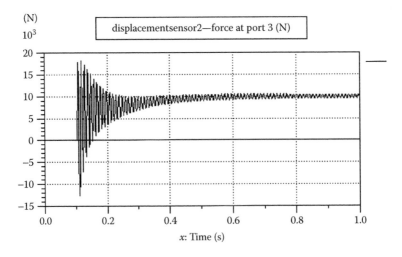

Figure 5.4.23 Force exerted by the actuator rod.

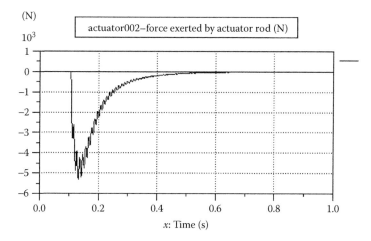

Figure 5.4.24 Force exerted by the damper rod.

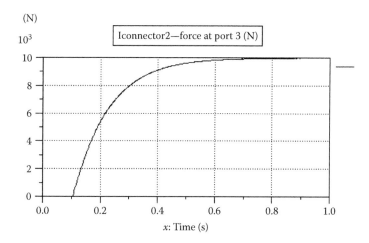

Figure 5.4.25 Force developed by the whole servomechanism at the output rod.

However, the force at the output of the servomechanism increases continuously to the final value of about 10,000 N (Figure 5.4.25).

5.4.4 Frequency response

The integration of the servomechanism in the airplane flight control system needs the study of its frequency response. This can be easily obtained by the transferometer block from the control library of Amesim (Figure 5.4.26).

The small value (1.2 Hz) of the cut frequency proves that the servomechanism behaves as low-pass filter for all types of input signals and for any external disturbance. This response quality is suitable for large airplanes [4].

An important role in the dynamics of the studied system plays the diameter of the damper orifice. Three common values of the orifice diameter d_o were studied: 0.5, 0.75, and 1.25 mm. The influence of the orifice size is important for diameters that are smaller than 0.75 mm (Figures 5.4.27 and 5.4.28).

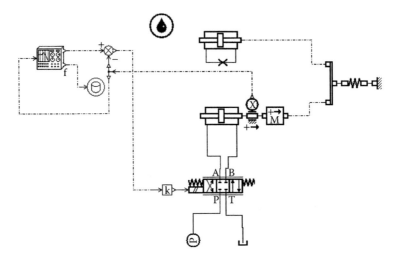

Figure 5.4.26 Simulation network for the frequency response.

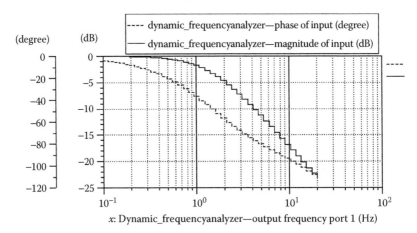

Figure 5.4.27 Servomechanism frequency response.

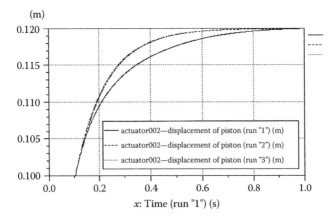

Figure 5.4.28 The influence of the damping orifice diameter on a small step input servomechanism response: The lower line corresponds to $d_o = 0.5$ mm.

5.4.5 Conclusion

The use of a hydraulic damper in parallel with the hydraulic cylinder improves the dynamics of the servosystem *without introducing a steady-state error*. The damping effect can be controlled from outside the damper by the opening of the by-pass orifice [3,4].

This *parallel damper* solution is used now in different similar applications due to the good dynamics and the small increase in the total price of the servomechanism. Different kinds of control algorithms are now developed to improve the dynamics [5].

5.5 Dynamics of the hydromechanical servomechanisms supplied at constant pressure

5.5.1 Applications of hydromechanical servomechanisms constant pressure supplied

From many reasons, the modern designers are using single-stage hydromechanical servomechanisms in the power stages of the fluid control systems instead of full electrohydraulic ones. The first reason is the physical security of the whole actuator, including the servovalve, the control and feedback levers, the filters, and so on. The force developed by the pressure difference between the cylinder ports can be transmitted to the load by the piston (Figure 5.5.1) or by the body of the cylinder (Figure 5.5.2). The force needed to control such a force amplifier is very small and in practice is supplied by a small electrohydraulic servomechanism, guided by a digital or analog controller.

The highest power multistage hydraulic control systems require a cascade of servomechanisms to cover forces of thousands of tones. Each stage has to be designed in close connection with the others to avoid stability and accuracy problems. This case can be encountered in the field of hydraulic turbines delivering 500–1000 MW per unit. Another important field using the high-power hydromechanical servomechanism is the navy one, the submarine control, and many others. A typical example can be found in the important field of the helicopters, where the power stage of the blades pitch control is a double

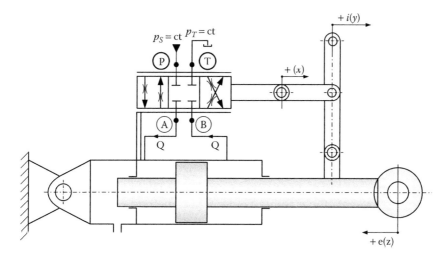

Figure 5.5.1 Single-stage moving piston hydraulic servomechanism.

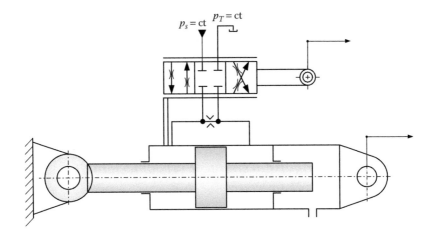

Figure 5.5.2 Single-stage moving body hydraulic servomechanism.

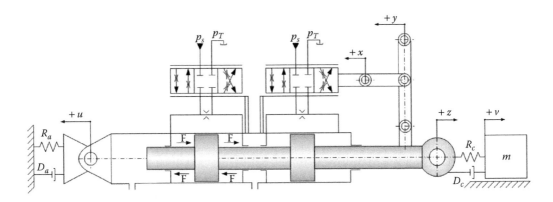

Figure 5.5.3 Dual-moving piston hydraulic servomechanism.

hydromechanical servomechanism including two independent power systems supplied by two independent pressure systems (Figure 5.5.3). The input lever role is complex: It makes the comparison between the input and the output joints displacement, delivering the error needed to actuate the spools of the two servovalves. In practice, this lever is called *error bar*. The complexity of the lever system significantly increases the cost of the servomechanism but offers the possibility of introducing a small electrohydraulic actuator connected to the automatic pilot. This problem can be avoided using a moving body feedback solution (Figure 5.5.2). This is the reason of starting any dynamic study of the servomechanisms by this elementary configuration in which the position feedback is implicitly achieved by the relative position of the valve spool and the cylinder body. This simple feedback solution was used by Marcel Guillon [1,2], when he designed the flight control system for the supersonic passenger airplane Concorde [3].

   This chapter presents the stability computation method of the dual moving piston hydraulic servomechanisms and the numerical simulation by Amesim of the dynamics of a single moving body hydraulic servomechanism. The numerical results, based on a set of real data, are interesting for any practical device using such force amplifier.

5.5.2   Mathematical modeling and dynamic analysis
of the hydraulic servomechanisms

The main difference between the dynamic behavior of a hydraulic servomechanism working in ideal conditions and the one working in real conditions comes from the finite stiffness of the mechanical structural connections of the body and the rod. The elastic adjacent components of the force amplifier can reduce both accuracy and stability but they can also introduce a small damping. The security of the complex modern hydraulic servomechanism requires double or even triple identical components supplied by independent constant pressure sources. For small mechanical inputs, the dynamic behavior of such a system can be performed using both the linear control system theory and the numerical simulation with high-performance digital simulators. All the computations are performed with real data taken from a fighter flight control system.

A dual (tandem) servomechanism (Figure 5.5.3) includes two servomechanisms, which actuate the load by a single rod. The body of the two servomotors serves as link with the helicopter structure. The two sections are supplied by two independent constant pressure sources at p_s. The servovalves are controlled with the same effect by a lever serving as error bar: It receives continuously the input signal (y) and the feedback one. The weight of input and output depends on the entire levers geometry.

The mathematical model of the studied system includes seven differential and algebraic equations [4–8]. For equal geometrical gains μ on the input way—y, and on the feedback way—z, the mechanical comparator has the simple following form:

$$x = \mu\left(y - z\right) - u \tag{5.5.1}$$

where:
   u is the common displacement of the two solitary bodies due to the elasticity of the
       link between the servomechanism and the airplane
   x is the displacement of the two spools

The structural stiffness R_a allows the displacement of the servovalves' sleeves in the same direction as the cylinders' bodies. This means that the control systems have two negative feedbacks.

The second equation is the servovalves' steady-state characteristics, written in the linearized form:

$$Q = K_{Qx}x - K_{QP}P \tag{5.5.2}$$

The overall pressure force on the output rod is about twice that of the force on a piston:

$$F = 2PA_p \tag{5.5.3}$$

The most difficult task of the servosystem is to drive an aerodynamic surface consisting of a concentrated mass, m. In this case, the motion equation of the load has the following elementary form:

$$F = m\ddot{v} \tag{5.5.4}$$

The pressure forces developed by the fluid on the two left internal covers act on the airplane structure, which can be described by the link stiffness, R_a, and the structural damping factor, D_a:

$$F = R_a u + D_a \dot{u} \qquad (5.5.5)$$

The pressure forces developed by the two pistons drive the load by a mechanical chain, which can be described from a dynamic point of view by the control stiffness, R_c, and the control damping, D_c:

$$F = R_c (z - v) + D_c (\dot{z} - \dot{v}) \qquad (5.5.6)$$

The continuity equation for a servovalve, a linear hydraulic cylinder elastically linked, and a damping orifice has the simple following form:

$$Q = A_p \dot{z} + A_p \dot{u} + K_l P + \frac{A_p}{2R_h^{(1)}} \cdot \frac{d}{dt} (2PA_p) \qquad (5.5.7)$$

This equation includes the axial body motion under pressure and the double hydraulic stiffness of the assembly:

$$R_h^{(2)} = R_h = 2R_h^{(1)} \qquad (5.5.8)$$

If the system is excited by inputs $y(t)$, the earlier system of equations has the following unknowns: x, z, u, v, Q, P, and F. All the initial conditions are considered nulls. The positioning systems have the following parameters: m, K_{Qx}, K_{QP}, A_p, m, R_a, D_a, R_c, D_c, K_l, and R_h.

The dynamic behavior of the above-defined system may be completely defined by the transfer function:

$$H(s) = \frac{z(s)}{y(s)} \qquad (5.5.9)$$

Laplace transform applied to previous equations, setting on null all the initial conditions, gives the following set of equations:

$$x = \mu(y - z) - u \qquad (5.5.10)$$

$$Q = K_{Qx} x - K_{QP} P \qquad (5.5.11)$$

$$F = 2PA_p \qquad (5.5.12)$$

$$F = ms^2 v \qquad (5.5.13)$$

$$F = u(R_a + D_a s) \qquad (5.5.14)$$

$$F = (z - v)(R_c + D_c s) \qquad (5.5.15)$$

$$Q = A_p s z + A_p s u + \frac{A_p}{R_h} sF + K_l P \qquad (5.5.16)$$

It is useful to introduce an auxiliary variable:

$$w = z - v \tag{5.5.17}$$

Equation 5.5.15 becomes:

$$F = w\left(R_c + D_c s\right) \tag{5.5.18}$$

The system output will be given by the relation:

$$z = v + w \tag{5.5.19}$$

In a first step, the Laplace transform of v, P, u şi w is expressed as functions of Laplace transform of F:

$$v = \frac{F}{ms^2} \tag{5.5.20}$$

$$P = \frac{F}{2A_p} \tag{5.5.21}$$

$$u = \frac{F}{R_a + D_a s} \tag{5.5.22}$$

$$w = z - v = \frac{F}{R_c + D_c s} \tag{5.5.23}$$

From Equations 5.20 and 5.23, a relation between the Laplace transform of F and z can be obtained:

$$F = z \frac{ms^2\left(R_c + D_c s\right)}{ms^2 + D_c s + R_c} \tag{5.5.24}$$

The two flow rate Equations 5.2 and 5.7 give a relation between y, z, and F:

$$K_{Qx}\mu y - K_{Qx}\mu z - \frac{K_{Qx}}{R_a + D_a s}F = \frac{F}{2A_p}K_P + A_p s z + A_p s \frac{F}{R_a + D_a s} + \frac{A_p s}{R_h}F \tag{5.5.25}$$

Here the coefficient K_P represents the influence of the hydraulic cylinder pressure drop:

$$K_P = K_{QP} + K_l \tag{5.5.26}$$

The previous relation shows the similar effect of the servovalve flow-pressure coefficient with the leakage coefficient of the stabilizing orifice placed between the cylinder chambers.

Equation 5.25 can be written in a new form containing the variables y, z şi F only:

$$K_{Qx}\mu y = \left(K_{Qx}\mu + A_p s\right)z + F\left(\frac{K_{Qx}}{R_a + D_a s} + \frac{K_P}{2A_p} + \frac{A_p s}{R_a + D_a s} + \frac{A_p s}{R_h}\right) \tag{5.5.27}$$

The transfer function results by eliminating the force F between Equations 5.24 and 5.27:

$$\frac{z}{y} = \frac{K_{Qx}\mu\left(ms^2 + D_c s + R_c\right)}{\left(K_{Qx}\mu + A_p s\right)\left(ms^2 + D_c s + R_c\right) + ms^2\left(R_c + D_c s\right)\left(\dfrac{K_{Qx}}{R_a + D_a s} + \dfrac{K_p}{2A_p} + \dfrac{A_p s}{R_a + D_a s} + \dfrac{A_p s}{R_h}\right)} \tag{5.5.28}$$

The final servomechanism transfer function has the following form:

$$\frac{z(s)}{y(s)} = \frac{b_3 s^3 + b_2 s^2 + b_1 s + b_0}{a_5 s^5 + a_4 s^4 + a_3 s^3 + a_2 s^2 + a_1 s + a_0} \tag{5.5.29}$$

in which

$$b_3 = m D_a \tag{5.5.30}$$

$$b_2 = m R_a + D_a D_c \tag{5.5.31}$$

$$b_1 = D_c R_a + D_a R_c \tag{5.5.32}$$

$$b_0 = R_a R_c \tag{5.5.33}$$

$$a_5 = \frac{m A_p D_a D_c}{K_{Qx}\mu R_h} \tag{5.5.34}$$

$$a_4 = \frac{m A_p\left(D_a + D_c\right)}{\mu K_{Qx}} + \frac{m K_p D_a D_c}{2\mu A_p K_{Qx}} + \frac{m A_p\left(D_c R_a + R_c D_a\right)}{\mu K_{Qx} R_h} \tag{5.5.35}$$

$$a_3 = m D_a + \frac{m D_c}{\mu} + \frac{m A_p\left(R_a + R_c\right)}{\mu K_{Qx}} + \frac{A_p D_a D_c}{\mu K_{Qx}} +$$
$$+ \frac{m K_p\left(D_c R_a + R_c D_a\right)}{2\mu A_p K_{Qx}} + \frac{m A_p R_a R_c}{\mu K_{Qx} R_h} \tag{5.5.36}$$

$$a_2 = m R_a + D_a D_c + \frac{A_p\left(R_a D_c + R_c D_a\right)}{K_{Qx}} + \frac{m R_c}{\mu} + \frac{m K_p R_a R_c}{2\mu A_p K_{Qx}} \tag{5.5.37}$$

$$a_1 = R_a D_c + D_a R_c + \frac{A_p R_a R_c}{\mu K_{Qx}} \tag{5.5.38}$$

$$a_0 = R_a R_c \tag{5.5.39}$$

The last stage of the study is to find the control system direct way transfer function. In Equation 5.5.25 written under the following form:

$$K_{Qx}\left[\mu(y - z) - \frac{F}{R_a + D_a s}\right] - K_p \frac{F}{2A_p} = A_p s z + A_p s \frac{F}{R_a + D_a s} + A_p s \frac{F}{R_h} \tag{5.5.40}$$

we introduce F, which is taken from Equation 5.5.24:

$$\frac{z}{y-z} = \frac{K_{Qx}\mu}{A_p s} \cdot \frac{ms^2 + D_c s + R_c}{\left(ms^2 + D_c s + R_c\right) + ms^2\left(R_c + D_c s\right)\left(\dfrac{K_{Qx}}{A_p s(R_a + D_a s)} + \dfrac{K_P}{2A_p^2 s} + \dfrac{1}{R_a + D_a s} + \dfrac{1}{R_h}\right)} \tag{5.5.41}$$

Finally, the direct way transfer function takes the following form:

$$\frac{z}{y-z} = \frac{1}{s} \cdot \frac{K_{Qx}}{A_p} 2A_p^2 R_h \left[mD_a s^3 + \left(mR_a + D_a D_c\right)s^2 + \left(R_a D_c + R_c D_a\right)s \right.$$

$$+ R_a R_c \bigg] \bigg/ \bigg\{ 2mA_p^2 D_a D_c s^4 + s^3 \left[2mA_p^2 R_h\left(D_a + D_c\right) + mR_h K_P D_a D_c \right.$$

$$+ 2mA_p^2\left(R_a D_c + R_c D_a\right)\bigg] + s^2 \left[2A_p^2 R_h\left(mR_a + D_a D_c\right) + 2mA_p R_h K_{Qx} D_c \right. \tag{5.5.42}$$

$$+ R_h K_P m\left(R_a D_c + R_c D_a\right) + 2mA_p^2 R_c\left(R_h + R_a\right)\bigg] + s\left[2A_p^2 R_h\left(D_c R_a \right.\right.$$

$$+ R_c D_a\bigg) + 2mA_p K_{Qx} R_h R_c + mK_P R_a R_c \bigg] + 2A_p^2 R_h R_a R_c \bigg\}$$

The terms of third and fourth order of this function can be neglected because they include the damper coefficients D_a and D_c. In practice, the following form of Equation 5.5.42 can be used with confidence:

$$H_a(s) = \frac{z}{\varepsilon} = \frac{K_\varepsilon}{s} \cdot \frac{c_2 s^2 + c_1 s + c_0}{d_2 s^2 + d_1 s + d_0} \tag{5.5.43}$$

Here:

$$K_\varepsilon = \frac{K_{Qx}\mu}{A_p} \tag{5.5.44}$$

$$c_2 = \frac{mR_a + D_a D_c}{R_a R_c} \tag{5.5.45}$$

$$c_1 = \frac{R_a D_c + R_c D_a}{R_a R_c} \tag{5.5.46}$$

$$c_0 = 1 \tag{5.5.47}$$

$$d_2 = m\left(\frac{1}{R_c} + \frac{1}{R_a} + \frac{1}{R_h}\right) + \frac{D_a D_c}{R_a R_c} + \frac{mK_{Qx} D_c}{A_p R_a R_c} + \frac{mK_P}{2A_p^2} \cdot \left(\frac{D_c}{R_c} + \frac{D_a}{R_a}\right) \tag{5.5.48}$$

$$d_1 = \frac{D_c}{R_c} + \frac{D_a}{R_a} + \frac{mK_{Qx}}{A_p R_a} + \frac{mK_P}{2A_p^2} \tag{5.5.49}$$

$$d_0 = 1 \tag{5.5.50}$$

Starting from the practical form of Equation 5.5.43, a simpler system transfer function can be derived:

$$H_{0a}(s) = \frac{H_a(s)}{1 + H_a(s)} \tag{5.5.51}$$

This way we get a realistic closed loop transfer function for the studied servosystem:

$$H_{0a}(s) = K_\varepsilon \frac{s^2 c_2 + s c_1 + 1}{s^3 d_2 + s^2 (d_1 + K_\varepsilon c_2) + s(1 + K_\varepsilon c_1) + K_\varepsilon} \tag{5.5.52}$$

If we neglect the fluid compressibility, the compliance and the damping of the mechanical structures, and the load inertia, the servomechanism behaves as a first-order element:

$$H_{0a}(s) \cong \frac{1}{\dfrac{s}{K_\varepsilon} + 1} \tag{5.5.53}$$

The constant time depends on three variables only:

$$T = \frac{1}{K_\varepsilon} = \frac{A_p}{\mu K_{Qx}} \tag{5.5.54}$$

5.5.3   Numerical study of the stability

The experience of the advanced aerospace companies in the field of flight control design has shown that the volume of any simulation session is strongly reduced by a preliminary numerical stability study. The use of the algebraic criteria is useful for small-order transfer functions. That is why the designer has to make an assessment of the different terms from the transfer function than to reduce its order. A practical example can sustain this procedure for a well-known servomechanism defined by the following parameters:

$$\mu = 0.4375; \ m = 500 \ \text{kg}; \ A_p = 1.046 \cdot 10^{-3} \ \text{m}^2; \ K_{Qx} = 8.256 \cdot 10^{-2} \ \text{m}^2/\text{s}$$

$$K_l = 6.177 \cdot 10^{-12} \ \text{m}^5/\text{Ns}; \ K_{QP} = 3.323 \cdot 10^{-12} \ \text{m}^5/\text{Ns}; \ K_P = 9.5 \cdot 10^{-12} \ \text{m}^5/\text{Ns}$$

$$R_h = 2 \cdot 3.648 \cdot 10^7 \ \text{N/m}$$

The following real experimental contact data are considered [7]:

$$R_a = 2.145 \cdot 10^7 \ \text{N/m}; \ R_c = 2.231 \cdot 10^7 \ \text{N/m}; \ D_a = 3.678 \ \text{Ns/m}; \ D_c = 4.188 \ \text{Ns/m}$$

The transfer function is defined by the following coefficients:

$$a_5 = \frac{m A_p D_a D_c}{K_{Qx} \mu R_h} = 3.056 \ \text{kg} \cdot \text{s} \tag{5.5.55}$$

$$a_{41} = \frac{m A_p (D_a + D_c)}{K_{Qx} \mu} = 1.139 \cdot 10^5 \ \text{kg}^2 \tag{5.5.56}$$

$$a_{42} = \frac{mK_P D_a D_c}{2A_p K_{Qx}\mu} = 9.682 \cdot 10^2 \ \text{kg}^2 \tag{5.5.57}$$

$$a_{43} = \frac{mA_p \left(D_c R_a + R_c D_a\right)}{K_{Qx}\mu R_h} = 3.41 \cdot 10^4 \ \text{kg}^2 \tag{5.5.58}$$

$$a_4 = a_{41} + a_{42} + a_{43} = 1.489 \cdot 10^5 \ \text{kg}^2 \tag{5.5.59}$$

$$a_{31} = mD_a = 1.839 \cdot 10^6 \ \text{kg}^2/\text{s} \tag{5.5.60}$$

$$a_{32} = \frac{mD_c}{\mu} = 4.786 \cdot 10^6 \ \text{kg}^2/\text{s} \tag{5.5.61}$$

$$a_{33} = \frac{mA_p \left(R_a + R_c\right)}{\mu K_{Qx}} = 6.336 \cdot 10^8 \ \text{kg}^2/\text{s} \tag{5.5.62}$$

$$a_{34} = \frac{A_p D_a D_c}{\mu K_{Qx}} = 4.46 \cdot 10^5 \ \text{kg}^2/\text{s} \tag{5.5.63}$$

$$a_{35} = \frac{mK_P \left(D_c R_a + R_c D_a\right)}{2A_p \mu K_{Qx}} = 1.08 \cdot 10^7 \ \text{kg}^2/\text{s} \tag{5.5.64}$$

$$a_{36} = \frac{mA_p R_a R_c}{\mu K_{Qx} R_h} = 0.95 \cdot 10^8 \ \text{kg}^2/\text{s} \tag{5.5.65}$$

$$a_3 = a_{31} + a_{32} + a_{33} + a_{34} + a_{35} + a_{36} = 7.464 \cdot 10^8 \ \text{kg}^2/\text{s} \tag{5.5.66}$$

$$a_{21} = mR_a = 1.072 \cdot 10^{10} \ \text{kg}^2/\text{s}^2 \tag{5.5.67}$$

$$a_{22} = D_a D_c = 1.54 \cdot 10^7 \ \text{kg}^2/\text{s}^2 \tag{5.5.68}$$

$$a_{23} = \frac{A_p \left(R_a D_c + R_c D_a\right)}{\mu K_{Qx}} = 4.977 \cdot 10^9 \ \text{kg}^2/\text{s}^2 \tag{5.5.69}$$

$$a_{24} = \frac{mR_c}{\mu} = 2.549 \cdot 10^{10} \ \text{kg}^2/\text{s}^2 \tag{5.5.70}$$

$$a_{25} = \frac{mK_P R_a R_c}{\mu} = 3.000 \cdot 10^{10} \ \text{kg}^2/\text{s}^2 \tag{5.5.71}$$

$$a_2 = a_{21} + a_{22} + a_{23} + a_{24} + a_{25} = 7.120 \cdot 10^{10} \ \text{kg}^2/\text{s}^2 \tag{5.5.72}$$

$$a_{11} = R_a D_c = 8.983 \cdot 10^{10} \ \text{kg}^2/\text{s}^3 \tag{5.5.73}$$

$$a_{12} = D_a R_c = 8.205 \cdot 10^{10} \ \text{kg}^2/\text{s}^3 \tag{5.5.74}$$

$$a_{13} = \frac{A_p R_a R_c}{\mu K_{Qx}} = 1.385 \cdot 10^{13} \ \text{kg}^2/\text{s}^3 \tag{5.5.75}$$

$$a_1 = a_{11} + a_{12} + a_{13} = 1.402 \cdot 10^{13} \ \text{kg}^2/\text{s}^3 \tag{5.5.76}$$

$$a_0 = R_a R_c = 4.785 \cdot 10^{14} \ \text{kg}^2/\text{s}^4 \tag{5.5.77}$$

$$b_3 = m D_a = 1.839 \cdot 10^6 \ \text{kg}^2/\text{s} \tag{5.5.78}$$

$$b_2 = m R_a + D_a D_c = 1.074 \cdot 10^{10} \ \text{kg}^2/\text{s}^2 \tag{5.5.79}$$

$$b_1 = D_c R_a + D_a R_c = 1.718 \cdot 10^{11} \ \text{kg}^2/\text{s}^3 \tag{5.5.80}$$

$$b_0 = R_a R_c = 4.785 \cdot 10^{14} \ \text{kg}^2/\text{s}^4 \tag{5.5.81}$$

The transfer function Equation 5.5.28 becomes:

$$\frac{z}{y} = \frac{1.839 \cdot 10^6 \cdot s^3 + 1.074 \cdot 10^{10} \cdot s^2 + 1.718 \cdot 10^{11} \cdot s + 4.785 \cdot 10^{14}}{3.056 \cdot s^5 + 1.489 \cdot 10^5 \cdot s^4 + 7.464 \cdot 10^8 \cdot s^3 + 7.120 \cdot 10^{10} \cdot s^2 + 1.402 \cdot 10^{13} \cdot s + 4.785 \cdot 10^{14}} \tag{5.5.82}$$

The frequency response corresponding to this transfer function (Figures 5.5.4 and 5.5.5) shows a normal dynamic behavior for the studied type of dual servomechanism:

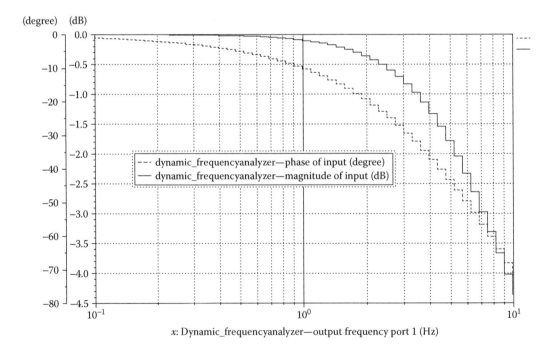

Figure 5.5.4 Frequency response given by the complete transfer function of a dual servomechanism.

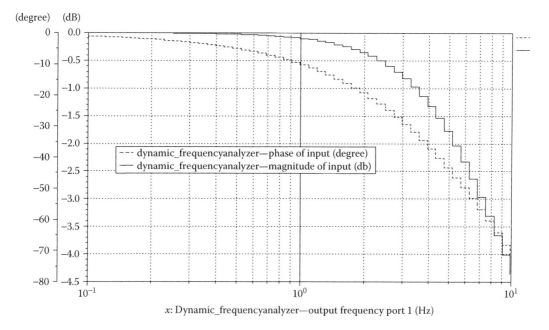

Figure 5.5.5 Frequency response given by the simplified transfer function of a dual servomechanism.

The numerical simulations and experimental validations lead to the practical form of the transfer function:

$$\frac{z}{y} = \frac{1.839 \cdot s^3 + 1.074 \cdot 10^4 \cdot s^2 + 1.718 \cdot 10^5 \cdot s + 4.785 \cdot 10^8}{7.464 \cdot 10^2 \cdot s^3 + 7.120 \cdot 10^4 \cdot s^2 + 1.402 \cdot 10^7 \cdot s + 4.785 \cdot 10^8} \tag{5.5.83}$$

From Figures 5.5.4 and 5.5.5, we can see the practical possibility of using the simplified transfer function without any risk.

5.5.4 The final result of the stability study

All the coefficients of the transfer functions are positive, and the stability condition is a simple one:

$$a_1 \cdot a_2 > a_3 \cdot a_0 \tag{5.5.84}$$

In the studied case, this condition is fulfilled, because

$$a_1 \cdot a_2 = 9.982 \cdot 10^{23} \ \mathrm{kg}^4/\mathrm{s}^5$$
$$a_3 \cdot a_0 = 3.571 \cdot 10^{23} \ \mathrm{kg}^4/\mathrm{s}^5 \tag{5.5.85}$$

Unhappily, the stability condition can be used to find an unknown quantity only. The most important dynamic parameters are K_{Qx} and K_d. It is interesting to find out the value of K_d

needed for stability, starting from a given flow gain K_{Qx}, which set mainly the servomechanism time response. The condition Equation 5.5.84 can be written in the form

$$\frac{a_3 \cdot a_0}{a_1 \cdot a_2} = \alpha < 1 \tag{5.5.86}$$

The practical values of α are included between 0.2 and 0.4. From the denominator of the transfer function results:

$$K_P = \frac{\alpha\left(m_1 + m_2 + m_3 + m_4 + m_5 + m_6\right) - \left(m_7 + m_8 + m_9 + m_{10}\right)}{n_1\left(1-\alpha\right) - n_2\alpha} \tag{5.5.87}$$

where:

$$m_1 = mR_a\left(R_aD_c + R_cD_a\right) \tag{5.5.88}$$

$$m_2 = D_aD_c\left(R_aD_c + R_cD_a\right) \tag{5.5.89}$$

$$m_3 = \frac{3A_pR_aR_cD_aD_c + A_p\left(R_a^2D_c^2 + D_a^2R_c^2\right) + mA_pR_a^2R_c}{K_{Qx}\mu} \tag{5.5.90}$$

$$m_4 = \frac{mR_c\left(R_aD_c + R_cD_a\right)}{\mu} \tag{5.5.91}$$

$$m_5 = \frac{A_p^2R_aR_c\left(R_aD_c + R_cD_a\right)}{K_{Qx}^2\mu^2} \tag{5.5.92}$$

$$m_6 = \frac{mA_pR_aR_c^2}{K_{Qx}\mu^2} \tag{5.5.93}$$

$$m_7 = mD_aR_aR_c \tag{5.5.94}$$

$$m_8 = \frac{mD_cR_aR_c}{\mu} \tag{5.5.95}$$

$$m_9 = \frac{A_pR_aR_c\left[m\left(R_a + R_c\right) + D_aD_c\right]}{K_{Qx}\mu} \tag{5.5.96}$$

$$m_{10} = \frac{mA_pR_a^2R_c^2}{K_{Qx}\mu R_h} \tag{5.5.97}$$

$$n_1 = \frac{mR_aR_c\left(R_aD_c + R_cD_a\right)}{2A_pK_{Qx}\mu} \tag{5.5.98}$$

$$n_2 = \frac{mR_a^2R_c^2}{2K_{Qx}^2\mu^2} \tag{5.5.99}$$

For a common value of $\alpha = 0.358$, the above-mentioned quantities have the following values:

$$m_1 = 1.843 \cdot 10^{21};\ m_2 = 2.647 \cdot 10^{18};\ m_3 = 1.497 \cdot 10^{23};\ m_4 = 4.383 \cdot 10^{21};$$

$$m_5 = 6.898 \cdot 10^{22};\ m_6 = 3.533 \cdot 10^{23};\ m_7 = 8.800 \cdot 10^{20};\ m_8 = 2.290 \cdot 10^{21};$$

$$m_9 = 3.034 \cdot 10^{23};\ m_{10} = 4.544 \cdot 10^{22};\ n_1 = 5.443 \cdot 10^{32};\ n_2 = 4.388 \cdot 10^{34}$$

Finally, we obtain $K_P = 9.5 \cdot 10^{-12}$ m^5/Ns. If $K_{QP} = 3.3 \cdot 10^{-12}$ m^5/Ns, the system needs a leakage coefficient $K_l = 6.2 \cdot 10^{-12}$ m^5/Ns. In geometrical terms, this means a sharp edge orifice of 0.35 mm, which is practically identical with the common ones. This result proves that Equation 5.5.86 can be used with confidence by the designers.

5.5.5 Experimental results

Systematic laboratory experiments were performed by the authors to validate the above-mentioned model [5,6,7]. A classical dual hydromechanical servomechanism used for controlling the blade pitch of a helicopter [8] was tested in the same conditions as that used in the earlier analytical study. The test bench hydromechanical diagram is presented in Figure 5.5.6. The lateral view of the test bench is presented in Figure 5.5.7.

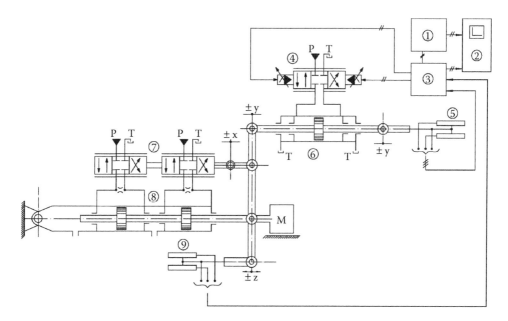

Figure 5.5.6 The test bench hydromechanical diagram: 1—universal analog signal generator (STANFORD); 2—data acquisition system NI-PXI; 3—analog servocontroller (Moog); 4—electrohydraulic high-speed servovalve (Moog 762); 5 and 9—dual track resistive displacement transducers (Penny & Giles); 6—high-speed low-friction hydraulic cylinder; and 7 and 8—dual servomechanism SAMM 7111A.

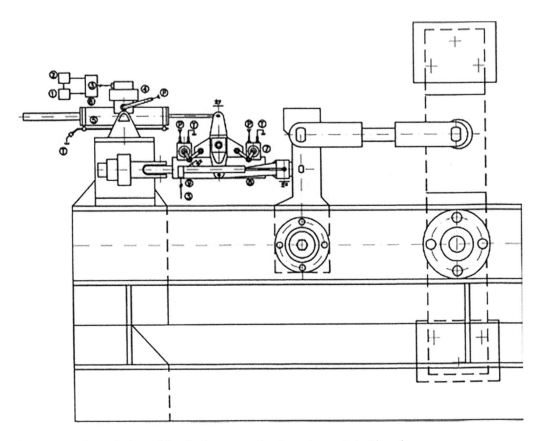

Figure 5.5.7 Lateral view of the dual servomechanisms dynamic test bench.

The experimental frequency response shows a good prediction of the dynamic performances. Two methods were used for exploring the common range of operating frequency with small input signals: (a) discrete input signal frequency variation and (b) continuous variation of the input signal to use the high-speed identification method developed in [9]. The results are presented in Figure 5.5.8. They confirm the possibility of using the linear analysis in the preliminary stages of the servomechanism design.

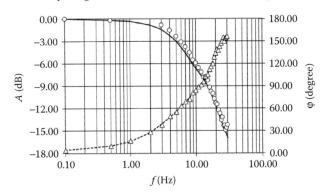

Figure 5.5.8 Experimental frequency response of a dual hydromechanical servomechanism: (a) continuous and dashed curve—magnitude and phase obtained by high-speed identification method and (b) circles (magnitude) and triangles (phase) obtained by point-by-point method.

5.5.6 Numerical simulation of a moving body servomechanism

The utility of the mathematical models developed for linear control systems is limited by a lot of nonlinearities existing in the design, manufacturing process, and operation in real conditions [10,11]. In practice, any linear analysis is followed by a long series of numerical simulations performed by different dedicated software. This chapter is devoted to an example of using Amesim for accomplishing the following phases of the design of a moving body servomechanism:

   a. Find the evolution of the main dynamic parameters during a transient generated by sine and step input signals.
   b. Study the influence of the main geometrical, hydraulic, and mechanical parameters on the dynamic performances.
   c. Identify the influence of the main design parameters on the frequency response applied at the input.

The simplest geometrical sleeve technology involves one or two rounds or shaped (trapeze) flow windows. This hole's geometry is used in many small-flow aerospace servomechanisms. A simple such case is presented in the simulation network shown in Figure 5.7, which contains all the dynamic components involved in transients. The main parameters of the system are as follows: piston diameter, 40 mm; rod diameter, 20 mm; radial piston clearance: 0.1 mm; contact length: 30 mm; flow control valve type: close center, four-way, zero underlap on all the control orifices; spool diameter: 10 mm; spool rod diameter: 5 mm; spool mass: 0.03 kg; jet angle: 69°; maximal flow coefficient: 0.7; critical flow number: 100; sleeve windows overall area: 0.8 or 2×0.8 mm²; load mass: 200 kg; coefficient of viscous friction of the load: 500 N/(m/s).

   Starting from the previous data, the following interesting information can be quickly obtained by numerical simulation with the network shown in Figure 5.5.9, for

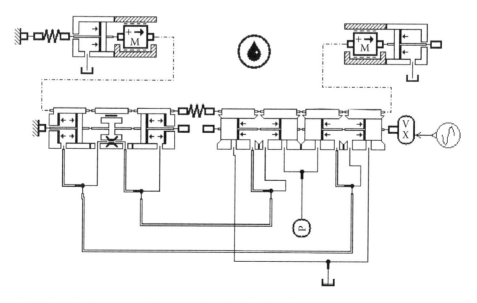

Figure 5.5.9 Simulation network for a moving body hydraulic servomechanism.

a given slow sine input signal (Figure 5.5.10), without writing any equation, but knowing all the important details:

a. The cylinder body displacement (Figure 5.5.11)
b. The cylinder body velocity (Figure 5.5.12)
c. The steady-state characteristics with no load (Figure 5.5.13)
d. Flow rate entering the flow control valve at the input port (Figure 5.5.14)
e. Pressure variations in the cylinder chambers (Figures 5.5.15 and 5.5.16)
f. Pressure evolution in the cylinder chambers for a small step input (Figure 5.5.17)
g. Small step input response (Figure 5.5.18)
h. Force on the cylinder body for a step input response (Figure 5.5.19)
i. Step input response for a double flow gain (Figure 5.5.20)

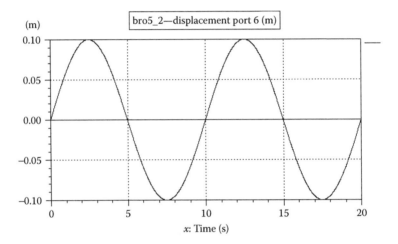

Figure 5.5.10 The spool displacement (input signal).

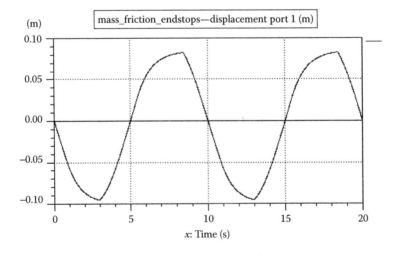

Figure 5.5.11 The cylinder body displacement (output signal).

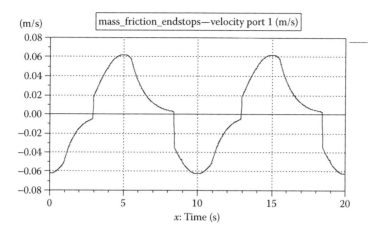

Figure 5.5.12 The cylinder body velocity.

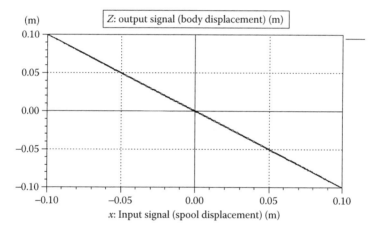

Figure 5.5.13 The steady-state characteristics (no load).

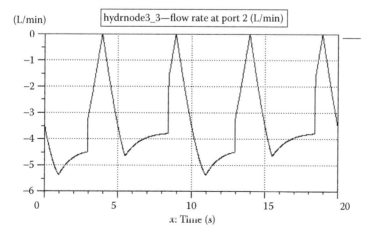

Figure 5.5.14 The flow rate at the input port.

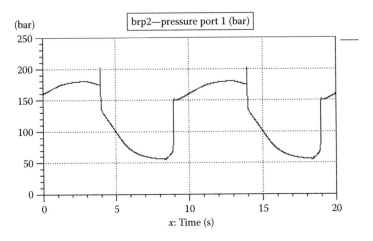

Figure 5.5.15 The pressure variation in the active chamber of the cylinder.

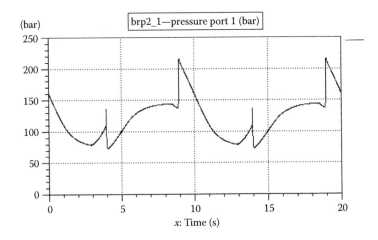

Figure 5.5.16 The pressure variation in the passive chamber of the cylinder.

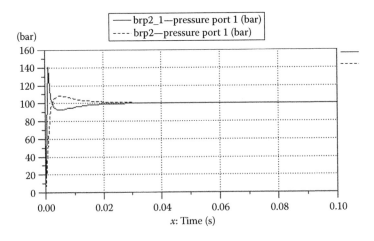

Figure 5.5.17 The pressure evolution in the cylinder chambers for a step input (1 mm).

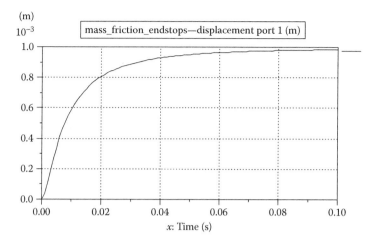

Figure 5.5.18 The step input response (1 mm).

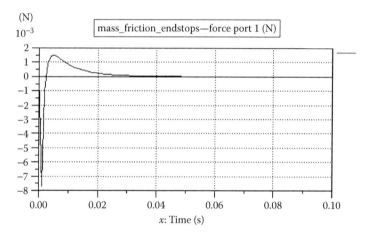

Figure 5.5.19 The force on the cylinder body for a step input response (1 mm).

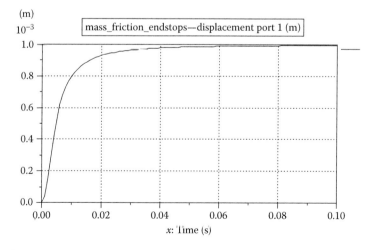

Figure 5.5.20 The step input response for a double flow gain.

The frequency response of the moving body servomechanisms needs a new simulation network (Figure 5.5.21). Both simple flow gain and double flow gain responses were investigated (Figures 5.5.22 and 5.5.23). As expected, the cut frequency is greater (from 15 to 20 Hz) for the double flow gain valve.

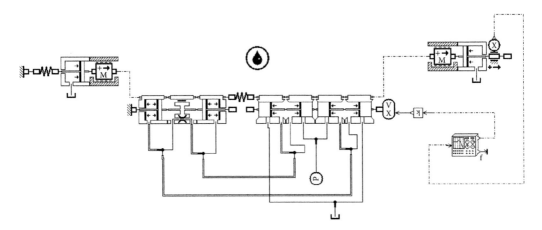

Figure 5.5.21 Simulation network for the servomechanism frequency response.

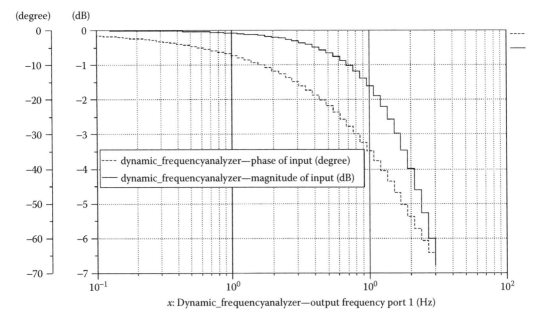

Figure 5.5.22 The frequency response of the basic servomechanism.

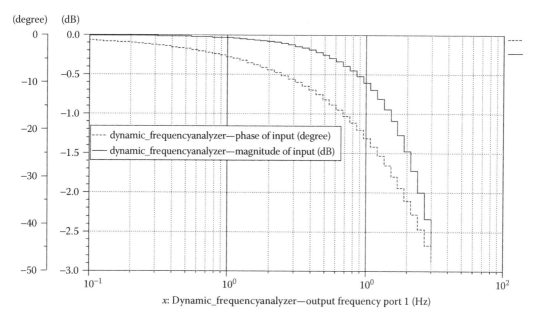

x: Dynamic_frequencyanalyzer—output frequency port 1 (Hz)

Figure 5.5.23 The frequency response of the servomechanism with two windows on each way of the valve sleeve.

5.5.7   Conclusion

This section gives the designers the possibility to compare the classical and the modern manners of study of the dynamic behavior of the hydraulic servosystems. Both methods are still used in the aerospace equipment companies, but the big gain of time offered by the numerical simulation methods and the possibility of virtual integration of the Amesim models in Hardware-in-the-Loop (HiL) simulation loops are more and more used in many innovative teams.

5.6   Improving the accuracy of the electrohydraulic servomechanisms by additional feedback

5.6.1   Problem formulation

This section contains a report on some theoretical and experimental researches aiming to create high-accuracy electrohydraulic digital servomechanisms used in high-tech applications as actuators, load simulators, flight control systems, dynamic test machines, and so on. Computational methods, control software, design problems, and experimental validation are shortly presented. The authors developed a new hardware and software solution to replace the old generation of two- or three-stage electrohydraulic servovalves and to achieve maximum flexibility of the testing programs.

Using one derivative feed forward path from the system input and one derivative path from the system output, the authors developed a hardware and software configuration that eliminates steady-steady error for step, ramp, and parabolic input. The Keithley ADwin PRO DSP high-performance industrial computer and AD basic programming language were used. Also, a configuration with PXI Controller from National Instruments and LabVIEW development environment were used as a very flexible combination. The experimental researches were carried out on a general-purpose low-friction actuator designed and built in the Fluid Power Laboratory from the University Politehnica of Bucharest. The same ideas were implemented in different kinds of high-performance applications as power units speed governors. The keywords of this section are as follows: digital electro-hydraulic control systems, proportional valves, and control algorithms.

5.6.2 Introduction

A classical electrohydraulic servomechanism (Figure 5.6.1) contains a position feedback and a P compensator only. The dynamic performance is limited by the load mass mainly. An effi-cient way of improving the overall damping is the use of an additional feedback of force, dif-ferential pressure or velocity, supplied by specific transducers (Figures 5.6.2 through 5.6.4).

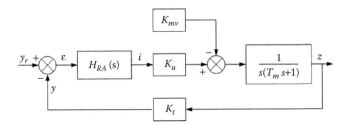

Figure 5.6.1 The structure of a classical electrohydraulic servomechanism.

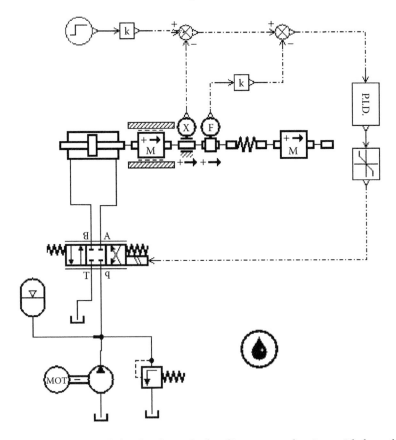

Figure 5.6.2 Simulation network for the electrohydraulic servomechanism with force feedback.

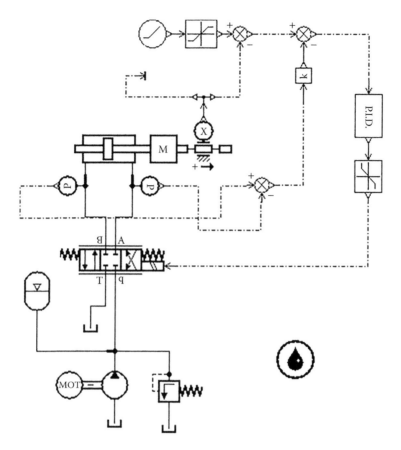

Figure 5.6.3 Electrohydraulic servomechanism with differential pressure feedback.

The transient force feedback (Figure 5.6.2) is more suitable for a general-purpose servo-mechanism loaded by a *random force* [1,2]. This hardware expensive additional feedback can be replaced by a software that is based on some additional derivative feedback suitable for high-speed digital servocontrollers. The general predictive control by additional state variables feedback can lead to good results even with low computation speed [3].

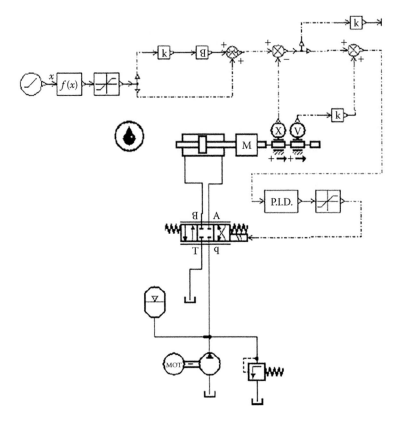

Figure 5.6.4 Simulation network for an electrohydraulic servomechanism with velocity correction.

5.6.3   *Mathematical modeling*

The basic set of equations describing the dynamic behavior of an electrohydraulic servo-system contains the following equations:

$$Q = K_{qi}i - K_{qp}P_m \tag{5.6.1}$$

$$Q = A_m \frac{dz}{dt} + aP_m + \frac{V_m}{4E}\frac{dP_m}{dt} \tag{5.6.2}$$

$$A_mP_m = M\frac{d^2z}{dt^2} + f\frac{dz}{dt} + F_0 \tag{5.6.3}$$

$$P_m = P_1 - P_2 \tag{5.6.4}$$

The simplest transfer functions can be obtained by neglecting the fluid compressibility:

$$H_m(s) = \frac{Z(s)}{I(s)} = \frac{K_m}{s(T_ms+1)} \tag{5.6.5}$$

$$H_{mv}(s) = \frac{Z(s)}{F_0(s)} = -\frac{K_{mv}}{s(T_ms+1)} \tag{5.6.6}$$

This simple dynamic description is valid by accepting the practical condition:

$$\frac{f \cdot K_{ce}}{A_m^2} \ll 1 \tag{5.6.7}$$

Here, the quantity:

$$T_m = \frac{MK_{ce}}{A_m^2} \tag{5.6.8}$$

is the equivalent time constant of the hydraulic motor, including the overall pressure drop influenced by the coefficient:

$$K_{ce} = a + K_{qp} \tag{5.6.9}$$

and

$$K_m = \frac{K_{qi}}{A_m} \; ; \; K_{mv} = \frac{K_{ce}}{A_m^2} \tag{5.6.10}$$

are the transfer factors on the two control channels. System performance can be derived by the aid of this model. A proportional compensator ensures a null steady-state error ($\varepsilon_{st} = 0$) for a step input signal, $y_r = 1(t)$ only. The steady-state error for a ramp input signal:

$$\varepsilon_{st} = \frac{1}{K_b} \tag{5.6.11}$$

depends on the overall speed gain

$$K_b = K_R K_t K_m \tag{5.6.12}$$

The high-level applications such as aerospace ones need the elimination of the steady-state error both for step, ramp, and parabolic input signals. This goal can be achieved by the aid of some derivative functional correction (Figure 5.6.5), which can be easily implemented by the aid of a high-speed industrial process computer (IPC) [1].

This new configuration is described by the following equation:

$$Z(s) = \frac{(K_{m1}K_R + K_{m1}T_{d1}s)Y_r(s) - K_{mv}F_0(s)}{T_m s^2 + (1 + K_{m1}T_{d2})s + K_{m1}K_R K_t} \tag{5.6.13}$$

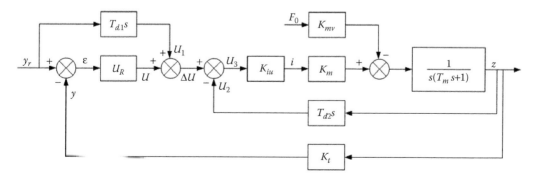

Figure 5.6.5 The structure of a servomechanism with derivative corrections.

The system follow error expressed as voltages becomes

$$\varepsilon(s) = Y_r(s) - Y(s) = Y_r(s) - K_t Z(s) \tag{5.6.14}$$

Hence,

$$\varepsilon = \frac{\left[T_m s^2 + \left(1 + K_{m1} T_{d2} - K_t K_{m1} T_{d1}\right)s\right] Y_r(s) + K_t K_{mv} F_0(s)}{T_m s^2 + \left(1 + K_{m1} T_{d2}\right)s + K_{m1} K_R K_t} \tag{5.6.15}$$

For a *P* compensator:

$$H_{RA}(s) = \frac{U(s)}{(s)} = K_R \tag{5.6.16}$$

and if we apply the condition:

$$T_{d1} = \frac{1 + K_{m1} T_{d2}}{K_t K_{m1}} \tag{5.6.17}$$

the input ramp steady-state error becomes null for $y_r(t) = t$. The parabolic input signal

$$y_r(t) = \frac{1}{2} t^2 \tag{5.6.18}$$

is followed by the steady-state error

$$\varepsilon_{st} = \frac{T_m}{2 K_{m1} K_t K_R} \tag{5.6.19}$$

Using the notation

$$a_1 = 1 + K_{m1} T_{d2} - K_t K_{m1} T_{d1} \tag{5.6.20}$$

a *PI* compensator

$$H_{RA}(s) = \frac{U(s)}{(s)} = K_R \left(1 + \frac{1}{T_i s}\right) \tag{5.6.21}$$

leads to the following equation:

$$\varepsilon(s) = \frac{(T_m s^2 + a_1 s) T_i s Y_r(s) + K_t K_{mv} T_i s F_0(s)}{T_m T_i s^3 + \left(1 + K_{m1} T_{d2}\right) T_i s^2 + K_{m1} K_R K_t T_i s + K_{m1} K_R K_t} \tag{5.6.22}$$

If $a_1 = 0$, the ramp input signal has no steady-state error, and the servomechanism rejects the force disturbances. At the same time, the system stability depends on the condition

$$T_i > \frac{T_m}{1 + K_{m1}T_{d2}} \tag{5.6.23}$$

which has to be always checked during the design process. A PID compensator can lead to a better dynamic performance, but the fine tuning is more delicate on the whole operation range.

A simple numerical simulation can prove the aforementioned considerations. For a typical servosystem:

$$A_m = 4 \cdot 10^{-3} \text{m}^2 \text{ } K_{qi} = 0{,}013 \text{ m}^3/\text{As}; K_{iu} = 5 \cdot 10^{-3} \text{ A/V}; a = 1.2 \cdot 10^{-11} \text{ m}^5/\text{Ns};$$

$$K_{qp} = 0.12 \cdot 10^{-11} \text{ m}^5/\text{Ns}; Z_{max} = 0.3 \text{ m}; K_t = \frac{100}{3} \text{ V/m}; M = 80 \text{ kg}$$

Figures 5.6.6 and 5.6.7, obtained by MATLAB®, show the advantages offered by the derivative corrections clearly.

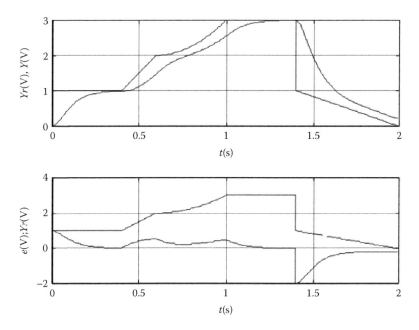

Figure 5.6.6 Response of an electrohydraulic servomechanism without corrections for step, ramp, and parabolic input signals: input and response (up); response and error (down).

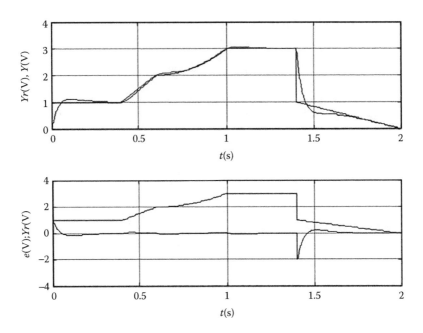

Figure 5.6.7 Response of an electrohydraulic servomechanism with corrections for step, ramp, and parabolic signal: input and response (up); response and error (down).

5.6.4   *Numerical simulation with Amesim*

The applications of the hydraulic servomechanism are so wide and different that any case needs a deep knowledge of the relation between the stability and accuracy. The digital compensators can solve complex dynamics problems but their implementation is a boundary domain, which requires a complex design team. In this frame, the preliminary use of Amesim for drastically reducing the number of preliminary models becomes very efficient. The logical structure of a position or force control system has to be deeply considered before any computation.

This final part of the hydraulic servomechanisms study shows the utility of such approach. Both numerical simulation and extensive experiments were performed in our fluid power lab.

One of the most important subjects concerning the structure of the servomechanisms is regarding the additional feedbacks. Using Amesim, we have investigated the utility of a velocity feedback. The results are presented in Figures 5.6.8 through 5.6.12. The steady

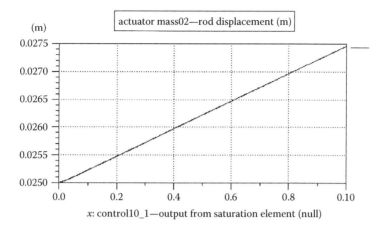

Figure 5.6.8 Steady-state characteristics of the servomechanism without velocity feedback.

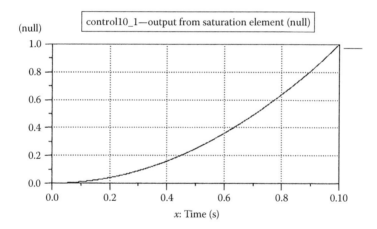

Figure 5.6.9 Parabolic input signal.

state of such a device is nearly a linear one (Figure 5.6.8). If we apply a parabolic input signal (Figure 5.6.9), the positive velocity feedback shortens the response time (Figure 5.6.10). This is a typical advantage for flight control systems. Figures 5.6.11 and 5.6.12 show the advantage of the velocity correction of the servomechanism dynamics by the extension of the work frequency clearly.

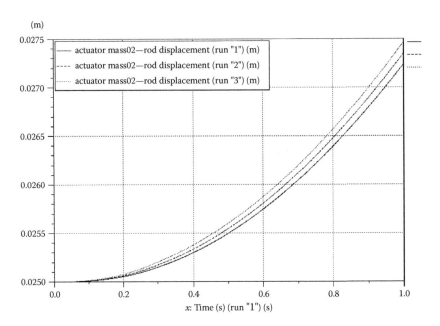

Figure 5.6.10 Rod displacement with different velocity feedbacks: positive, null, and negative.

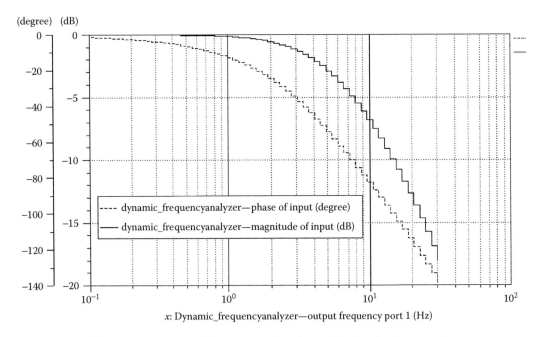

Figure 5.6.11 Frequency response of the servomechanism without velocity correction.

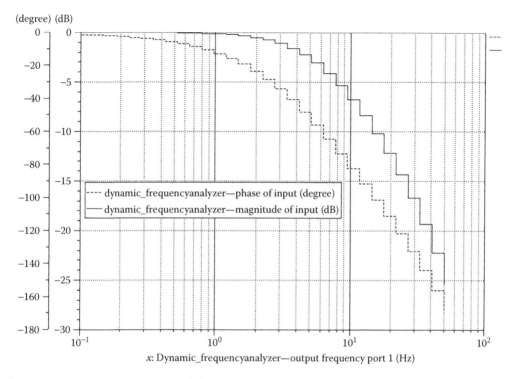

x: Dynamic_frequencyanalyzer—output frequency port 1 (Hz)

Figure 5.6.12 Frequency response of the servomechanism with velocity correction.

5.6.5   *Experimental validation of the theoretical developments*

The best solution for fast Real-Time applications is to place a dedicated CPU close to the signal source and therefore having dedicated resources for the purpose of processing this data. Only this structure gives the ability of exact response times with predictable delays. If the intelligence is not localized and dedicated but centered on a host PC platform; all calculations are under the control of PC's operating system, Windows, and its available resources. As a consequence, there can be no guarantees for response times to either an external event interrupt or an internal timer interrupt. Furthermore, processes executed based on timer feedback will become erratic, at best, due to inconsistencies in the system timer. However, Windows offers comfortable user interfaces, multitasking functionality, and great possibilities for network functionality. In order to take advantages of a Windows environment and run fast and stable Real-Time processes, it is useful to use the ADwin family of products. ADwin Real-Time systems are complete process controllers with ana-log and digital I/O, a local CPU, and local memory [6–8].

ADwin systems use DSPs, which guarantee response times as little as 0.5 μs to an inter-rupt while maintaining complete software stability, even in a Windows environment. Since the local processor handles the process control and/or data acquisition, the PC processor is free to run a user interface program, for example, a man–machine interface with data visu-alization, user input, data storage, and so on, without regard for the effect which the user front-end software has on PC resources. As the programs run on the processor of the ADwin board, up to 10 processes can run simultaneously on one CPU with priorities assigned where required. Processes can interact by exchanging parameters and data. Thus each process has its own independent timing, but it is also possible to exchange parameters and data with other running processes. Finally, it is possible to develop complex applications with processes, which interact more or less. This is the case of multiple missiles launcher, aerospace gunnery, ground defenses, and other special applications.

The experimental researches were carried out on a full-scale general-purpose load simulator and on a test bench designed and built in the Fluid Power Laboratory from the University Politehnica of Bucharest. The test bench diagram is presented in Figure 5.6.13. The digital compensators included in the overall loops are proportional or PI ones, but any other type such as fuzzy can be used. A Stanford signal generator supplies the input signal of the system. The use of an analog proportional compensator offers not only a good dynamic but also leads to a steady-state error, which strongly depends on the actuator load. This fact was systematically checked by numerical simulation and extensive experiments. A very simple control system cannot solve the problem of the accuracy, even for low dynamic requirements. The main characteristics of the test bench are as follows: IPC-type ADwin-Pro (Keithley), servosolenoid valves with integrated amplifier OBE (Bosch), analog speed/position servocontroller AVPC (Bosch), high-speed actuator or heavy hydraulic cylinder, inductive position transducers (Penny & Giles), and an industrial computer IPC connected by an Ethernet interface with an IPC. The dual control system structure designed for high-speed complex applications is presented in Figure 5.6.14 [7].

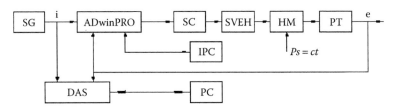

Figure 5.6.13 Test bench diagram: SG—signal generator; SVEH—high-speed proportional valve; SC—analog servocontroller; IPC—industrial process computer; DAS—data acquisition system; i—input signal; e—output signal; PT—position transducer; and ADWin PRO-DSP industrial process computer.

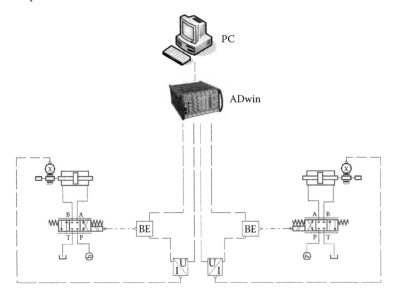

Figure 5.6.14 Dual test bench diagram.

The theoretical results are found in good agreement with the experimental ones. The improvement of the dynamic performance of the low-speed servosystems for a sine input signal can be identified in Figures 5.6.15 and 5.6.16. The overall sine input amplitude was great enough to point out the improvement introduced by the control algorithm. The sampling period of the Real-Time control was 10 ms, good enough for a maximum speed of about 45 mm/s. The tuning parameters of the digital error amplifier were correlated with the tuning parameters of the analog amplifiers included in the systems. Some restrictions were introduced by the numerical derivative velocity corrections. A good dynamic needs a close correlation of the T_{d1} and T_{d2} parameters by theory aid.

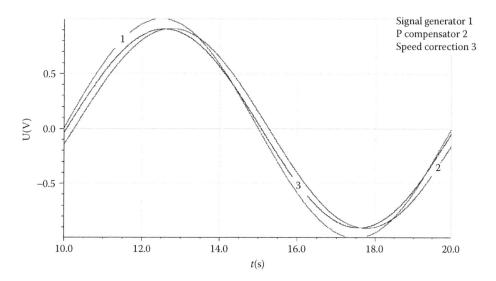

Figure 5.6.15 Numerical simulation by Amesim of sine input response.

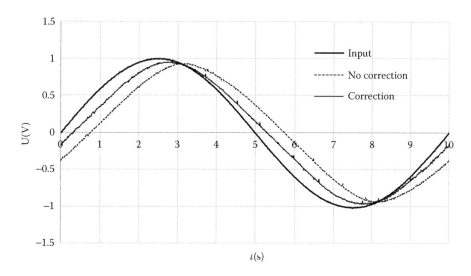

Figure 5.6.16 Typical experimental sine input response with correction and without correction.

5.7 Modeling, simulation, and experimental validation of the synchronized electrohydraulic servomechanisms

5.7.1 Practical problem formulation

Many technical systems include heavy or wide metal structures driven by two or more hydraulic cylinders. Typical examples from these categories can be found in the field of the water gates and the floodgates of the hydropower stations. Many other technical systems such as complex cranes, rolling mill, and so on demand the synchronous control of two or more hydraulic motors. Usually, the driving forces are supplied by hydraulic cylinders that are working in heavy environmental, cinematic, and random dynamic conditions. Under these situations, the position-control accuracy needs a high-speed response flow control system. This performance can be achieved by high-speed industrial servovalves or by high-speed industrial servopumps.

This chapter is mainly devoted to the dynamic study of the hydraulic control systems of the water gates included in the sluice of the Iron Gate I Hydropower Station. This important component of the hydropower system has two basins, which are divided into two heads of about 33 m by the aid of a two-section water gate of 900 t (Figure 5.7.1). Two heavy hydraulic cylinders are moving vertical to the two sections of the gate, thereby covering important friction random forces that can tilt and lock the metal sections with a width of 33 m and weight of 450 t. The main condition that has to be accomplished by the

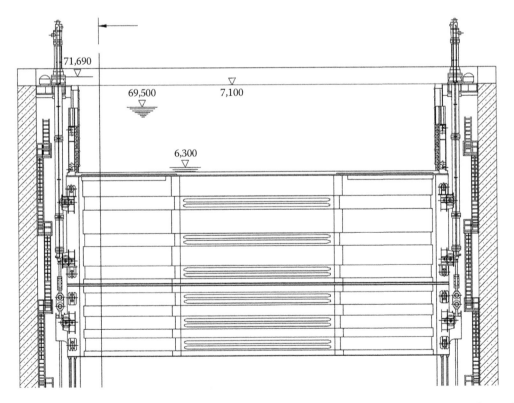

Figure 5.7.1 Upper section of the water gate from the intermediate section of the two-stage sluice of Iron Gates I Hydropower Station.

driving system is to move the sections horizontally to respect the narrow gaps between the guiding railway system and the concrete walls of the motion space. Two cylinders with pistons of 600 mm and rods of 220 mm are supplied by two electrohydraulic servopumps of maximum 355 L/min at the nominal pressure of 350 bar. The stroke of the cylinders is 16 m under a nominal load of 450 t.

The two hydraulic cylinders are supplied by independent circuits. The cylinder rods are synchronized by different methods according to the flow needed to move the gate. The small cylinders can be controlled by industrial servovalves. The high diameters and strokes need the control by servopumps. Both systems are studied in this section to match the cinematic demands and to limit the energy consumptions.

The cylinders include redundant position transducers (analog or digital) connected to industrial controllers, which can be loaded with different control lows. The hydraulic supply circuits include pressure relief valves and security systems controlled by the synchronization error. The mechanical integrity of the gates is also preserved by a stress-monitoring system included in a complex supervisory control and data acquisition (SCADA) system. Such a system was implemented in 2000 by the authors in the Iron Gate I water gate sited between the two stages of the Romanian sluice. All the computation was performed by ANSYS, and the hardware and software implementation of the stress sensors was guided by the theoretical results.

This successful application led to the conclusion that the whole driving system design has to be performed by numerical simulation to find the best control low of the motion process and to know the main hydraulic, mechanic, and electric process parameters.

Another study target is to find the effect of different failure events on the system components and to supply the strength computation programs with realistic data about the stress introduced by the failure of the driving system.

In this section, all the numerical simulations were performed by Amesim. The numerical application was performed for the plain gate of intermediate head of the Iron Gates I lock, which is a critical equipment.

5.7.2 Dynamics of the synchronization systems with servopumps

The response time of the modern industrial swash plate pumps is small enough for efficiently reacting in the case of the common errors that occur in the driving system supplied with well-filtered oil and sited in the common temperature range (Figure 5.7.2).

The simplest mathematical model of the synchronizing system with servopumps (Figure 5.7.3) includes two hydraulic cylinders loaded by half of the gate weight and by disturbing forces generated by different factors such as frictions, guiding roller jam, and so on. The gate motion direction is controlled by two high-flow synchronized proportional valves. Two position transducers from the new generation dedicated to high rod diameters are sending information to the error stage of the overall controller. This one sends the reference signals to the servopumps to eliminate the error [2–5]. The servopumps pressure is limited by two proportional two-stage pressure relief valves.

One of the cylinders is loaded by a sudden external periodical force of 500,000 N (Figure 5.7.4).

The efficiency of the synchronization system can be assessed by examining the rods path, (Figure 5.7.5) from the pressure variations into the cylinder chambers (Figures 5.7.6 and 5.7.7), but for such a long stroke cylinder (18 m) the best information is supplied by the synchronization error offered by the system controller (Figure 5.7.8).

Figure 5.7.2 The servopump station of the intermediate watergate of Iron Gate I sluice.

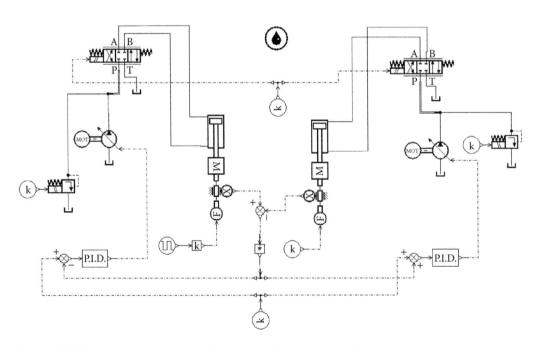

Figure 5.7.3 Simulation network for the synchronizing system with servopumps.

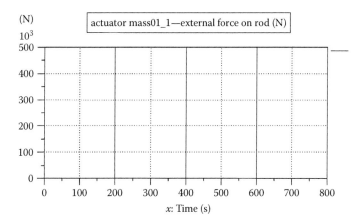

Figure 5.7.4 External periodical force applied to the left cylinder.

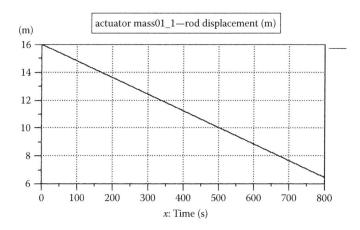

Figure 5.7.5 Left cylinder (no.1) rod displacement in 800 s.

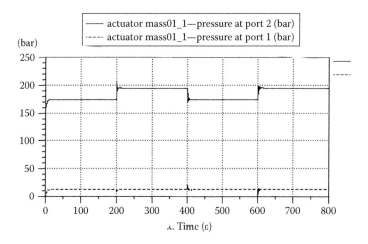

Figure 5.7.6 Pressure variation in the first cylinder-active chambers.

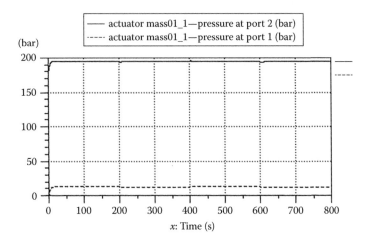

Figure 5.7.7 Pressure variation in the second cylinder chambers.

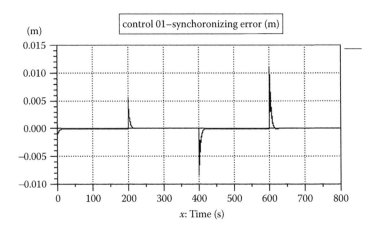

Figure 5.7.8 Synchronization error given by the system controller.

The initial actuation system of the water gate allows stroke differences between the two cylinder rods of about 10 cm. This condition is fulfilled without any opening of the pressure relief valves. This means a maximum efficiency of about 85% for each branch of the driving system (a good one). It is useful to examine the rod velocity variation of the two cylinders (Figures 5.7.9 and 5.7.10). The rod speed of the disturbed cylinder remains in narrow limits (−8–+7 cm/s) for an average of about 12 mm/s.

The rod speed of the second cylinder has a smaller variation: −2–2 mm/s (Figure 5.7.10).

The pressure difference between the two servopumps during the forces shocks applied to the left cylinder (Figure 5.7.11) remains under 20 bar (10% from the optimal pressure of 200 bar).

The reference flows of the servopumps are setup from the beginning at 50% from the nominal ones (Figures 5.7.12 and 5.7.13).

The rejection of the force of 500 kN applied as a disturbance to the left cylinder needs a variation of about 6% from the reference setup of the servopump (50% from the nominal flow).

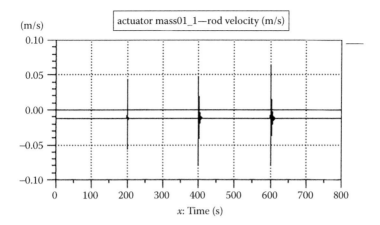

Figure 5.7.9 First cylinder rod velocity variation.

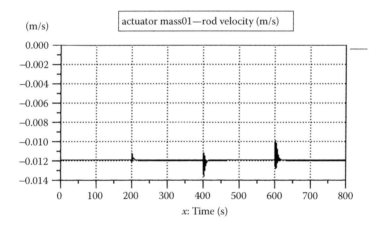

Figure 5.7.10 Second cylinder rod velocity variation.

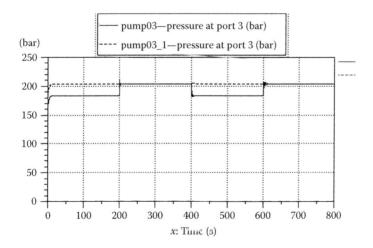

Figure 5.7.11 Servopumps' pressures during the gate motion with periodical shocks.

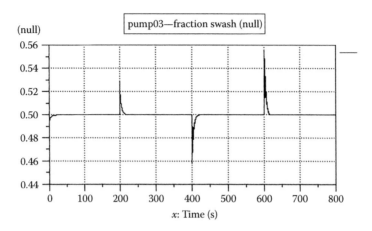

Figure 5.7.12 First pump swash plate fraction variation.

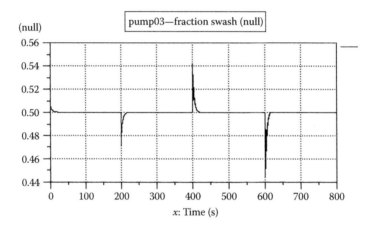

Figure 5.7.13 Second pump swash plate fraction variation.

The overall conclusion regarding the use of the servopump for the synchronization is a positive one: The accuracy of the control is good enough for such a load (900,000 kg) and the energy consumption is maintained at a minimum value.

5.7.3 Dynamics of the synchronization systems with industrial servovalves

The synchronization of the common hydraulic cylinders can be realized directly with the aid of two good enough servovalves. The simulation network from Figure 5.7.14 is similar to the previous one, but the price is much smaller in spite of the relatively large amount of energy lost by passing and heating all the flows through the narrow channels from the servovalves. This kind of hydraulic diagram is used in many industrial or civil systems by assembling the cylinder, the servovalve, and the position transducer into one unit called *servocylinder* (Figure 5.7.15). The controller task becomes more simple by changing information by CAN-like process field bus (PROFIBUS).

Figures 5.7.16 through 5.7.23 describe the synchronization process of two hydraulic cylinders of 250/100 mm with a stroke of 1 m. The high-speed industrial servovalves of

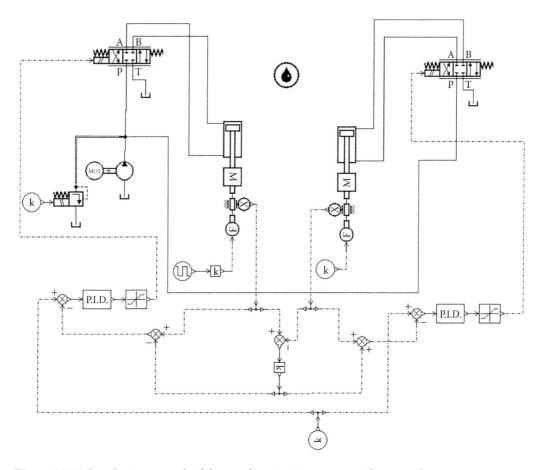

Figure 5.7.14 Simulation network of the synchronization system with servovalves.

Figure 5.7.15 Electrohydraulic servocylinder for hydraulic gates. (From The Bosch Rexroth Corporation is the manufacturer of the biggest hydraulic cylinders with piston rods using Ceramax Engineered Coatings, www.boschrexroth.com.)

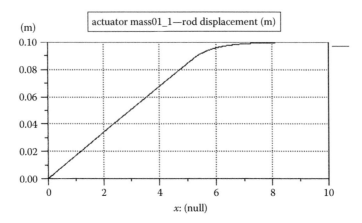

Figure 5.7.16 First cylinder rod displacement.

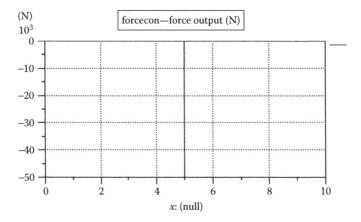

Figure 5.7.17 External force applied to the first cylinder.

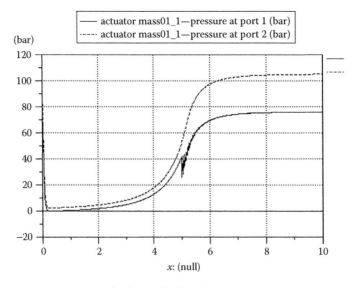

Figure 5.7.18 Pressure variation in the first cylinder chambers.

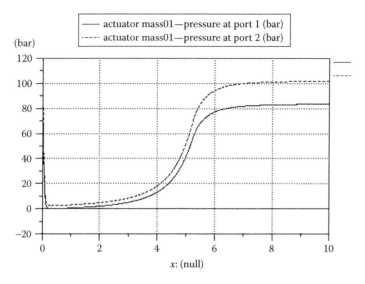

Figure 5.7.19 Pressure variation in the second cylinder chambers.

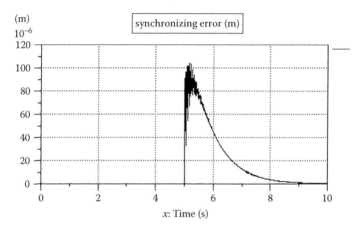

Figure 5.7.20 Position synchronization error.

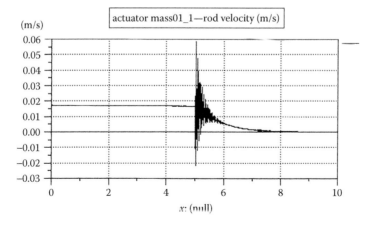

Figure 5.7.21 First cylinder rod velocity variation.

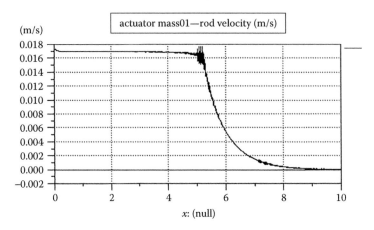

Figure 5.7.22 Second cylinder rod velocity variation.

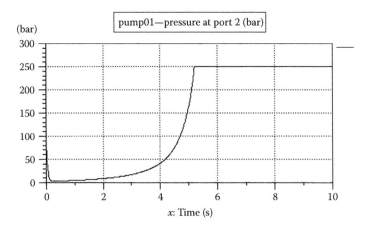

Figure 5.7.23 Pump delivery pressure variation during the motion process.

100 L/min supplied at 250 bar are moving a mass of 1000 kg. A sudden asymmetric load of 50,000 N generates a maximum error of 100 μm, which is eliminated in 5 s.

5.7.4   *Experimental identification of the relation between the synchronizing error and the maximum effort introduced in the structure*

A good example of the practical importance of the hydraulic cylinders' synchronization is offered by the study of the correlation between the stress generated into a big metal structure by the velocity difference between two long hydraulic driving cylinders. The authors studied the stress and the displacements of the old water gate deserving the sluice of the Iron Gate I Hydropower Station sited on the Danube river. After 30 years of service, a lot of cracks appeared on the different components of the gate [8].

A complete strength analysis was performed by ANSYS program for the lower section of the gate. All the categories of loading forces were considered including the weight, the water pressure difference, and the synchronization errors, which can introduce big efforts into gate sections. The driving system includes two hydraulic cylinders having a stroke of 16 m. The elastic rods are spherically jointed with the lower section of the gate.

The mechanical structure is sliding on the dam surface on strong free rollers, which are not guiding laterally the gate. Under these circumstances, the vertical motion of the gate becomes vital. The vertical displacement generated by the gate mass is shown in Figure 5.7.24, with a maximum value of 3 mm, checked by a laser theodolite from Leica Corporation.

The pressure difference between the two faces of the gate reaches 17.4 mm, precisely measured by the same device (Figure 5.7.25).

The authors set up a SCADA system in the gate including nine stress-integrated transducers produced by HBM and calibrated by them in the Fluid Power System Laboratory. Figure 5.7.26 shows the evolution of the stress in the measurement points,

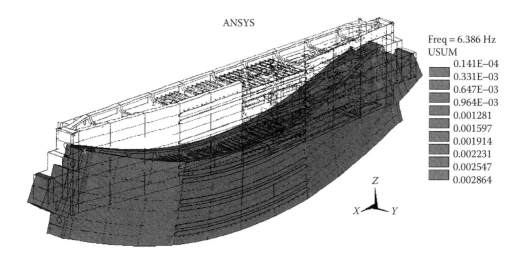

Figure 5.7.24 The shape of the lower section of the water gate under the metal weight.

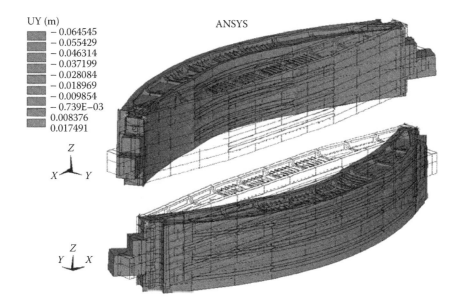

Figure 5.7.25 Gate deformation under the water pressure force.

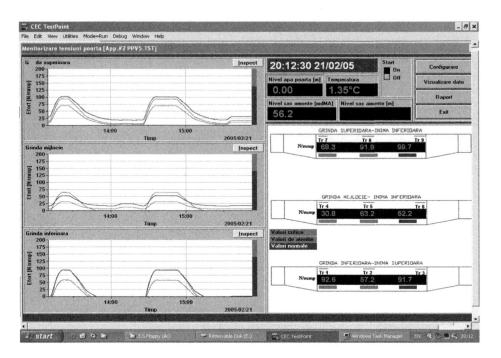

Figure 5.7.26 Screenshot from the SCADA system for two complete cycles of lockage.

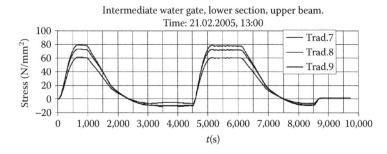

Figure 5.7.27 Stress measurement in the intermediate water gate, lower section, upper beam; the maximum stress was found at the end of the beam; and the minimum stress was found always at one end of the beam (99.7 N/mm²) for a synchronization error of 100 mm.

detailed in Figure 5.7.27. The stress recorded during two gate motion cycles was found in good agreement with the predicted ones by ANSYS. The maximum value of the tension is not exceeding 100 N/mm², much lower than that of the yield stress indicated in the original gate project (360 N/mm²).

The stress recorded during two gate motion cycles was found in good agreement with the predicted ones by ANSYS. The maximum value of the tension is not exceeding 100 N/mm², much lower than that of the yield stress indicated in the original gate project (360 N/mm²).

The biggest value of the tension is associated with a synchronization error of about 100 mm, detected during the gate motion with a speed of about 12 mm/s.

5.7.5   Conclusion

The modern synchronization systems of the hydraulic cylinders can satisfy high requirements under the condition of using high-quality equipment, keeping the oil as clean as possible by constant recirculation through tank filters and using high-quality industrial controllers with redundant properties. At the same time, the fine tuning of the error amplifiers can be very simple by the aid of a realistic numerical simulation of the whole system. A huge number of application of civil and power engineering such as lock barriers for coastal protection, hydroelectric power plants, [6,7] or locks are important applications of this computational technology using Amesim.

Section 1: Bibliography

1. IMAGINE, S.A. *Advanced Modelling and Simulation Environment, Release 4.2.1. User Manual.* Roanne, France, 2005.
2. Lebrun, M., Claude, R. How to create good models without writing a single line of code. *Fifth Scandinavian International Conference on Fluid Power,* Linköping, Sweden, 1997.
3. Lebrun, M. EHA's model reduction using activity indexes. *Recent Advances in Aerospace Hydraulics Conference,* I.N.S.A. Toulouse, France, 2004.
4. Vasiliu, D. Researches on the servo pumps and the servo motors of the hydrostatic transmissions. PhD Thesis, University Politehnica of Bucharest, Bucharest, *Romania,* 1997.
5. Mare, J.-C.H., Cregut, S. Electro hydraulic force generator for the certification of a thrust vector actuator. *Recent Advances in Aerospace Hydraulics Conference,* I.N.S.A. Toulouse, France, 2001.
6. Vasiliu, N., Călinoiu, C., Vasiliu, D., Ofrim, D., Manea, F. Theoretical and experimental researches on a new type of digital electro hydraulic speed governor for hydraulic turbines. *1st International Conference on Computational Methods in Fluid Power Technology,* Melbourne, Victoria, Australia, 2003.
7. Vasiliu, N., Vasiliu, D. *Electro Hydraulic Servomechanisms with Two Stages DDV for Heavy Load Simulators Controlled by ADWIN.* Recent Advances in Aerospace Hydraulics Conference, I.N.S.A. Toulouse, France, 2004.
8. Vasiliu, N., Vasiliu, D. *Fluid Power Systems, Vol. I.* Technical Publishing House, Bucharest, (In Romanian), 2005.
9. https://www.dspace.com/en/inc/home/products/systems/ecutest.cfm#

Section 2: Bibliography

1. Dardac, L.T., Vasiliu, N., Călinoiu, C. *Electrohydraulic Variable Valve Train System for Diesel Engines.* Romanian Patent No. 126878, 2013.
2. Feher, S.I. Researches on the electrohydraulic variable valves systems. PhD Thesis, University POLITEHNICA of Bucharest, Bucharest, Romanian, 2013.
3. Kehtarnavaz, N. *Digital Signal Processing System Design: LabVIEW-Based Hybrid Programming.* Academic Press Elsevier, San Diego, CA, 2008.
4. Vasiliu, N., Vasiliu, D. et al. *Engineering Innovation in Fluid Power Systems by IT.* State Project 12-104, National Research Program, Bucharest, Romanian, 2013.
5. Vorih, J.M., Egan, J.F. 2001. *Lost Motion Valve Actuation System.* Patent US6321701 B1, Diesel engine retarders, Delaware.
6. Grünwald, B. 1980. *Internal Combustion Engines Computation and Design.* E.D.P., Bucharest, (in Romanian).
7. ISO10770-1-2009. *Hydraulic Fluid Power—Electrically Modulated Hydraulic Control Valves. Part 1. Test methods for four-way directional control valves,* Geneva, Switzerland, 2009.
8. MOOG *Electrohydraulic Valves—Applications, Selection Guide, Technology, Terminology, Characteristics.* East Aurora, New York, 2010.
9. National Instruments Corporation. *LabVIEW 13,* Manu, TX, 2013.

10. Robert Bosch GmbH, *EHVS*. FPNI Forum, Bad Ischl, Austria, 2004.
11. Robert Bosch GmbH. *Diesel-Engine Management, 4th edition*, Wiley, *Chichester, England*, 2005.
12. Sturman Industries. *The Sturman Digital Engine*, San Diego, CA, 2008.
13. www.sturmanindustries.com
13. www.ni.com
14. www.parker.com
15. www.boschrexroth.com
16. www.lmsintl.com
17. www.moog.com
18. www.dspace.com
19. www.mathworks.com

Section 3: Bibliography

1. LMS INTERNATIONAL. *Advanced Modeling and Simulation Environment*, Release 13 User Manual, Leuven, *Belgium*, 2013.
2. Mare, J.-C.H. Modelling of aerospace actuation systems. *European AMESim Conference*, Strasbourg, France, 2006.
3. Mouvand, S. Advances and applications of AMESim aviation system simulation. *LMS Aerospace Competence Center*, Internal Report, LMS International, 2011.
4. http://www.thebanque-pdf.com/fr_commandes-de-vol-a380.html
5. http://blascheck.franck.free.fr/IMG/pdf/TD_4_Airbus_A380.pdf
6. http://www.sujets-de-concours.net/sujets/ccp/2005/mp/si.pdf

Section 4: Bibliography

1. Xavier Le tron. *A380 Flight Controls Overview*. Presentation at Hamburg University of Applied Sciences, AIRBUS, 2007.
2. https://www.ac-paris.fr/serail/jcms/s1_942790/ee2-lg-clg-paul-valery-portail
3. Merritt, H.E. *Hydraulic Control Systems*. John Wiley & Sons, New York, , 1967.
4. Guillon, M. *L'asservissement Hydraulique et Electrohydraulique*. Dunod, Paris, 1972.
5. Ursu, I., Tecuceanu, G., Toader, A., Calinoiu, C. Switching neuro-fuzzy control with antisaturating logic. Experimental results for hydrostatic servoactuators. *Proccedings of the Romanian Academy, Series A*, Vol. 12, Number 3/2011, pp. 231–238, ISSN 1454-9069.

Section 5: Bibliography

1. Guillon, M. *L'asservissement hydraulique et électrohydraulique*. Dunod, Paris, France, 1972.
2. Guillon, M., Thoraval, B. Commande et asservissement hydrauliques et electrohydraulique. *Technique et documentation—Lavoisier*, Londres, New York, 1992.
3. http://www.heritageconcorde.com/
4. Merritt, H.E. *Hydraulic Control Systems*. John Wiley & Sons, New York, 1967.
5. Vasiliu, D. Researches on the transients from the servopump and servomotors of the hydrostatic transmissions. PhD Thesis, University Politehnica of Bucharest, Bucharest, Romania, 1987.
6. Călinoiu, C. Contributions on the dynamic Identification of the Power Systems. PhD Thesis, University Politehnica of Bucharest, Bucharest, Romania, 1988.
7. Vasiliu, N., Vasiliu, D., Catană, I., Theodorescu, C. *Hydraulic and Pneumatic Servomechanisms, Vol. I*, Politehnica Press House, Bucharest, Romanian, 1992.
8. SAMM 7111A, *Technical Documentation*, ISSY-LES-MOULINEAUX, Paris, *Ile-de-France*, 1976.
9. Călinoiu, C., Vasiliu, N. *Experimental Identification of the Electrohydraulic Servomechanisms by variable Frequency Test Signals*. Scientific Bulletin of the University politehnica of Timişoara, Tom 44 (58), Mechanical Series, Timişoara, Romania, 1999.

10. Vasiliu, N., Vasiliu, D. *Fluid Power Systems. Vol. I.* Technical Press Bucharest, Romania, 2005.
11. Ursu, I., Popescu, T., Vladimirescu, M., Costin, R.D. *On Some Linearization Methods of the Generalized Flowrate Characteristics of the Hydraulic Servomechanisms.* Revue Roumaine des Science Techniques, Serie de Mecanique Appliquee, Tome 39, Romanian Academy, 1994.

Section 6: Bibliography

1. Catana, I., Călinoiu, C., Vasiliu, N. *High Speed Electrohydraulic Servosystems,* Scientific Bulletin of the University Politehnica of Timisoara, Tome 44, Timisoara, Romania, 1999.
2. Lebrun, M., Claude, R. How to create good models without writing a single line of code. *Fifth Scandinavian International Conference on Fluid Power,* Linköping, Sweden, 1997.
3. Lebrun, M. *EHA's Model Reduction using Activity Indexes.* Recent Advances in Aerospace Hydraulics, I.N.S.A. Toulouse, France, 2004.
4. Mare, J.-C.H. Dynamics of the electrohydraulic rotary servomechanisms, PhD Thesis, I.N.S.A. Toulouse, France, 1994.
5. Mare, J.C., Cregut, S. *Electrohydraulic Force Generator for the Certification of a Thrust Vector Actuator.* Recent Advances in Aerospace Hydraulics, I.N.S.A. Toulouse, France, 2001.
6. http://www.adwin.de/index-us.html.
7. Vasiliu, N., Călinoiu, C., Vasiliu, D. *Electrohydraulic Servomechanisms with Two Stages DDV for Heavy Load Simulators Controlled by ADWIN.* Recent Advances in Aerospace Hydraulics, I.N.S.A. Toulouse, France, 2004.
8. Vasiliu, N., Vasiliu, D. *Fluid Power Systems, Vol. I.* Technical Publishing House Bucharest, Bucharest, Romania, 2005.
9. Vasiliu, N., Călinoiu, C., Vasiliu, D. Modeling, simulation and identification of the electro-hydraulic speed governors for hydraulic turbines by AMESIM. *European AMESIM User Conference,* 4th edition, 2006.
10. Vasiliu, D. Researches on the electrohydraulic servopump and servomotors, PhD Thesis, University politehnica of Bucharest, Romania, 1997.
11. Vasiliu, N., Antonescu, I. Hydrostatic transmission for aircraft weapons loaders. AEROTEH Research Report, 1998.
12. Vasiliu, D., Călinoiu, C., Vasiliu, N. *Governing Hydraulic Turbines by Servo Solenoid Valves— Numerical Simulation and Test,* 2000 ESS CONFERENCE, The Society for Computer Simulation, Hamburg, Germany, 2000.
13. IMAGINE, S.A. *Advanced Modelling and Simulation Environment, Release 4.3.0 User Manual,* Roanne, France, 2006.
14. BOSCH. *Servo Solenoid Valves,* Technical Specification 13/2, Automation Technology, Stuttgart, Germany, 1999.

Section 7: Bibliography

1. https://www.boschrexroth.com/en/xc/products/product-groups/industrial-hydraulics/cylinders/large-hydraulic-cylinders/products (Hydraulic cylinders with piston rods using Ceramax Engineered Coatings).
2. Vasiliu, N., Vasiliu, D., Catană, I., Theodorescu, C. *Hydraulic and Pneumatic Servomechanisms. Vol. I,* University politehnica of Bucharest Press, Bucharest, Romania, 1992.
3. Vasiliu, D. Researches on the transients from the servopump and servomotors of the hydro-static transmissions. PhD Thesis, University politehnica of Bucharest, Bucharest, Romanian, 1997.
4. Călinoiu, C. Contributions on the dynamic identification of the fluid power systems. PhD Thesis, University politehnica of Bucharest, Bucharest, Romanian, 1998.
5. Călinoiu, C., Vasiliu, N. Experimental identification of the electrohidraulic servomechanisms by variable frequency test signals. Scientific Bulletin of the University politehnica of Timișoara, Tom 44 (58), Mechanical Series, Timișoara, Romania, 1999.

6. Vasiliu, D., Calinoiu, C., Vasiliu, N. *Governing Hydraulic Turbines by Servo Solenoid Valves— Numerical Simulation and Test*. ESS CONFERENCE, The Society for Computer Simulation, Hamburg, Germany, 2000.
7. Vasiliu, N., Calinoiu, C., Vasiliu, D., Ofrim, D., Manea, F. Theoretical and experimental researches on a new type of digital electro hydraulic speed governor for hydraulic turbines. *1st International Conference on Computational Methods in Fluid Power Technology, Methods for Solving Practical Problems in Design and Control*, Melbourne, VA, 2003.
8. Cazanacli, C. Researches on the predictive maintenance of hydromechanical equipments from hydropower stations. PhD Thesis, University politehnica of Bucharest, Bucharest, Romanian, 2005.

chapter six

Numerical simulation of the automotive hydraulic steering systems

6.1 Numerical simulation and experimental identification of the car hydraulic steering systems

6.1.1 Steady-state behavior of an open-center flow valve

Being regarded as force or torque amplifier, which reproduces the input to output with various gains, the steering systems are used everywhere for *motion control*: from cars, trucks, heavy vehicles, and agricultural machines, to airplanes, missiles, navy, submarines, and so on. The static and the dynamic requirements of these machines are very different, leading to different structures, both for reaching specific static and dynamic performances and for different stability criteria. The stability is strongly influenced by the load type, the interaction between the load and the hydraulic motor [4], the dynamic behavior of the flow rate in the presence of pressure waves, and so on. The main parameter that decides the stability of a hydraulic servomechanism is the flow gain around the null region of the control valve. From this point of view, an open-center flow control valve is the worst case but for the applications requiring the *automatic recovery of the neutral position* of the input and the output components, it is the only one possibility. The main practical case from this point of view can be found in the automotive field (Figures 6.1.1 and 6.1.2).

Similar to any other control system, the hydraulic servomechanisms have to be studied from accuracy and stability points of view. A realistic approach starts with a realistic mathematical modeling, taking into consideration the nonlinear components of the systems and their linearized form. The main target of this study is to determine the quantitative influence of the design parameters over the stability and the accuracy of these systems. First, it was considered the study of unsteady phenomena associated with liquid flows through the components of the servomechanism, using the continuity equation with its specific form related to the unsteady hydraulic power systems. The obtained equation was applied to the open-center valve—hydraulic cylinder subsystem. Another two necessary equations were studied: (1) the equation of motion for the piston and (2) the equation of the comparison system. The steady-state characteristics were linearized for a small interval around the hydraulic null position of the flow valve. This operating interval reveals the main important nonlinearity of the servomechanism. By joining all these equations together, one may obtain a system of equations, which describes the dynamic working regime of the hydraulic open-center servomechanisms supplied with constant flow.

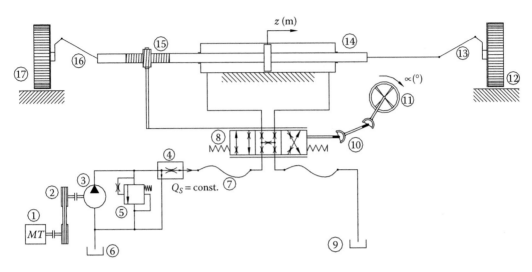

Figure 6.1.1 Hydraulic diagram of an automotive steering system: 1—thermal engine; 2—pump pulley; 3—gear or sliding vanes pump; 4—three-way flow control valve; 5—pressure relief valve; 6 and 9—free surface oil tank; 7—hoses; 8—four-way three position open-center rotary valve; 10—steering column; 11—steering wheel; 12 and 17—orientable front tire; 13 and 16—steering power link; 14—hydraulic cylinder; and 15—rack-pinion gear.

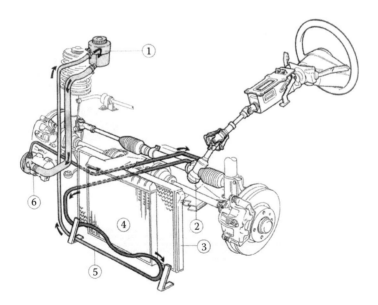

Figure 6.1.2 Overall view of a classical steering unit: 1—oil tank; 2—steering servomechanism; 3—engine cooler; 4—air intercooler; 5—steering oil cooler; and 6—oil pump with flow rate valve and pressure relief valve. (Source: Volvo)

Regarding the theory of control systems, the hydraulic flow valves are amplifiers having the spool displacement as input variable and the flow rate as output variable. Their behavior is described by the following equation:

$$F(p_S, x, Q_M, P_M) = 0 \qquad (6.1.1)$$

To supply the servomechanism with constant flow, one may need a supply system containing a special volumetric pump, including a three-way flow control valve and a pressure safety valve for limiting the pressure in the supply line of the servomechanism. The main important feature of the constant flow-supplied servomechanisms is that their flow valves always have a narrow hydraulic open center. An exploded view of the rotary valve is shown in Figure 6.1.3.

The main details of the rotary valve are presented in Figures 6.1.4 and 6.1.5. The flow area variation has three domains indicated in Figure 6.1.6.

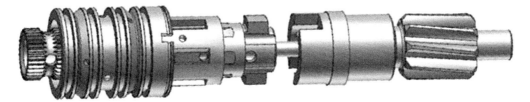

Figure 6.1.3 Expanded view of a typical rotary valve [6].

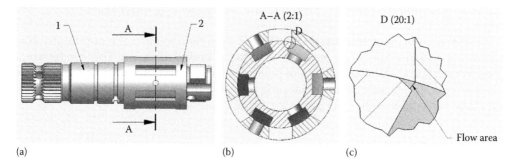

(a)   (b)   (c)

Figure 6.1.4 Main sections of a rotary valve. (a) Lateral view of the spool and the sleeve of the rotary valve; (b) Main section of the rotary valve; (c) The flow area between spool and sleeve in a neutral position of the steering wheel.

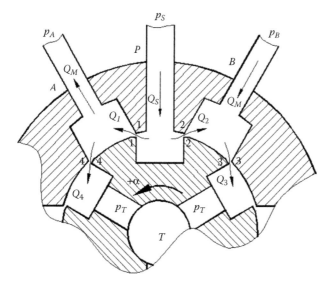

Figure 6.1.5 Fluid flow into an open-center rotary spool valve used in power steering systems.

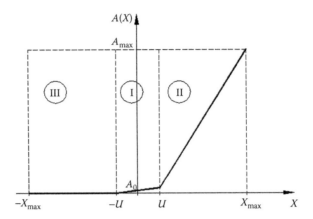

Figure 6.1.6 Flow area domains as functions of the relative motion of the spool and the sleeve: I-underlap domain; II and III-overlap.

Figure 6.1.7 offers a practical design of the rotary steering valve used in a car steering box.

The behavior of an open-center flow valve is strongly nonlinear. An automotive open-center valve operates most of the running time in the underlap region. This can be studied in close connection with the possibility of building a linear dynamic model for such a valve.

When operating inside the underlapping zone, the characteristic equation of the open-center flow valve domain has the following form:

$$Q_M = Q_M(x, P_M) = c_d \cdot b \cdot (U + x) \cdot \sqrt{\frac{p_S - P_M}{\rho}} - c_d \cdot b \cdot (U - x) \cdot \sqrt{\frac{p_S + P_M}{\rho}} \qquad (6.1.2)$$

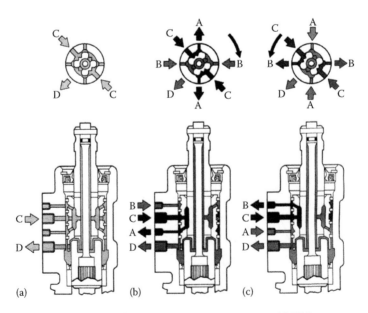

Figure 6.1.7 Rotary spool valve used in power steering systems: (a) Valve not actuated, (b) right turning, and (c) left turning.

Due to the constant flow supply, the pressure inside the supply line depends on the valve opening and on the pressure drop on the hydraulic motor. From the equation of the supply flow,

$$Q_S = c_d \cdot b \cdot (U+x) \cdot \sqrt{\frac{p_S - P_M}{\rho}} + c_d \cdot b \cdot (U-x) \cdot \sqrt{\frac{p_S + P_M}{\rho}} \qquad (6.1.3)$$

it is possible to find the supply pressure of the servomechanism $p_s = p_S(x, P_M)$ necessary for the numerical simulation based on the nonlinear model:

$$p_S = \frac{-b + \sqrt{b^2 - a \cdot c}}{a} \qquad (6.1.4)$$

where a, b, and c are functions of P_M and x [5]. The flow supplied to the servomechanism, Q_S, from Equation 6.1.3, may be obtained by computing the flow consumed by the valve, when operating in the neutral position:

$$Q_S = Q_S\big|_0 = 2 \cdot c_d \cdot b \cdot U \cdot \sqrt{\frac{p_{S0}}{\rho}} = \text{const.} \qquad (6.1.5)$$

In Equation 6.1.5, p_{S0} is the value of the supply pressure of the valve when working in neutral position. In order to obtain the linear form of the steady-state regime characteristic, one must develop in Taylor series Equation 6.1.2, around hydraulic null:

$$Q_M - Q_{M0} = \Delta Q = \left(\frac{\partial Q_M}{\partial x}\right)\Bigg|_0 \cdot \Delta x + \left(\frac{\partial Q_M}{\partial P_M}\right)\Bigg|_0 \cdot \Delta P_M + \dots \qquad (6.1.6)$$

Using the notations K_1 and K_2 for the partial derivatives $\left(\partial Q_M / \partial x\right)\big|_0$ $\left(\partial Q_M / \partial P_M\right)\big|_0$, respectively, and neglecting the terms containing high-order partial derivatives, one may obtain the linear form of the characteristic equation of the valve for the steady-state regime:

$$Q_M = K_1 \cdot x + K_2 \cdot P_M \qquad (6.1.7)$$

where:

$$K_1 = \left(\frac{\partial Q_M}{\partial x}\right)_0 = \frac{Q_S}{U} \qquad (6.1.8)$$

$$K_2 = \left(\frac{\partial Q_M}{\partial P_M}\right)_0 = -\frac{Q_S}{2 \cdot p_{S0}} \qquad (6.1.9)$$

are the valve coefficients.

6.1.2   Continuity equation for the flow control valve—hydraulic linear motor subsystem

Taking into consideration the subsystem containing a 4/3 open-center rotary flow valve and a double-ended hydraulic cylinder, the continuity equation becomes:

$$Q_M = K_l \cdot P_M + A_p \cdot \dot{z} + \frac{A_p^2}{R_h} \cdot \frac{dP_M}{dt} \qquad (6.1.10)$$

Here

$$R_h = 2\frac{\varepsilon}{V_0}A_p^2 \tag{6.1.11}$$

is the hydraulic stiffness, and Q_M represents the average ports flow rate:

$$Q_M = \frac{Q_A + Q_B}{2} \tag{6.1.12}$$

6.1.3   Motion equation of the piston of the hydraulic cylinder

The pressure force, F_p, must counteract an elastic type force, F_e, the friction force, F_f and the inertia force:

$$m_p\ddot{z} = F_p - F_e - F_f \tag{6.1.13}$$

Here:

$$F_p = A_p \cdot P_M \tag{6.1.14}$$

$$F_e = K_e \cdot z \tag{6.1.15}$$

and m_p is the equivalent actuated mass, reduced to the cylinder rod. For the friction force, one has to consider two components: (1) a static one, F_{fs} and (2) a viscous one, F_{fv} [4,5]:

$$F_{fs} = F_{fs0} \cdot sign(\dot{z}) \tag{6.1.16}$$

$$F_{fv} = K_{fv} \cdot \dot{z} \tag{6.1.17}$$

6.1.4   Equation of the following error

The following error of the servomechanism results from the comparison of the input variable—angular displacement of the spool measured from the hydraulic neutral position, α [degree], and the output variable—linear displacement of the cylinder rod, z [m]. The effective error is the rotary flow valve opening, x [m], considered to be the relative displacement between the spool and the sleeve of the valve, and measured on their contact diameter:

$$x(\alpha,z) = \mu_\alpha \cdot \alpha - z \cdot \mu_z \tag{6.1.18}$$

The final computational model of the servomechanism is shown in Figure 6.1.8. The model of the load is a simple one to detect mainly the influence of the underlap on the dynamic behavior of the control system. The real underlap of the rotary valve was found by the aid of a test bench used in the Fluid Power Laboratory of the University Politehnica of Bucharest (Figures 6.1.9 and 6.1.10).

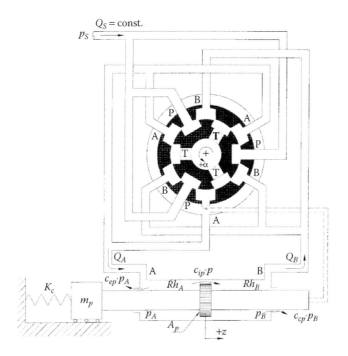

Figure 6.1.8 Computational scheme of the servomechanism.

Figure 6.1.9 Top level of the underlap experimental identification.

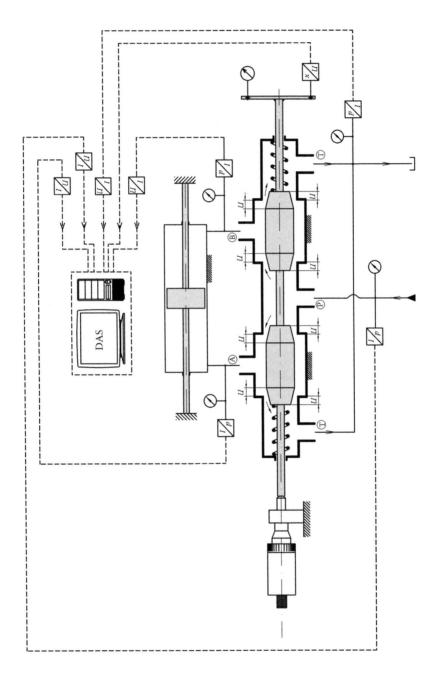

Figure 6.1.10 The overall diagram of the underlap test bench.

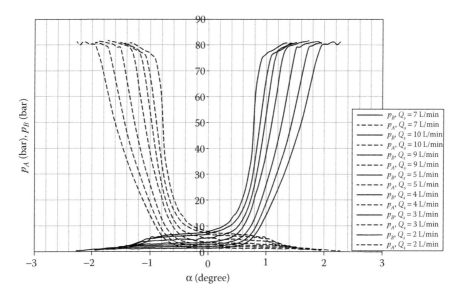

Figure 6.1.11 Pressure variations in the locked cylinder chambers as function of the twist angle of the torsion bar and the supply flow.

Experimental pressure variations in the locked cylinder chambers as function of the twist angle of the torsion bar and the supply flow are presented in Figure 6.1.11. The minimum insensitivity of the rotary valve (less than 1°) results for the nominal flow of the pump: about 10 L/min.

The small flows enlarge the insensitivity angle, explaining the option of some manufacturers of introducing a by-pass valve in the steering case for reducing the pump flow entering in the valve [6]. The explanation of this option is connected to the increase in the damping effect of the steering box for high speeds on highways. The steering systems supplied by electrical driven pumps are adapting the pump speed to the car velocity from the same reason.

6.1.5 Numerical simulation

The results of the numerical simulations performed in SIMULINK®—MATLAB® environment are useful under the condition to keep the following error inside the range of the underlap. Otherwise, the simulations cannot reveal the influence of the design and operation parameters over the dynamic behavior of the servomechanisms.

The real model is strongly nonlinear and needs a high-power simulator that contains predefined blocks for any kind of nonlinear component as the tire friction with the road.

The simulations were performed for the next typical data corresponding to a small car power steering system, tested on an experimental rig [6,7]:

$Q_s = 4.5$ L/min; $p_{S0} = 2$ bar; $A_p = 1.26 \cdot 10^{-3}$ m²; $b = 0.04$ m; $U = 0.106 \cdot 10^{-3}$ m; $R_h = 11.3 \cdot 10^6$ N/m;

$m_p = 70$ kg; $\mu_\alpha = 1.57 \cdot 10^{-4}$ m/rad; $\mu_z = 1.06$; $K_e = 25 \cdot 10^3$ N/m; $F_{fs0} = 100$ N, $K_{fv} = 3.10^4$ Ns/m.

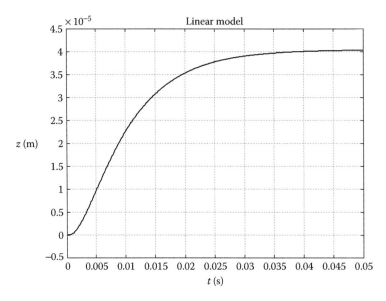

Figure 6.1.12 The response of the servomechanism to a small step input (45 μm).

A typical response of the servomechanism to a step input signal (41 μm) is presented in Figure 6.1.12. The value of the amplitude of the signal was maintained inside the underlapping interval of the valve opening. In this interval, the response of the system is critical from the system stability point of view. The value of the amplitude of the signal was maintained inside the underlapping interval of the valve opening. The time constant of the response (about 17 ms) is small enough for any automotive application.

6.1.6   Conclusion

The fast response of the system and the quick rejection of the external perturbations are both stated by the relatively high value of the K_1 coefficient of the flow valve, which is specific to this type of valves; the value of this parameter is twice larger than its value in the case of the critical center valves [1–3].

For a more comprehensive and accurate approach, it is necessary to study the influence of the valve microgeometry on the static and dynamic performances of the servomechanism.

Another important issue, which has to be considered for further studies, is the interaction between the servomechanism and the integrating system. It is necessary to link the mathematical model of the servomechanism with the mathematical model specific to the integrating dynamic system, using adequate software such as Amesim [8].

The main practical problem of the studied system is the rejection of the flutter induced by the car steering wheels. The new concepts based on the eliminate of the valve effect for high car speed are solving this problem by the aid of a small electrohydraulic valve controlling a small conical clutch sited between the spool and the sleeve. This way, all the system becomes a true hydraulic damper [9,10].

6.2   Modeling and simulation of the hydraulic power steering systems with Amesim

6.2.1   Nonlinear analysis of a steering system dynamics for a linear input

A linear dynamic analysis of a system offers a lot of *virtual* tools but the *distance* between the virtual image and the reality can hide a lot of real practical problems that need a procedure of *cut and try*, and time and money consumption. One of the classical case of this type is found in the field of the power steering shudder, which has a small amplitude vibration that can be felt in the passenger compartment while steering a vehicle in low-speed maneuvers. The frequency of these vibrations is practically found within the range of 10–40 Hz or more. The shudder can be eliminated step by step by changing the different components design, but this time-consuming approach does not allow a deep insight look into the nature of the physical generator of the events. Analyzing this problem by pencil and paper is practically impossible due to the complex nature of the system component interactions. The main practical case from this point of view can be found in the automotive field (Figure 6.2.1).

A hydraulic pump is connected to a rotary *following* valve with pipes and hoses. A pinion-rack converts the rotation of the steering column into a linear displacement of the front axle. The simulation model considers that the pump delivers a flow rate proportional to the engine speed and that a three-way flow control valve reduces this flow to a constant one (similar to the one in Chapter 5). The hydraulic line from this device to the rotary valve and from the rotary valve to the tank is a combination of pipes, hoses, and orifices. In the frame of the IMAGINE vision [1–10], the hoses are modeled with a lumped parameter submodel with compressibility of oil and expansion of line walls. Pipes are modeled with additional assumptions and they take pipe friction and fluid inertia into account. All Amesim line submodels are able to predict air release and cavitation effect. In this case, it is vitally important that density and bulk modulus are totally consistent to achieve rigorous mass conservation. The core of the simulation network is the bridge of the four hydraulic variable orifices created between the rotary spool and the rotary sleeve having the end coupled with the torsion bar (Figures 6.2.2a and b). This one detects the positioning error of the servosystem. The bridge is included in the simulation network as a super component (not available for the simulation user).

Other two super components are complex entities: (1) the hydraulic cylinder and (2) the tire (Figure 6.2.3).

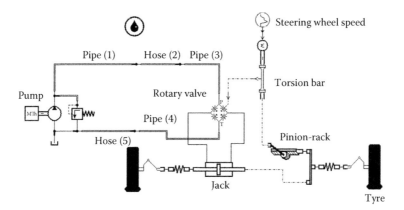

Figure 6.2.1 Amesim simulation model of a hydraulic power steering.

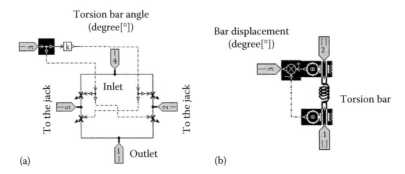

Figure 6.2.2 Complex super components of the system: (a) the hydraulic control orifices bridge and (b) the control error sensor.

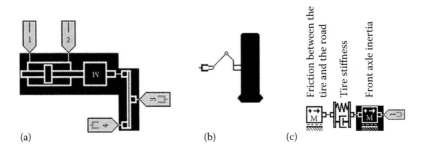

Figure 6.2.3 Other super components of the system: (a) hydraulic cylinder with hydraulic and mechanical connections; (b) the tire with the steering rod; and (c) front axle parameters.

Some important quantities define the practical case studied in this chapter as follows:

- Spring stiffness of the torsion bar: 1.5 Nm/degree
- Double rod jack piston diameter: 28 mm
- Rod diameter: 12 mm (both sides)
- Length of stroke: 500 mm
- Total mass moved in jack: 3 kg
- Viscous friction coefficient: 5000 N/(m/s)
- Geometrical radius of the pinion: 7 mm
- Spring rate of the direction rod: 5.0×10^6 N/m
- Coulomb friction force between the tire and the road: 2000 N
- Stiction force between the tire and the road: 2000 N
- Tire lateral stiffness: 200000 N/m
- Damper rating: 20000 N/(m/s)
- Mass of the front axle: 60 kg
- Oil density: 850 kg/m³
- Bulk modulus: 17000 bar
- Absolute viscosity: 51 cP
- Absolute viscosity of air/gas: 0.02 cP
- Air/gas content: 0.1%
- Polytrophic index of air/gas/vapor content: 1.4 [−]

A deep understanding of the whole system dynamic behavior needs three types of input signals in angular position: (1) a linear periodical one, (2) a sine one, and (3) a step-in velocity one without recovering the initial position. These complex investigations follow the world experience in systems identification [13–20]. Some new ideas, patented by the authors [10,11,12], still need such researches to be applied in practice.

6.2.2   *Study of a linear periodical input steering process*

A periodical constant velocity of 20 rev/min is applied to the steering column (Figures 6.2.4 and 6.2.5). The steering wheel angular position has a linear periodical variation with an amplitude of 270°. The rotary valve opens and the pressure in the left piston chamber increases to help the rotation of the wheel. The torque at the free end of the torsion bar has a periodical variation of 2 Nm around the average of 7 Nm.

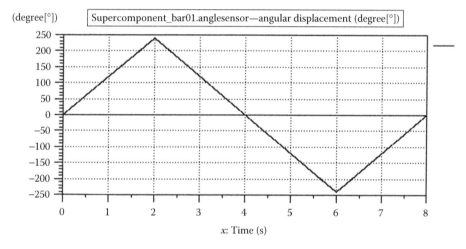

Figure 6.2.4 Linear periodical input steering angle.

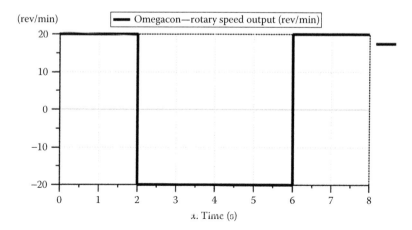

Figure 6.2.5 Periodical input angular velocity.

The corresponding twist angle follows by 1.2° of the torque high-frequency variations. Figures 6.2.6 through 6.2.8 present some preliminary results regarding the input signal and its effect.

While plotting the pressure from the hydraulic cylinder chambers (Figures 6.2.9 and 6.2.10), we notice a lot of oscillations between 43 and 72 bar. Performing a fast Fourier transform analysis, we have access to the frequency of these oscillations: approximately 31 Hz.

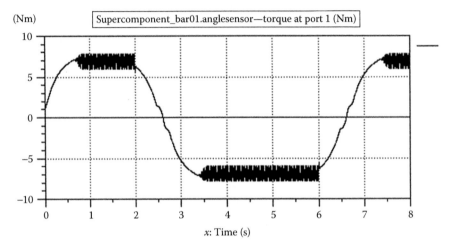

Figure 6.2.6 Torque at the free end of the torsion bar for a linear steering angle.

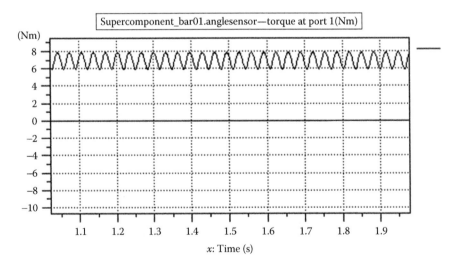

Figure 6.2.7 Detail of the torque.

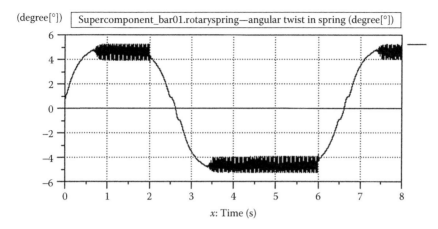

Figure 6.2.8 Angular twist variation of the torsion bar.

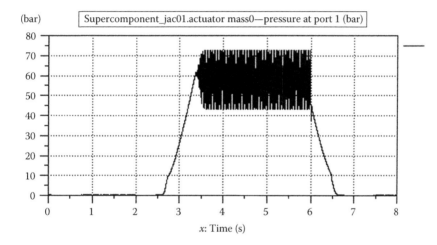

Figure 6.2.9 Pressure variation in the first port of the hydraulic cylinder.

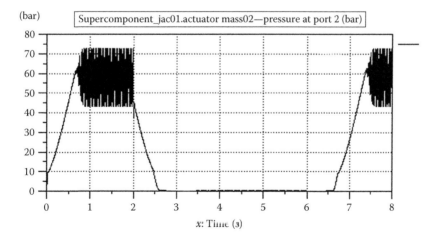

Figure 6.2.10 Pressure variation in the second port of the hydraulic cylinder.

The rod displacement of the hydraulic cylinder seems to be a linear one (Figure 6.2.11), with a small shudder that comes from the interaction between the tires and the rod. The external force applied on the rod of the hydraulic cylinder (Figure 6.2.12) reaches 3500 N (the biggest force from the system). The system applied to the tire reaches 2000 N (Figure 6.2.13), and the small *slips* needed to follow the path are not exceeding 300 N, small enough to avoid a strong discomfort of the passengers.

The overall force applied to the tire-front steering axle has near the same aspect (Figure 6.2.14).

The main conclusion of this first series of numerical simulations concerns the vibrations generated during any change of direction in the whole hydromechanical system. At the first sight, another kind of steering law may change the situation, but the next series of simulation will show the contrary.

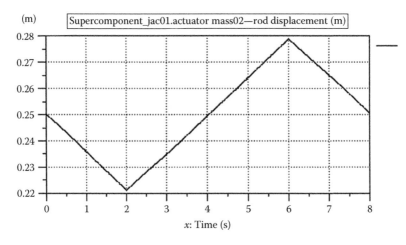

Figure 6.2.11 Hydraulic cylinder rod displacement in a cycle of steering.

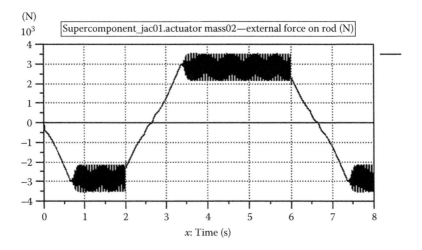

Figure 6.2.12 External force applied on the rod of the hydraulic cylinder.

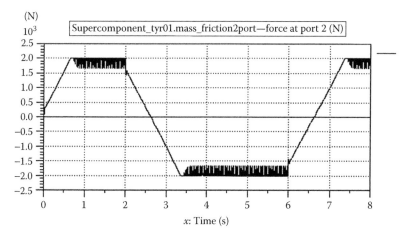

Figure 6.2.13 Force applied to the tire during the steering.

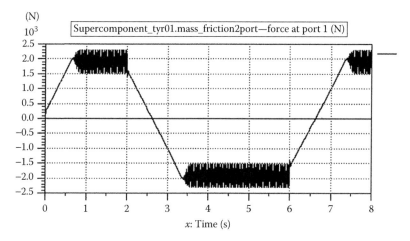

Figure 6.2.14 Force applied to the steering axis of the wheels.

6.2.3 Study of the sine periodical input process

A common test in the automotive industry is called *Moose Test* and combines a sudden sine turn with the action of the Electronic Stability Program. From 2012, this kind of handle has to be performed by any car manufactured or sold in European Union (UE). In order to study the behavior of a hydraulic power steering in such a situation, the input steering speed (Figure 6.2.15) is a cosine curve, which is obtained by applying a similar torque to the torsion bar (Figure 6.2.16). The maximum torque value is a normal one (average between 6 and 8 Nm), but for half of the turning time it has high-frequency variations with a frequency of about 31 Hz (Figure 6.2.17).

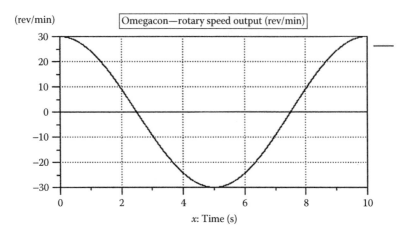

Figure 6.2.15 Steering wheel rotary speed.

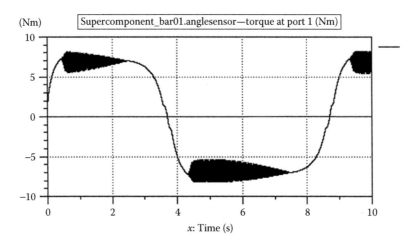

Figure 6.2.16 Torque applied to the steering wheel in a cycle of turn.

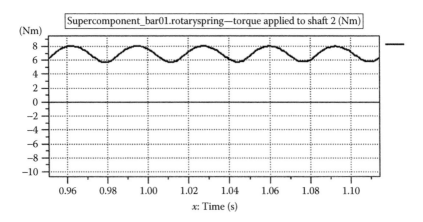

Figure 6.2.17 Detail of the torque applied to the steering wheel.

The steering wheel of the car follows a complete sine curve with an amplitude of about 290° (Figure 6.2.18).

The pressures in the two chambers of the hydraulic cylinder have strong vibrations (Figures 6.2.19 and 6.2.20).

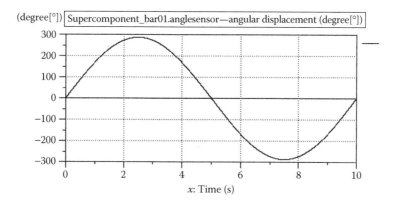

Figure 6.2.18 The angular displacement of the steering wheel (maximum 290°).

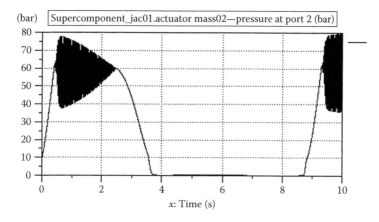

Figure 6.2.19 Pressure variation at port 1 of the hydraulic cylinder (maximum 80 bar).

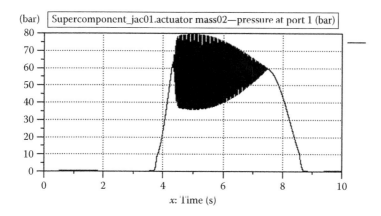

Figure 6.2.20 Pressure variation at port 2 of the hydraulic cylinder (maximum 80 bar).

The flows entering from the hydraulic cylinder ports have also strong variations (Figure 6.2.21).

The following graphs (Figure 6.2.22 through 6.2.31) reveal the evolution of different dynamic, cinematic, hydraulic, and mechanical parameters during the typical steering process of a car.

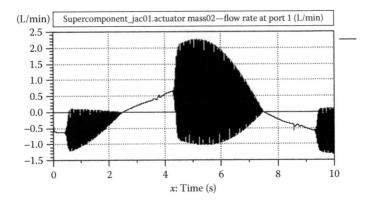

Figure 6.2.21 Flow rate at port 1 of the hydraulic cylinder.

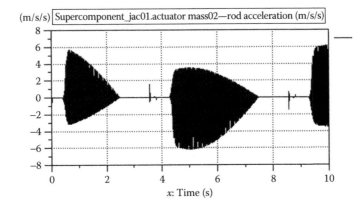

Figure 6.2.22 Rod acceleration of the hydraulic cylinder (maximum 6 m/s/s).

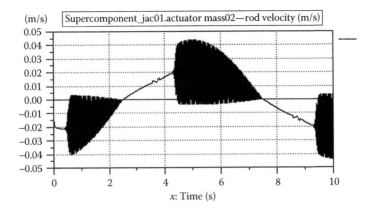

Figure 6.2.23 Rod velocity of the hydraulic cylinder (maximum 42 mm/s).

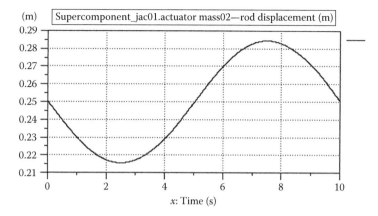

Figure 6.2.24 Rod displacement during the steering wheel revolution (maximum ±33 mm).

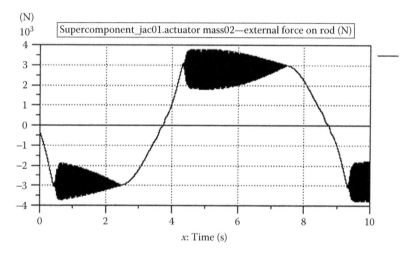

Figure 6.2.25 External force applied to the cylinder rod (maximum ±3800 N).

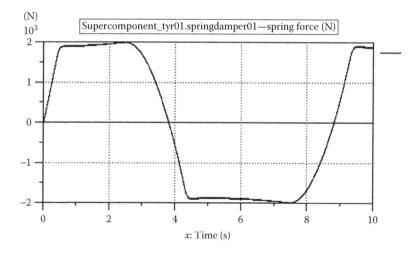

Figure 6.2.26 Spring force in tire (maximum ±2000 N).

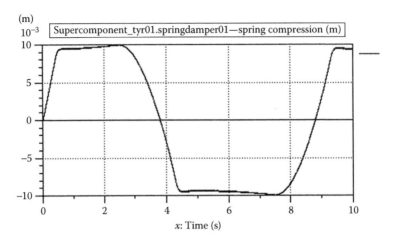

Figure 6.2.27 Spring compression in tire (maximum ±10 mm).

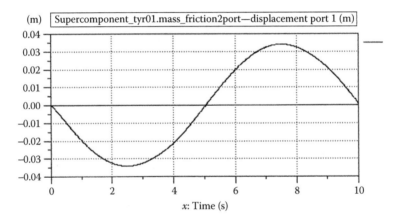

Figure 6.2.28 Displacement of the tire center (maximum ±33 mm).

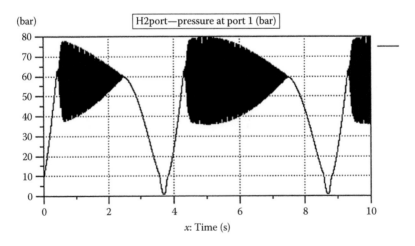

Figure 6.2.29 Pump delivery pressure evolution during the car turn.

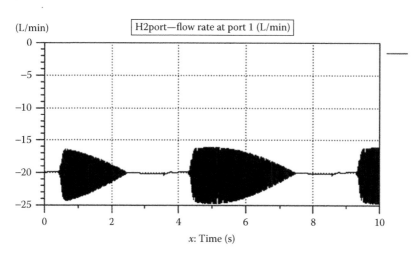

Figure 6.2.30 Pump flow rate variation.

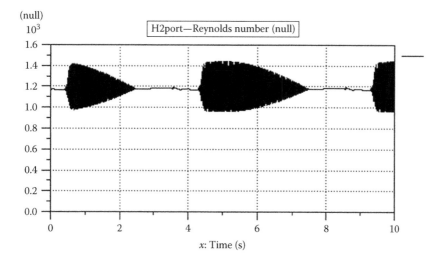

Figure 6.2.31 Reynolds number variation in the pump delivery line.

6.2.4 Conclusion

The main conclusion regarding the simulation of a hydraulic power steering is the complexity of the phenomena occurring inside a direction bridge of a car. At a first glance, the small underlap of the rotary control valve cannot generate a lot of transient in the system. A deep view inside the process shows the contrary conclusion: Even a sudden turn of the steering wheel generates finally strong pressure variations in the supply system, and the only component that can reduce the characteristic noise is the insertion of a medium elastic hose in the delivery line of the pump. The reduction of the underlap leads to a great sensitivity that requires a high level of attention for the driver. The connection between the pump and the tank by a small servovalve, used by Volvo Corporation on the Servotronic steering system is very useful for high speed, which is increasing in a significant manner of the driving comfort and the danger of sudden change of the track on the highways.

6.3 Researches on the electrohydraulic steering systems of the articulated vehicles

6.3.1 Defining precision agriculture

This section presents the structure and the performance of a new electrohydraulic steering system for articulated mobile equipments, which can be remotely controlled via GPS. The system performances were identified using a forestry-articulated tractor. The front chassis was locked, leaving free the back one to rotate on a circular raceway. The tractor steering system was identified by the aid of a rotary hydraulic motor controlled in closed loop by a servovalve. Then, the steering unit was replaced by a servovalve. The system feedback was supplied by a position transducer attached to one of the hydraulic cylinders. The new system performance is found good enough for tractor's remote control. Precision farming or precision agriculture is about making farming processes more accurate, efficient, and sustainable through the use of advanced satellite navigation systems (such as GPS) and Information Technology (IT).

Precision farming is raising the agriculture to a new level: Doing the right thing, in the right place, in the right way, at the right time in the field. Apart from satellite navigation systems, precision farming makes use of a number of other key technologies [1]. They include automated steering systems, which take over specific driving tasks such as auto—steering, overhead turning, following field edges, and overlapping of rows. On tractors, combine harvesters, maize harvesters, and other simulate vehicles, there is often a need for electrically actuated steering to make automatic GPS-controlled steering possible. Manual steering with variable ratio is an often-wanted feature to improve productivity and driver comfort. For this purpose, Sauer–Danfoss Company has developed [2] a combined steering unit and electrohydraulic steering valve named OSPE (Figure 6.3.1a and b): OSP for normal manual steering wheel-activated steering (Figure 6.3.2) and E for electrohydraulic steering activated by electrical input signal either from GPS or vehicle controller or from steering wheel sensor (SASA) for variable steering ratio (Figure 6.3.3).

In variable steering mode, the electrohydraulic valve part adds flow to the metered-out flow from the steering unit part of the OSPE. This has built-in safety function in the form of a *cut off valve*, which makes unintended steering from electrohydraulic valve part impossible. So OSPE is the right steering element first of all to build up steering system with very high safety level and hence to be able to fulfill legislations demands such as demands in EU Machinery Directive 2006/42/EC. In cases where space does not allow room enough for OSPE, an ordinary OSP nonreaction steering unit combined with the electrohydraulic in-line steering valve is an alternative offered with the same safety functions as OSPE.

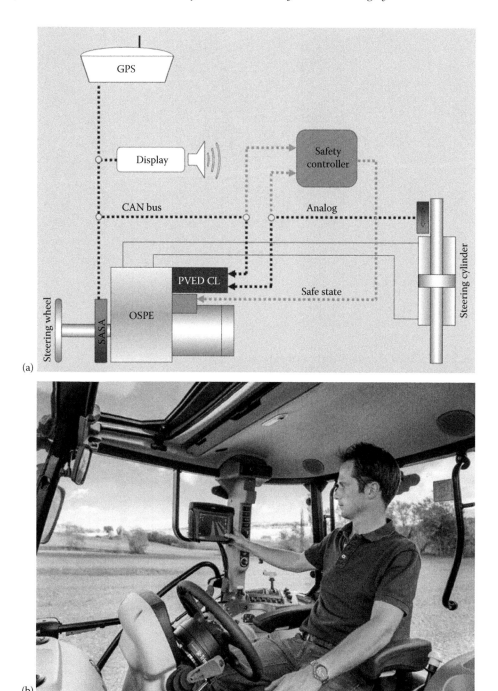

Figure 6.3.1 (a) Hybrid steering system for mobile equipment and (b) modern digital electrohydraulic steering systems. (Source: Sauer–Danfoss)

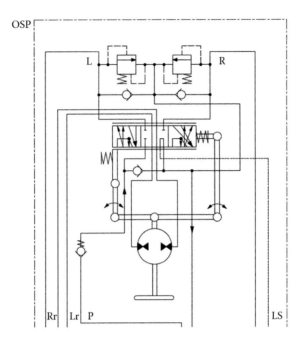

Figure 6.3.2 Modified manual steering wheel activated steering (OSP).

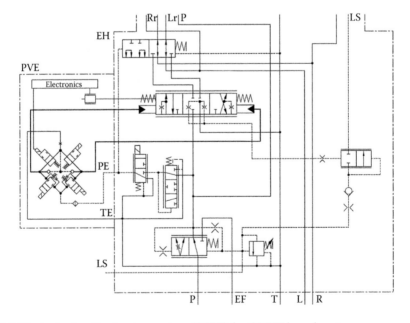

Figure 6.3.3 Electrohydraulic steering system (OSPE) for ORBIT-based system.

6.3.2 Options for a new hybrid steering system

There are some initial decisions to make to design a hybrid steering system for any kind of mobile equipment. First of all, the new system has to allow the human operator to drive himself the equipment anytime, that is, to have the highest priority. Second, the new control system has to be stable and to preserve the accuracy of the previous one under any operation condition. The electrical component strongly depends on the electric power supply, increasing real problems from the security point of view. Some electrical power supply systems can reduce the risk of the electrical systems failure. The main problem is to use the same steering cylinders with two kinds of hydraulic metering valves. Both of them need to have closed center or have to be isolated upon the request of the driver by an additive directional valve. The relative small fuel consumption introduced by an open-center Orbitrol hydraulic steering units can be preserved by driving the input with an electric actuator such as a stepping motor with 100–200 pulses/revolution. This option needs a special design of the stepping motor to insert it into a steering column. On the other hand, the use of an open-center flow valve leads to a poor tracking ability that increases with the steering wheel speed [3]. A typical steady-state diagram of such a hydromechanical feedback steering unit shows a small force capability for high turning speeds. The modern product range has to cover applications of all types of mobile equipment. This chapter presents the structure and the performance of a new electrohydraulic steering system for articulated mobile equipments that are remote controlled. The new concept was designed and tested in the Fluid Power Systems of the University Politehnica of Bucharest to point out the accuracy and the stability of a control solution based on a dual input (digital and analog) integrated axis controller (IAC)-R Rexroth servovalve (Figure 6.3.4) and a single input on board electronics (OBE) analog input servovalve of the same size.

The CAN bus type of *controller area network* is best suited for multipoint, long-range cabling in high electromagnetic interferences areas where analog feedback signals may fail.

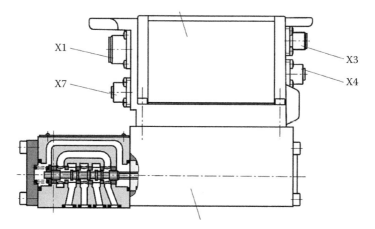

Figure 6.3.4 High-response valve with integrated digital axis controller (IAC-R) and field bus interface. (Source: Rexroth)

This new type of control is already used in military, automotive, and aerospace simulating platforms or other systems where precision is very important. The system performances were identified by the aid of a dedicated test bench [4], shown in Figure 6.3.5a, and using the forestry articulated tractor (Figure 6.3.5b) designed as prototype by the research department of the former company *Progresul* from Braila [5]. The front semichassis was locked, leaving free the back one to rotate on a circular raceway (Figure 6.3.6).

(a)

(b)

Figure 6.3.5 (a) Test bench for testing high-response valve with integrated digital axis controller (IAC-R) and field bus interface in the Fluid Power Laboratory of University Politehnica of Bucharest and (b) overall view of the articulated full hydraulic tractor with front semitrucks locked set up in the Fluid Power Laboratory of U.P.B.

Figure 6.3.6 Back semitruck with light contact pressure guiding rollers.

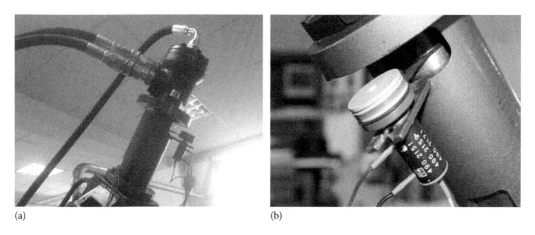

(a)   (b)

Figure 6.3.7 Hybrid electrohydraulic steering servosystem driven by an ORBIT motor in close loop: (a) driving system view and (b) rotary feedback potentiometer driven by the hydraulic motor shaft.

The tractor steering system was identified by the aid of a rotary hydraulic motor (Figure 6.3.7) controlled in closed loop by a servovalve constant pressure supplied (Figure 6.3.8). Then the steering amplifier was replaced by a servovalve, both analog and digital.

The system feedback was supplied by a position transducer (Figure 6.3.9) attached to one of the hydraulic cylinders (Figure 6.3.10).

The control valve of the Orbitrol hydraulic steering unit is normally designed with an open center for reducing both fuel consumption and steering mechanical shocks [6]. The *price* of these gains is a big backlash that increases with the input frequency and the loss of the available pressure drop for high speeds (Figure 6.3.11).

Figure 6.3.8 Fluid power supply unit with electrohydraulic proportional servovalve (OBE-Bosch) set up in the Fluid Power Laboratory of U.P.B.

Figure 6.3.9 Hydraulic steering cylinder with double track position transducer. (*Penny & Gilles*)

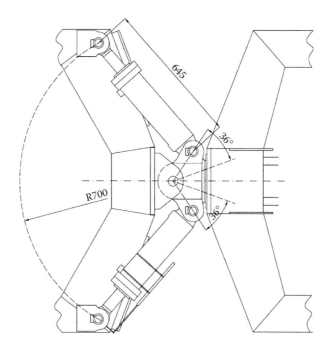

Figure 6.3.10 Plain view of the steering hydraulic cylinders.

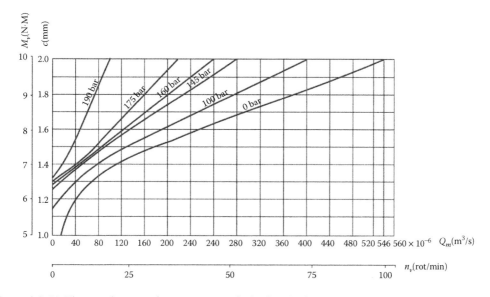

Figure 6.3.11 The steady-state characteristics of a hydraulic feedback steering unit [6].

6.3.3   *Numerical simulation*

The Amesim model of the steering system is shown in Figure 6.3.12. The main results of the simulations are presented in Figures 6.3.13 through 6.3.20. They show a good following capacity for a slow vehicle.

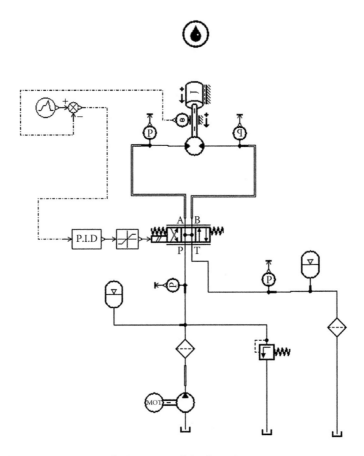

Figure 6.3.12 Steering system simulation network in Amesim.

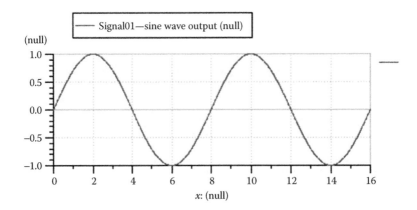

Figure 6.3.13 Sine input signal: $f = 0.125$ Hz.

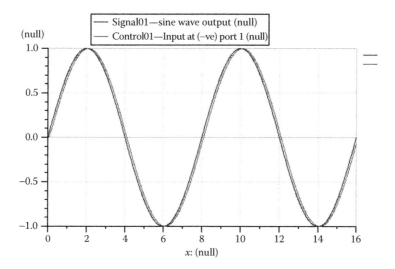

Figure 6.3.14 Input and output signals for $f = 0.125$ Hz.

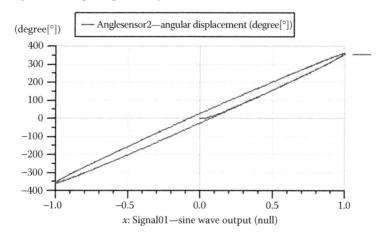

Figure 6.3.15 System steady-state characteristics for $f = 0.125$ Hz.

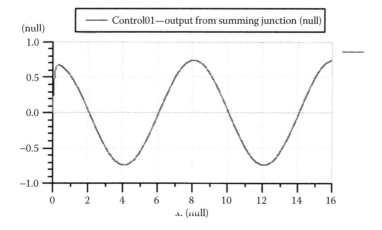

Figure 6.3.16 Error signal for $f = 0.125$ Hz.

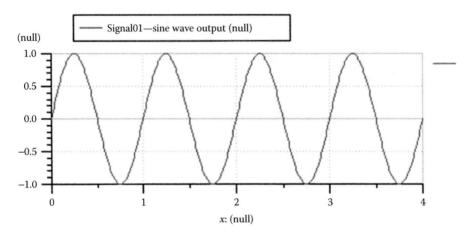

Figure 6.3.17 Input signal with $f = 1$ Hz.

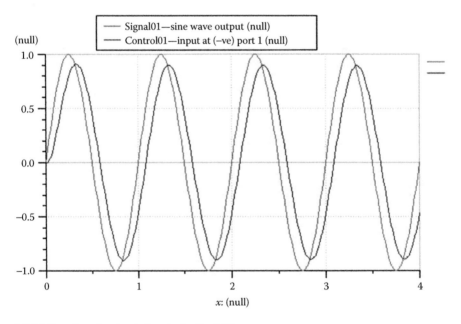

Figure 6.3.18 Input and output signals for $f = 1$ Hz.

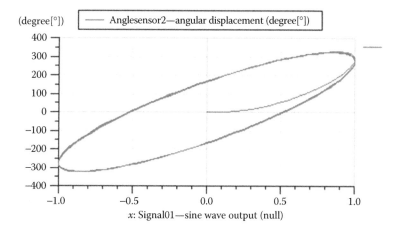

Figure 6.3.19 Steady-state characteristics for $f = 1$ Hz.

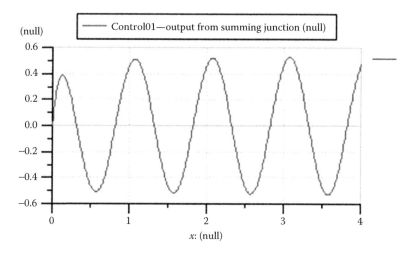

Figure 6.3.20 Error for $f = 1$ Hz.

6.3.4   *Experimental results*

Two series of experiments were performed using the drive of the original steering unit of the tractor and then replacing it by a direct close loop control of the back semitruck. The main results of the investigations are presented in the following figures (Figures 6.3.21 through 6.3.28).

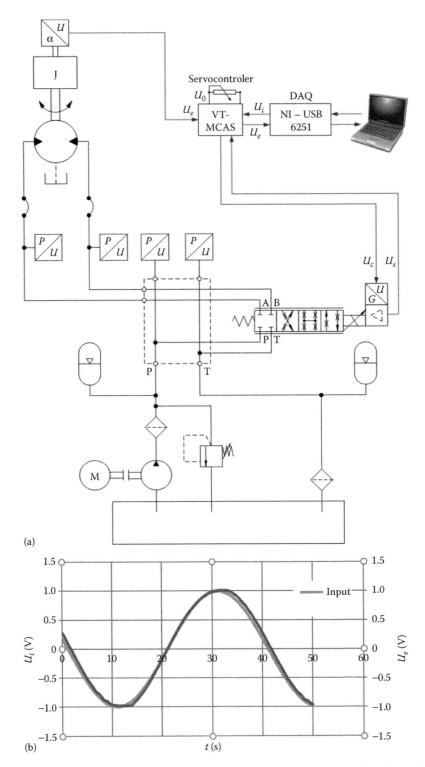

(a)

(b)

Figure 6.3.21 (a) Hydraulic diagram of the drive system of the steering unit [4], [12] and (b) steering system response when Orbitrol unit is driven by a hydraulic motor ($U_i = 1$ V; $f = 0.025$ Hz).

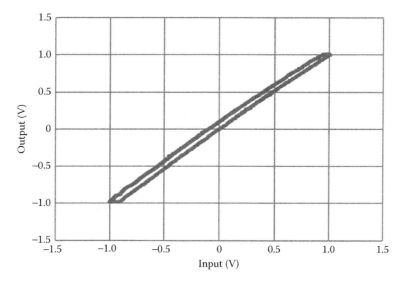

Figure 6.3.22 Steering system static characteristics when Orbitrol unit is driven by a hydraulic motor ($U_i = 1$ V; $f = 0.025$ Hz).

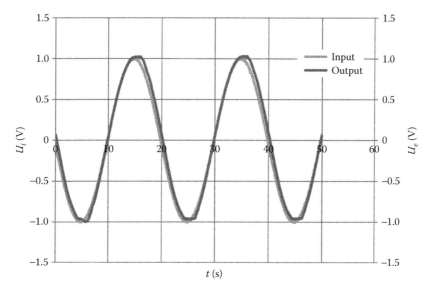

Figure 6.3.23 Steering system response when Orbitrol unit is driven by a hydraulic motor ($U_i = 1$ V; $f = 0.05$ Hz).

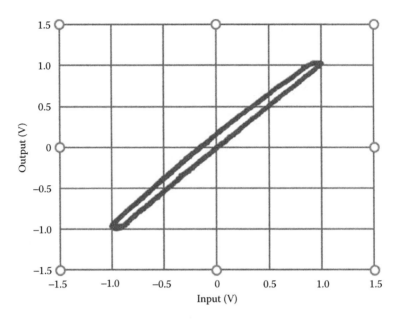

Figure 6.3.24 Steering system steady-state characteristics when Orbitrol unit is driven by a hydraulic motor ($U_i = 1$ V; $f = 0.05$ Hz).

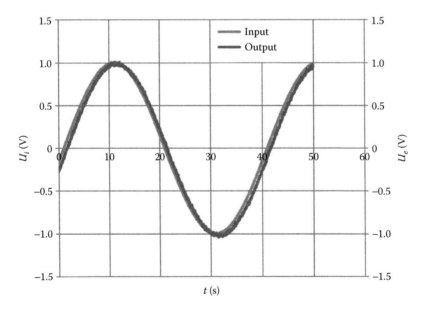

Figure 6.3.25 Steering system steady-state characteristics when cylinders are driven by a servovalve ($U_i = 1$ V; $f = 0.025$ Hz; $p_s = 100$ bar).

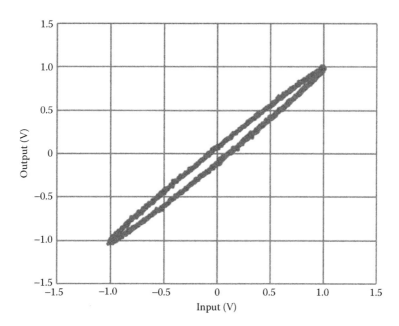

Figure 6.3.26 Steering system steady-state characteristics when cylinders are driven by a servo-valve in closed loop ($U_i = 1$ V; $f = 0.025$ Hz; $p_s = 100$ bar).

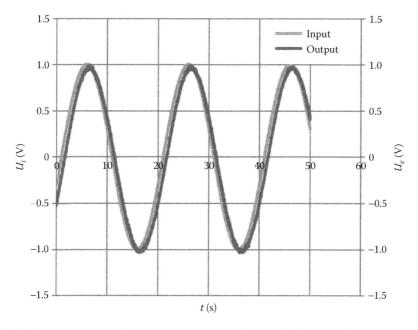

Figure 6.3.27 Steering system frequency response when cylinders are driven by a servovalve ($U_i = 1$ V; $f = 0.05$ Hz; $p_s = 100$ bar).

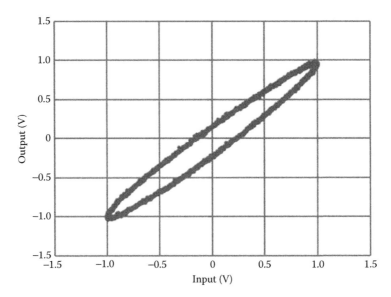

Figure 6.3.28 Steering system steady-state characteristics when cylinders are driven by a servo-valve in closed loop ($U_i = 1$ V; $f = 0.05$ Hz; $p_s = 100$ bar).

The control valve of the Orbitrol hydraulic steering unit is normally designed with open center for reducing both fuel consumption and steering mechanical shocks. The opposite of these gains is a big backlash that increases with the input frequency.

The direct control of the steering angle by a servovalve needs a constant pressure supply but avoids the backlash. However, the hysteresis increases with the input frequency. The steering process quality can be improved by increasing the pressure supply, reducing the length of the hoses, and the servovalve gain around the null. This last way can be easier applied by software tools using a digital servovalve from the same family [7–11].

6.3.5   Conclusion

The new test bench can offer a better understanding of the digital servovalve characteristics, response times, and EMI susceptibility. The new system performance is found good enough for tractor's remote control. Future projects and modern applications, where flexibility and accuracy are needed, will be supported by developing redundant and more EMI resistant digital servovalves with field-programmable gate array logic control unit and optical communications.

The steering process quality can be improved by increasing the pressure supply, reducing the length of the hoses, and the servovalve gain around the null. This last way can be easily applied by software tools using a digital servovalve from the same family [7–19]. In any case, a preliminary simulation with Amesim is very useful for predicting the system performance.

Section 1: Bibliography

1. Blackburn J.F., Reethof G., Sharer J. *Fluid Power Control.* John Wiley & Sons Inc., Cambridge, MA, 1960.
2. Merritt, H.E. *Hydraulic Control Systems,* John Wiley & Sons Inc., New York, 1967.

3. Viersma T.J. *Analysis Synthesis and Design of Hydraulic Servo Systems and Pipelines*, Elsevier Scientific Publishing Company, Amsterdam, the Netherlands, 1980.
4. Armstrong-Helouvry, B. Stick-slip arising from stribeck friction. *Proceedings of the IEEE International Conference on Robotics and Automation*, Cincinnati, OH, May 13–18. IEEE Computer Society Press, Los Alamitos, CA, 1990.
5. Mare, J.-C. *L'apport de la simulation pour la synthese de la commande adaptive d'un actionneur electro-hydraulique*. Journee Automatique Assistee par Ordinateur 2 AO92 ESIEE, Paris, 1992.
6. Vasile L. Researches on the dynamics of the automotive steering systems. PhD Thesis, University POLITEHNICA of Bucharest, 2005 (in Romanian).
7. Vasiliu, N., Vasiliu et al. *Fluid Power Systems – Fundamentals and Applications, Vol.1*, Technical Press House Bucharest, 2005 (in Romanian).
8. Power Steering System Studied with AMESim. Technical Bulletin n°107, IMAGINE, Roanne, 2004.
9. www.zf.com
10. www.volvo.com

Section 2: Bibliography

1. Lebrun, M. Digital simulation of stick-slip friction. SIMBOND, 1987.
2. IMAGINE. Power Steering System Studied with AMESim. Technical Bulletin n°107, Roanne, 2001.
3. IMAGINE. The hydraulic component design library. Technical Bulletin n°108, Roanne, 2001.
4. IMAGINE. AMESim in the Automobile Industry. Technical Bulletin n°106, 2001.
5. IMAGINE Numerical Challenges Posed by Modeling Hydraulic Systems, Technical Bulletin n°114, Roanne, 2002.
6. Alirand, M., Lebrun, M., Richards, C.W. *Front Wheel Vibrations: A Hydraulic Point of View— Models and First Results*. SAE 2001 World Congress Detroit, MI, March 5–8, 2001.
7. Schuster, M. Simulation of power steering systems using AMESim. *European AMESim Users' Conference*, Paris, 2000.
8. Birsching, J., Two-dimensional modeling of a rotary power steering valve. *SAE 2001 World Congress* Detroit, MI, March 5–8, 2001.
9. Vasile, L.N., Modeling and simulation of the hydraulic power steering. PhD Thesis, University POLITEHNICA of Bucharest, 2005 (in Romanian).
10. Vasiliu, N., Manea, I., Fulop, E. Hydraulic servomechanism with mechanical feedback. Romanian Patent no. 87155, 1983.
11. Vasiliu, N., Vasiliu, D., Gheorghiu, D. Rotary Hydromechanical Servomechanism. Romanian Patent no. 87234, 1983.
12. Vasiliu, N., Vasiliu, D., Manea, I. A new type of hydraulic rotary servomechanism. *IFAC Workshop on Trends in Hydraulic and Pneumatic Components and Systems*, Chicago, IL, November 8–9, 1994.
13. Călinoiu, C., Vasiliu N., Vasiliu D., Catană I. *Modelling, Simulation and Experimental Identification of the Hydraulic Servomechanisms*. Technical Press House Bucharest, 1998 (in Romanian).
14. Landau, I.D. *System Identification and Control Design*. Prentice-Hall, New York, 1990.
15. Oprean, I.M. *Modern Automotive Engineering*. Romanian Academy Press House, 2003. (in Romanian).
16. Radeş, M. *Dynamic Methods for Identification of the Mechanical Systems*. Academy Press House, Bucharest, 1979 (in Romanian).
17. Wylie, E., Streeter, V. *Fluid Transients*. McGraw-Hill Book Company, New York, 1983.
18. Armstrong-Helouvry, B. Stick-slip arising from stribeck friction. *Proceedings of the 1990 IEEE International Conference on Robotics and Automation*, Cincinnati, OH, May 13–18, 1990. IEEE Computer Society Press, Los Alamitos, CA, 1990.
19. Viersma, T.J. Suppression of pressure fluctuations in pipelines supplying hydraulic servosystems. *IFAC Symposium on Pneumatic and Hydraulic Components and Instruments in Automatic Control*, Warsaw, Poland, 1980.

20. Mare, J.- Ch. Dynamics of the Electrohydraulic Rotary Servomechanisms. PhD Thesis, I.N.S.A. Toulouse, 1994.
21. www.volvo.com
22. www.zf.com
23. www.siemens.com

Section 3: Bibliography

1. Agro-Mechatronics Sensor Based Process Control in Agriculture, ATB Potsdam-Bornim, 2009.
2. SAUER-DANFOSS. OSPE Steering Valve Technical Information 11968682, 2010.
3. Vasiliu D. Researches on the servo pumps and the servo motors of the hydrostatic transmissions. PhD Thesis, University POLITEHNICA of Bucharest, 1997 (in Romanian).
4. Ganziuc Al. *Electrohydraulic Servomechanisms with High Electromagnetic Imunity*, PhD Thesis, University POLITEHNICA of Bucharest, 2013 (in Romanian).
5. Aramă Şt., Daşchievici V., Axinte G. Forestry Hydraulic Articulated Tractor, Brăila, 1984.
6. Vasiliu, N., Vasiliu, D., Gheorghiu, D. Rotary hydromechanical servomechanism. Romanian Patent no. 87234, 1983.
7. Johnson, J.L. *Design of Electrohydraulic Systems for Industrial Motion Control*, Parker Hannifin Corporation, Cleveland, OH, 1995.
8. BOSCH - Automation Technology. Servo Solenoid Valves. Technical Specification 13/2, Stuttgart, Germany, 1999.
9. Pfeiffer, O., Ayre, A., Keydel, C. *Embedded Networking with CAN and CANopen*, RTC Books, San Clementem, CA, 2003.
10. Manring N.D. *Hydraulic Control Systems*, Wiley, New York, 2005.
11. Köhler, Th. Safer landings start with smarter servo valves, Moog, 2011.
12. Irimia, P.C. Power management of the articulated tractors, PhD Thesis, University POLITEHNICA of Bucharest, 2015 (in Romanian).
13. www.cema-agri.org
14. www.boschrexroth.com
15. www.eaton.com
16. www.moog.com
17. www.parker.com
18. www.lmsintl.com
19. www.fluidpower.net

chapter seven

Modeling, simulation, and identification of the hydrostatic pumps and motors

7.1 Numerical simulation of a single-stage pressure compensator

7.1.1 Structure of the servopumps

This section contains the results of a systematic research on the steady-state behavior, and the transients occurring in a basic hydrostatic system containing a servopump. The optimum structure of such a device working in given conditions is found step by step, the synthesis stages are alternating with the analysis ones in order to establish some general synthesis criteria, and for creating some images of the design parameters influence on the real-system dynamic behavior. [1,2]. The dynamic computation of the servopumps is based on the experimental researches carried out by the authors [3] on the control force of the variable displacement axial piston pumps. The analysis takes into account the pressure feedback, the rigid or spring mechanical feedback, or the electrical position feedback. The theoretical models are developed by SIMULINK [4] or Amesim [5,6]. The experimental tests were performed by the aid of LabVIEW software from National Instruments (NI) Corporation in the frame of a national certified laboratory set up by the aid of Parker Hannifin Corporation [7].

The past two decades pointed out some major changes in the industrial hydraulic control systems architecture: (a) the constant displacement hydraulic pumps were more and more replaced by variable displacement pumps for saving energy and increasing the systems performance, (b) the inflexible analog servocontrollers were replaced by versatile industrial process computers (IPC), and (c) the *delicate* servovalves were replaced by reliable proportional valves (Direct Drive Valves—DDV).

The IPC's large computational facilities have created the possibility of using sophisticated control algorithms such as fuzzy and adaptive ones according to the specific system requirements. Next, the high-speed electrohydraulic position servosystems were included in the variable displacement pumps and motors. These machines became high-quality servopumps and servomotors, successfully used in open or closed circuits in order to save a big amount of energy by high-speed displacement control. More than this, the new technology of *secondary control* electrohydraulic driving systems generated the possibility to save energy in complex systems without limiting the overall performance. The key to all these achievements is the performance of the displacement-control system. The high-speed electrohydraulic servosystems match these requirements, but their synthesis needs a deep understanding of the operating conditions, and a wide investigation of the stability.

For the moment, the hydromechanical servopumps and servomotors still cover the needs of a wide range of industrial applications, but the next generation of hydraulic control systems will be an electrohydraulic one.

This chapter contains a short report on the researches carried out by the authors on the structure, stroking force, modeling, simulation, and experimental performance identification for two important categories of axial piston servopumps: (1) bent axis and (2) swashplate one. The pressure compensator was considered in both cases because of the large practical interest in limiting the overflow supplied by the pump at the maximum operating pressure.

In any hydrostatic transmission, saving energy means first to match the pump flow with the hydraulic motor speed demand without opening any system relief valve. A two-stage pressure compensator (Figure 7.1.1) needs a small flow to reduce the pump displacement in order to keep the supply pressure at a set value. Such a compensator is always used in constant pressure supply systems such as fly control servomechanisms or electrohydraulic dynamic load simulators. The stroking force of the common servopumps requires the use of a servomechanism to set the pump flow at a given value (Figure 7.1.2). The pressure setting can be introduced by a remote pressure valve

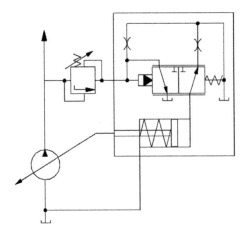

Figure 7.1.1 Standard two-stage pressure compensator.

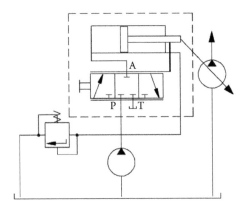

Figure 7.1.2 Flow control system by a hydromechanical servomechanism.

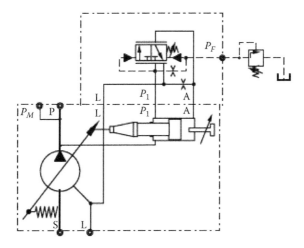

Figure 7.1.3 Remote pressure compensator of a swashplate pump.

(Figure 7.1.3). Both low and upper delivery pressures can be controlled by a Load Sensing compensator (Figure 7.1.4) using small permanent pressure drop on a variable hydraulic resistance. Pressure compensators can be remote controlled by an electrohydraulic proportional pressure valve (Figure 7.1.5). By the aid of an electrical position feedback given by a low voltage displacement transducer (LVDT) and an electrohydraulic proportional flow valve (DDV), the pump flow can be controlled by the aid of an industrial process computer (IPC) (Figure 7.1.6). The last generation of the electrohydraulic servopumps combines all the above-mentioned functions into an intelligent and fully flexible unit (Figure 7.1.7).

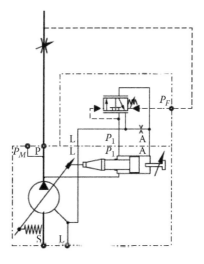

Figure 7.1.4 Load sensing flow control of a swashplate pump.

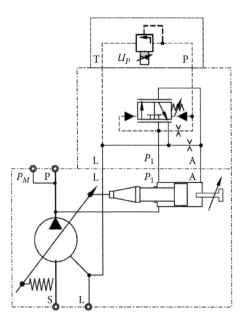

Figure 7.1.5 Electrohydraulic pressure compensator for a swashplate pump.

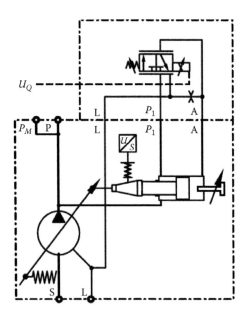

Figure 7.1.6 Electrohydraulic flow control system for a swashplate pump.

The use of a pressure transducer and a displacement gives the possibility to compensate the pump horsepower [1]. The last generation of DSP enables the servocontrollers to match all the peculiarities of the electrohydraulic proportional valves [2,3].

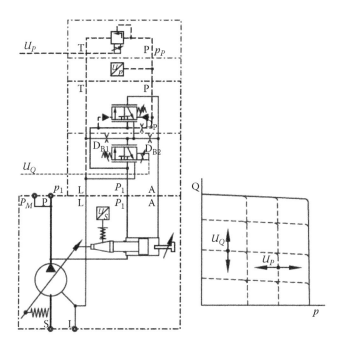

Figure 7.1.7 Electrohydraulic servopump with pressure compensator and horsepower compensator.

7.1.2 Dynamics of a single-stage pressure compensator

A good way to understand the dynamic behavior of a two-stage pressure compensator is to study the transient of a single-stage one (Figure 7.1.8) controlling a bent axis pump. A sudden decrease in the hydraulic motor flow generates an overpressure that opens the pressure sequence valve. The output flow needs an overpressure to pass through the

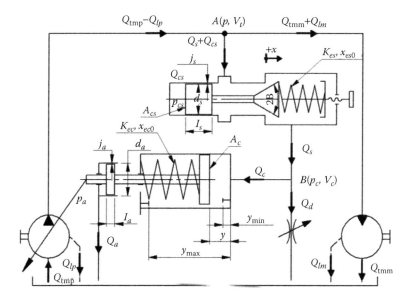

Figure 7.1.8 Computational model of a single-stage pressure compensator.

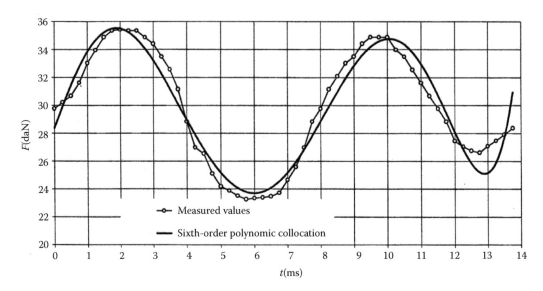

Figure 7.1.9 High-frequency component of the stroking force of a bent axis pump for n = 600 rev/min; p = 100 bar; z = 10.45 mm; and F_m = 392.3 N.

variable hydraulic resistance. A fraction of the valve flow enters the hydraulic cylinder and decreases the pump displacement.

The stroking force of a bent axis pump (Figure 7.1.9) has two components: (1) a steady-state one, proportional to the average delivery pressure and (2) a high frequency one, depending on the shaft speed and the piston number [4]. Each change in delivering the piston number generates a sine wave of stroking force.

The mathematical model of the system contains the following equations:

1. Pump control characteristics:

$$Q_{tmp} = \omega_p K_D \left(\sin \alpha_{max} - \frac{y}{R_m} \right) \tag{7.1.1}$$

where the quantity:

$$K_D = 0.25 z R D^2 = \frac{D_{p\,max}}{\sin \alpha_{max}} \tag{7.1.2}$$

is the pump constant.

2. Continuity equation in the fluid volume filling the space between pump, motor, and valve:

$$\dot{p} = \frac{\varepsilon}{V_t} \left(Q_{tmp} - Q_{tmm} - Q_{lpm} - Q_s - Q_{cs} \right) \tag{7.1.3}$$

3. Continuity equation in the volume filling the space between valve, stroking cylinder, and variable hydraulic resistance:

$$\dot{p}_c = \frac{\varepsilon}{V_c}\left(Q_s - Q_d - Q_c\right) \tag{7.1.4}$$

4. Valve poppet motion equation:

$$\ddot{x} = \frac{1}{m_s}\left(F_{cs} - F_{es} - F_{es0} - F_{hs}\right) \tag{7.1.5}$$

5. Stroking piston motion equation:

$$\ddot{y} = \frac{1}{m_c}\left(F_{cc} + F_b - F_{ec} - F_{ec0} - F_{ac}\right) \tag{7.1.6}$$

The main parameters of a typical compensator are the following: $V_{pmax} = 55$ cm^3/rev; $\alpha_{max} = 25°$; $\sin\alpha_{max} = 0.4226$; $n = 1500$ rev/min; $R_m = 75$ mm; $d_c = 60$ mm; $A_{cc} = 2.827 \times 10^{-3}$ m^2; $K_{ec} = 87000$ N/m; $m_c = 20$ kg; $y_{max} = 30$ mm; $y_{ec0} = 27.8$ mm; $l_a = 7$ mm; $d_a = 16$ mm; $A_a = 2.01 \times 10^{-4}$ m^2; $j_a = 0.157$ mm; and $\eta_{vp} = 0.95$ for $p = 30$ MPa.

The pressure sequence valve is a standard one: $d_s = 6 \times 10^{-3}$ m; $A_{cs} = 2.827 \times 10^{-5}$ m^2; $K_{es} = 25000$ N/m; $m_s = 17.6 \times 10^{-3}$ kg; $\beta = 30°$; $l_s = 6.0 \times 10^{-3}$ m; $j_s = 0.06 \times 10^{-3}$ m; $x_{es0} = 0.0113$ m for $p_{s0} = 100$ bar; $x_{es0} = 0.0169$ m for $p_{s0} = 150$ bar, $x_{es0} = 0.0226$ m for $p_{s0} = 200$ bar; $c_{ds} = 0.61$; $c_{vs} = 0.98$; $\rho = 900$ kg/m^3; $\eta = 2.72 \times 10^{-2}$ Ns/m; $\varepsilon = 7 \times 10^8$ N/m^2; $V_t = 1.0 \times 10^{-3}$ m^3; $V_c = 0.25 \times 10^{-3}$ m; $\varepsilon/V_t = 7 \times 10^{11}$ N/m^5; and $\varepsilon/V_c = 2.8 \times 10^{12}$ N/m^5.

The SIMULINK block diagram for a single-stage pressure compensator is shown in Figure 7.1.10. If the high-frequency component of the stroking force is neglected, the

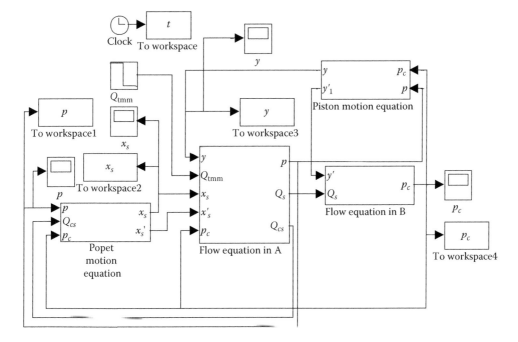

Figure 7.1.10 SIMULINK block diagram for a single-stage pressure compensator.

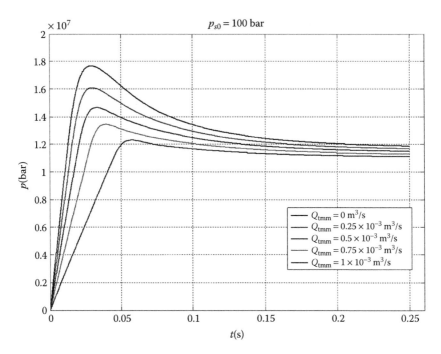

Figure 7.1.11 Single-stage pressure compensator response for different step inputs (sudden negative change of the hydraulic motor flow).

compensator response for different step inputs (sudden reduction of the hydraulic motor flow) is always a smooth one as shown in Figure 7.1.11. The pressure overshot can reach 50% from the final steady-state pressure value. The recovery time strongly depends on the size of the variable hydraulic resistance and is short enough for a wide class of industrial applications such as mobile hydraulics.

7.1.3 Dynamics of two-stage pressure compensators

The single-stage pressure compensators have a poor dynamics, especially during the recovery of the maximum displacement after a sudden servopump pressure overshot. They are used especially in predictable periodic operating cycles. The two-stage pressure compensators (Figure 7.1.12) include a standard pressure sequence valve, well damped, which controls the opening of a three-way flow valve against a spring. The flow valve supplies a single-effect hydraulic cylinder that reduces the pump displacement quickly, or recovers the maximum displacement of the servopump in a short time due to the great control area of the spool (Figure 7.1.13). The piston impact speed with the case is limited by a hydraulic one-way damper.

The dynamics of the above-mentioned pressure-control system can be found by integrating a set of differential equations composed by

- The flow continuity equation in the node (A) sited between the pump, the pressure valve, and the hydraulic motor (Figure 7.1.12)
- The flow continuity equation in the node (B) sited between the output port of the pressure valve, the control chamber of the three-way flow control valve, and the hydraulic resistance D1 and the input port of the flow control valve

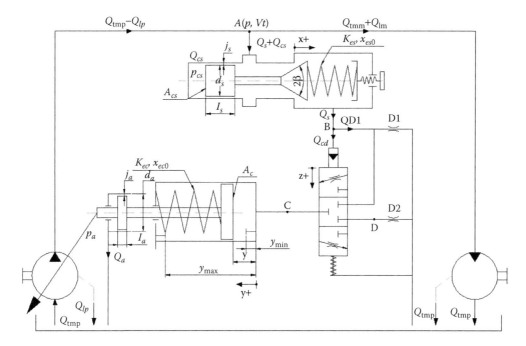

Figure 7.1.12 The hydraulic scheme of a two-stage pressure compensator.

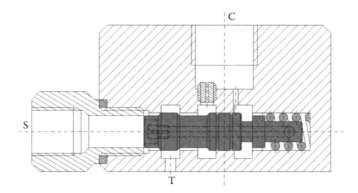

Figure 7.1.13 The flow control valve of a two-stage pressure compensator of a bent axis pump Brueninghaus A2V55: S—pressure valve output port; C—hydraulic cylinder control port; and T—tank port.

- The flow continuity equation in the node (C) placed between the flow control valve output port and the hydraulic cylinder
- Motion equation of the pressure valve poppet
- Motion equation of the compensator-stroking piston
- Motion equation of the spool of the flow control valve

The mathematical model needs many simplifying assumptions regarding the behavior of the hydraulic resistances, different friction phenomena, and so on. A global image of the transient generated by a sudden rise of the hydraulic motor load can be seen in Figure 7.1.14 obtained for a servopump Brueninghaus A2V55 [8]. Three transients generated by the

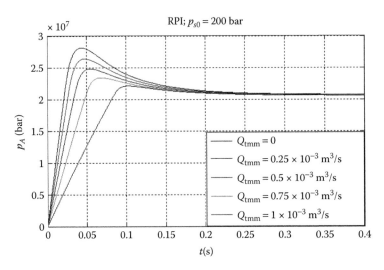

Figure 7.1.14 The influence of the hydraulic motor's sudden flow changes on the pump delivery pressure (sequence valve cracking pressure $p_{s0} = 200$ bar).

opening of the sequence pressure valve have been studied: $p_{s0} = 100, 150,$ and 200 bar. The numerical simulations were performed for the following set of common definition parameters:

$V_{pmax} = 55\,\text{cm}^3/\text{rev}; \alpha_{max} = 25°; \sin\alpha_{max} = 0.4226; n = 1500\,\text{rev}/\text{min}; R_m = 75\,\text{mm}; d_c = 60\,\text{mm};$
$A_{cc} = 2.827 \times 10^{-3}\,\text{m}^2; K_{ec} = 87000\,\text{N}/\text{m}; m_c = 20\,\text{kg}; y_{max} = 30\,\text{mm}; y_{ec0} = 27.8\,\text{mm}; l_a = 7\,\text{mm};$
$d_a = 16\,\text{mm}; A_a = 2.01 \times 10^{-4}\,\text{m}^2; j_a = 0.157\,\text{mm}; \eta_{vp} = 0.95$ for $p = 300\,\text{bar}; d_s = 6.0 \times 10^{-3}\,\text{m};$
$A_{cs} = 2.827 \times 10^{-5}\,\text{m}^2; K_{es} = 25000\,\text{N}/\text{m}; m_s = 17.6 \times 10^{-3}\,\text{kg}; \beta = 30°; l_s = 6.0 \times 10^{-3}\,\text{m};$
$j_s = 0.06 \times 10^{-3}\,\text{m}; x_{es0} = 0.0113\,\text{m}$ for $p_{s0} = 100\,\text{bar}; x_{es0} = 0.0169\,\text{m}$ for $p_{s0} = 150\,\text{bar}; x_{es0} = 0.0226\,\text{m}$
for $p_{s0} = 200\,\text{bar}; c_{ds} = 0.61; c_{vs} = 0.98; \rho = 900\,\text{kg}/\text{m}^3; \eta = 2.72 \times 10^{-2}\,\text{Ns}/\text{m}; \varepsilon = 7 \times 10^8\,\text{N}/\text{m}^2;$
$V_t = 1.0 \times 10^{-3}\,\text{m}^3; V_c = 0.25 \times 10^{-3}\,\text{m}; \varepsilon/V_t = 7 \times 10^{11}\,\text{N}/\text{m}^5; \varepsilon/V_c = 2.8 \times 10^{12}\,\text{N}/\text{m}^5;$
$\omega_p = 157\,\text{rad}/\text{s}; D_{pmax} = 8.753 \times 10^{-6}\,\text{m}^3/\text{rad}; K_D = 2.071 \times 10^{-5}\,\text{m}^3/\text{rad}; K_{dcs} = 4.8 \times 10^{11}\,\text{Ns}/\text{m}^5;$
$K_s = 2.71 \times 10^{-4}\,\text{kgm}; z_{ed0} = 6.0 \times 10^{-3}\,\text{m}; K_{ed} = 1.0 \times 10^4\,\text{N}/\text{m}; m_d = 0.070\,\text{kg}; K_{ad} = 4000\,\text{Ns}/\text{m};$
$K_{hd} = 17.593 \times 10^{-3}\,\text{m};$ and $A_{cd} = 1.539 \times 10^{-4}\,\text{m}^2.$

At the first sight, the dynamics of the two-stage pressure compensator is good enough for common industrial bent axis pumps operating with small oil volumes ($V_t = 1 \times 10^{-3}\,\text{m}^3$). However, the small inertia of the pressure compensators set up on the swashplate pumps explains the very short response time of this widespread type of pumps, in spite of their severe oil-filtering conditions.

7.1.4   Conclusion

The authors regarded the above-mentioned research field in close connection with other scientific and technical domains trying to create a global image of the mathematical modeling, numerical simulation, experimental identification, and performance tests chain when the simulation is performed by SIMULINK. The theoretical aspects are connected in a practical manner, starting from the needs of the design, trying to draw a straight line between theory and practice. The next chapter treats the same problem by Amesim [6,7], offering a global image of the available tools for solving engineering problems in the field of the fluid-power systems.

7.2 Dynamics of two-stage pressure compensator for swashplate pumps

7.2.1 Structure of the swashplate pumps with pressure compensator

From the beginning of the use of the swashplate pumps (1936), the fluid-power system designers included in the pump case at least a pressure relief valve. The optimal hydraulic structure includes a *pressure compensator* that cuts the external flow rate keeping the internal pump flow rate at a very low value when the fluid pressure overcome a top value *avoiding the opening of the system pressure relief valve*. The remaining flow rate includes the one needed to lubricate all the internal bearings and a very small part for supplying the compensator. This option saves a significant amount of energy in any fluid-power system that works intermittently or cyclically, and avoids overheating of the oil and the loss of their lubrication properties. The most important practical application is found in the aerospace field: All the main pumps first fill the hydraulic accumulators needed for safety reasons and then cut the external flow rate, remaining in standby. The only requirement for using the swashplate pumps regards the high level of oil cleaning needed for all the hydrostatic bearings working inside (at least slippers and flow valve plate). The modern industrial fluid-power systems have adopted this solution for any type of variable displacement pump, taking into account the gain of overall efficiency, and the possibility of integrating the electrohydraulic displacement control by industrial servovalves, and digital servocontrollers, integrated in the overall control systems.

This chapter is devoted to the numerical simulation of the dynamics on the hydromechanical pressure compensators that are included in any modern servopump together with the hydraulic or electrohydraulic servosystem for the pump's flow rate control.

Figure 7.2.1 shows a virtual view in SOLIDWORKS CAD environment of the servopump PV046R1K1T1NUPG that supplies the electrohydraulic servovalves test benches from the Fluid Power System Laboratory of the Politehnica University of Bucharest, Romania [1–3].

The hydraulic scheme of the servopump (Figure 7.2.2) includes a swashplate pump (Figure 7.2.3), an electrohydraulic proportional servovalve, a hydromechanical pressure

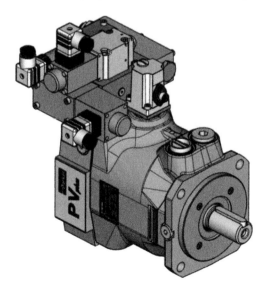

Figure 7.2.1 CAD view of a complete electrohydraulic servopump PARKER PV Plus (SOLIDWORKS representation).

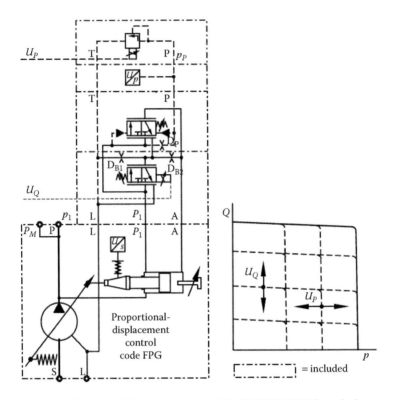

Figure 7.2.2 Hydraulic diagram of the servopump PV046R1K1T1NUPG, including pressure, flow, and power control systems.

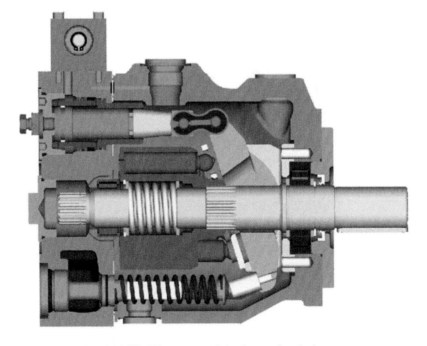

Figure 7.2.3 Cut view of a PARKER PV pump with hydromechanical pressure compensator.

compensator piloted by an electrohydraulic proportional pressure valve, a differential hydraulic cylinder, an LVDT measuring the piston displacement, and a pressure transducer. The three control systems allow the precise tuning of the pump operation parameters and the automatic limiting of the pump power consumption by the aid of a special P–Q digital servocontroller, compatible with any common industrial controller. The next Figures 7.2.4 through 7.2.6 show the design peculiarities of the pump, which reduces the noise, flow rate pulsations, and reduces the response time for any digital settings of the flow rate, pressure, and power overflow. The swashplate has two hydrostatic bearings supplied by the piston slippers passing over it. The PTFE plated lining and the hydrostatic bearings are eliminating the static friction between the swashplate and the pump case. Consequently, the response time of the pump for any input occurs in a very short time (between 80 and 200 ms).

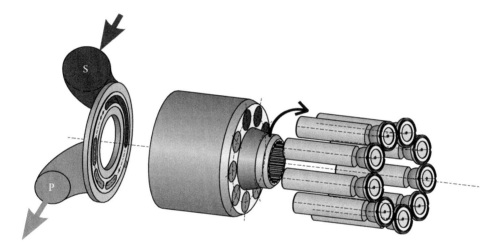

Figure 7.2.4 Swash pumps active components: S—suction port and P—delivery port.

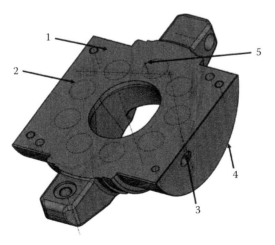

Figure 7.2.5 Swashplate with hydrostatic bearings: 1—swash plane surface; 2—hydrostatic shoes projections; 3—channels for supplying the two hydrostatic bearings of the swashplate; 4—cylindrical sliding surfaces; and 5—hydrostatic bearings supply orifice.

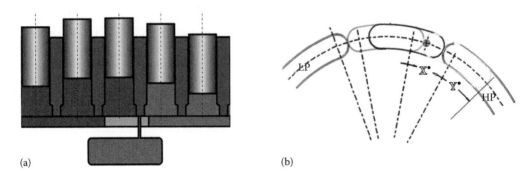

Figure 7.2.6 Reducing the pump noise with an intermediate chamber: (a) position of the damping chamber and (b) position of the orifice connecting the damping chamber with the cylinder sited in dead outer point.

The plate valve has no classical dumping notches, but a small orifice (Figure 7.2.6) that connects the cylinder to an additional chamber linked to the pump delivery port. The piston passing by the dead outer point is first connected to this chamber; the pressure rising time in the cylinder is longer, and the hydraulic noise and vibrations are diminished.

The sensitivity of the modern pressure compensator is very high because the differential hydraulic cylinder that drives the swashplate against a spring is directly connected through an annular orifice to the pump delivery port (Figure 7.2.7). The control spool edge is chamfered and the overlap of the control orifice is very small. The hydraulic diagram of the direct-action pressure compensator is shown in Figure 7.2.8a together with the steady-state characteristics. An exploded realistic cut view of the device included in the servo-pump PV46 is shown in Figure 7.2.8b.

The small chamber of the stroking hydraulic cylinder is always connected to the pump delivery port keeping the pump displacement at the maximum value. When pressure rises over the level set up by a very rigid spring ($K_e \sim 121.000$ N/m), the greater area chamber is connected to the pump suction port by a small orifice D_B. The control

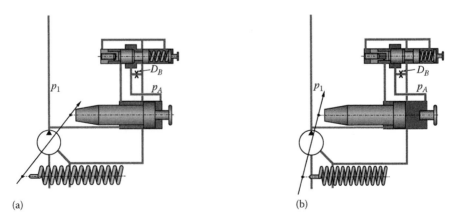

Figure 7.2.7 Direct-action pressure-compensator operation principle: (a) low delivery pressure and (b) high delivery pressure.

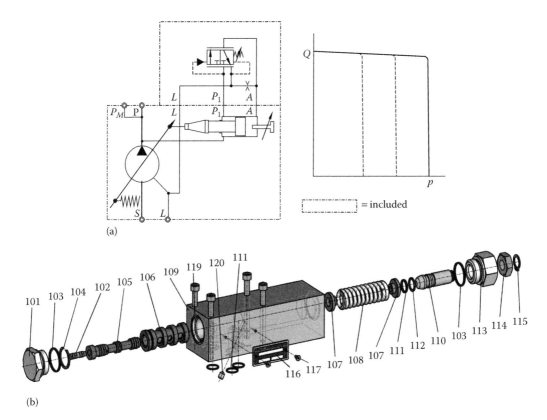

Figure 7.2.8 (a) Hydraulic diagram of a direct-action pressure compensator and the steady-state characteristics of the servopump. (b) Exploded view of a typical direct-action pressure compensator.

area of the spool is lower than the full section area, and the internal control chamfer is supplied by a small orifice. Due to this structure, the response time of the pressure compensator is very low, and the overall damping is good enough for avoiding the pump-load vibrations.

The best dynamics of the pressure limiting system is obtained using the integrated conical pilot valve with cylindrical damper (Figure 7.2.9).

The dynamics study of pressure compensated servopumps needs the complete knowledge of their geometry and of all pressure, inertia, elastic, and other forces involved in the displacement control. Between the above-mentioned forces, the most uncertain one is the overall force exerted by the hydrostatic shoes on the swashplate (Figure 7.2.10a). The finite number of pistons generates a strong alternative force during any passing of a piston from the suction phase to the delivery phase (Figure 7.2.10b). The torque variation on the swashplate is very important: for a PV46 pump with 9 pistons of 17 mm diameter that are delivering the fluid at 350 bar, 4 delivering pistons generate a force of 31750 N, and 5 pistons generate a force of 39750 N. The difference of 8000 N generates a high-frequency torque applied on the swashplate with a variable position around the tilting line. This torque cannot be neglected in the computation of the displacement control systems [4,5].

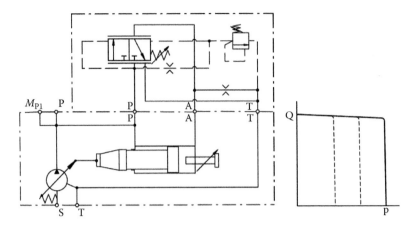

Figure 7.2.9 Standard pressure control system with integrated pilot valve.

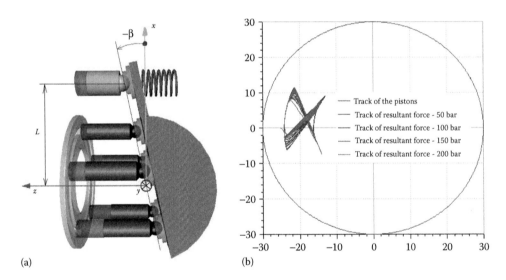

(a)   (b)

Figure 7.2.10 Computational elements for the control system dynamics: (a) reference system of the swashplate and (b) the path of the pressure center of the tilting force.

7.2.2 Numerical simulation of the dynamic behavior

Figure 7.2.11 contains a basic Amesim model for a swashplate pump with a typical direct-action pressure compensator, including both the mechanical inertia of the swashplate and the barrel inertia, the set of pistons, the stroking piston, and the control valve dynamics [6–9]. The numerical simulations were performed for the start sequence of the servopump, considering the relative area of the variable orifice sited in the output port of the servopump as the input signal (Figure 7.2.12), finding the swashplate-tilting

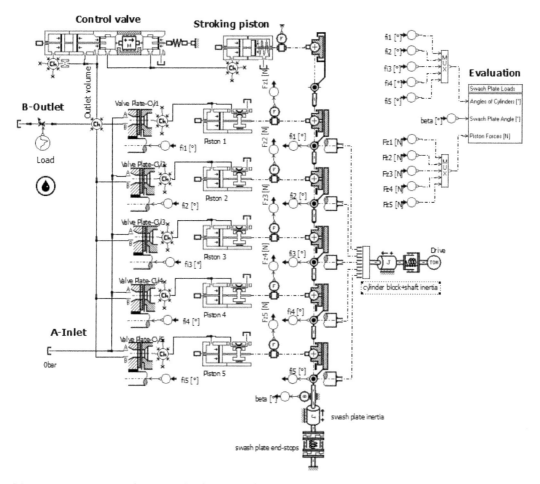

Figure 7.2.11 Amesim basic model for a swashplate axial piston pump with mechanical pressure compensator.

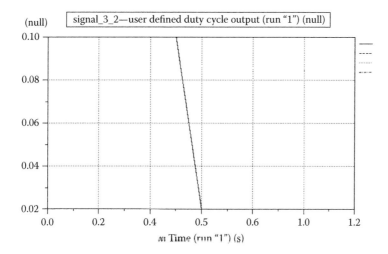

Figure 7.2.12 The input signal applied to the output port of the servopump.

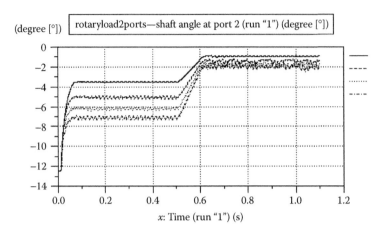

Figure 7.2.13 The swashplate tilting angle variation.

angle (Figure 7.2.13), pump delivery pressure variation (Figure 7.2.14), the pump flow rate (Figure 7.2.15), the control pressure variation (Figure 7.2.16), and the control piston force variation (Figure 7.2.17).

In order to find the main dynamic parameters variation for the two-stage pressure compensator included in the servopump PV046R1K1T1NUPG, which supplies the electrohydraulic servovalves test benches from the Fluid Power System Laboratory of the Politehnica University of Bucharest, Romania, the two-stage compensator type PVCMAMCN1 (Figure 7.2.9) was considered, keeping the same input signal connected

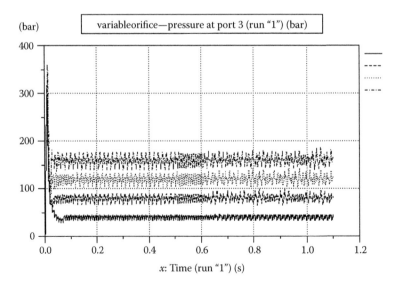

Figure 7.2.14 Pump delivery pressure variation.

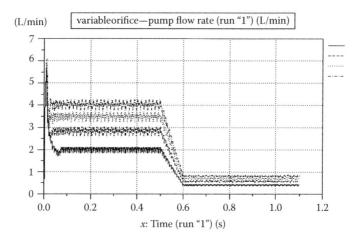

Figure 7.2.15 Pump flow rate variation.

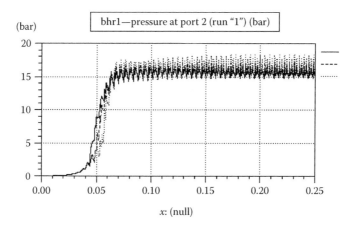

Figure 7.2.16 Control pressure variation.

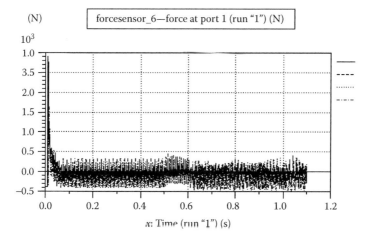

Figure 7.2.17 Control force applied to the swashplate variation.

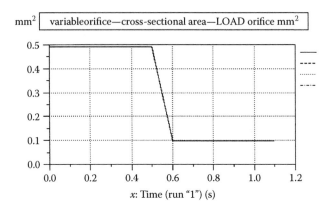

Figure 7.2.18 Variation of the cross-sectional area of the load orifice.

to the load orifice area from Figure 7.2.18. The results are shown in Figures 7.2.19 through 7.2.25.

The main practical conclusions of the simulations are the following:

1. The pump start sequence takes 80 ms only; the stroking piston has a cyclic movement with small amplitude and high frequency (shaft frequency multiplied by piston number).
2. The compensator flow consumption is relatively low.
3. The pump pressure has a high-frequency variation with relatively small amplitude.
4. The response time of the two types of pressure compensators is near the same; the use of an independent pilot pressure valve with damped conical poppet makes the cracking pressure of the compensator more precise.

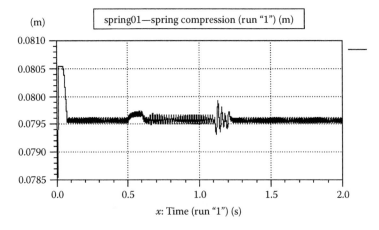

Figure 7.2.19 Pressure-compensator spool compression variation.

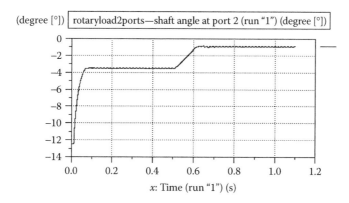

Figure 7.2.20 The swashplate-tilting angle variation.

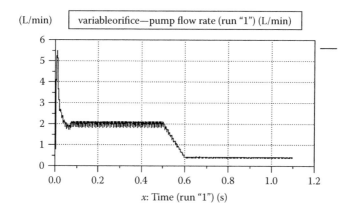

Figure 7.2.21 Pump flow rate variation.

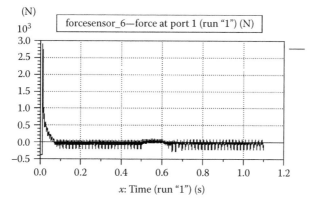

Figure 7.2.22 Swashplate-tilting force variation.

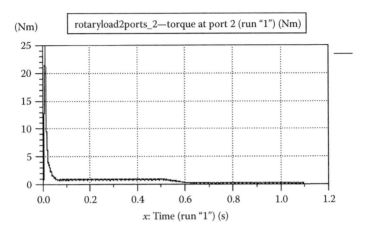

Figure 7.2.23 Pump shaft torque variation.

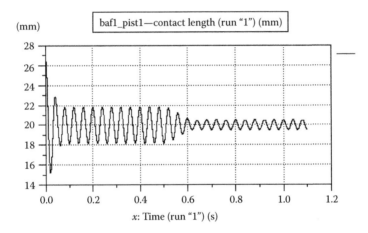

Figure 7.2.24 Variation of the contact length between the pistons and the cylinder.

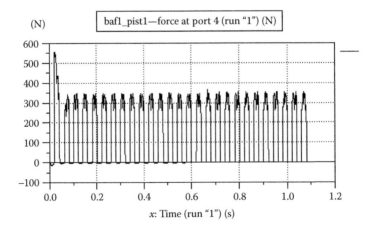

Figure 7.2.25 Variation of the force applied to a piston.

7.2.3 Experimental researches

The experimental researches were carried out in order to validate the mathematical model created in Amesim for the swashplate axial pistons servopumps with pressure compensators [6–8]. The researches were performed using the test bench for testing the electrohydraulic servovalves (Figures 7.2.26 through 7.2.30). A lot of transients were created by different means: sudden closure of a high dynamics two-stage pressure relief valve controlled by an electrohydraulic bypass, sudden closure of a high-speed electrohydraulic servovalve, and so on.

The 37 kW hydraulic power pack (Figure 7.2.29) supplies with high-purity mineral hydraulic fluid (Level 6 according to NAS 1683) all the tested equipments, which complies to all the mechanical, electrical, thermal, and electromagnetical compatibility international requirements. The servopump of the power pack (PV046R1K1T1NUPG) includes an electrohydraulic flow control system, a hydromechanical two-stage pressure compensator, and an electrohydraulic combined pressure and power governor included into a digital servocontroller P–Q–L (PARKER HY11-3243/K). The oil temperature is maintained in close limits by a digitally controlled cooler. The pressure and the flow-rate pulsations are reduced by a high-capacity hydraulic membrane accumulator.

Figure 7.2.26 Overall view of the hydraulic power supply and the servovalve dynamic test bench.

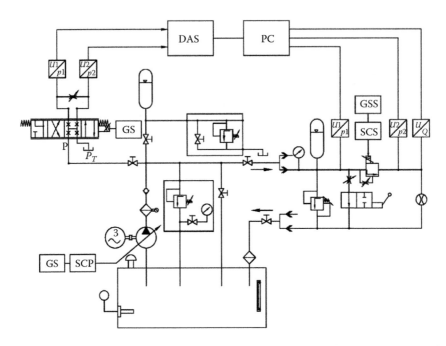

Figure 7.2.27 Hydraulic scheme of the servovalve dynamic test bench from Fluid Power Laboratory of the Politehnica University of Bucharest, Romania.

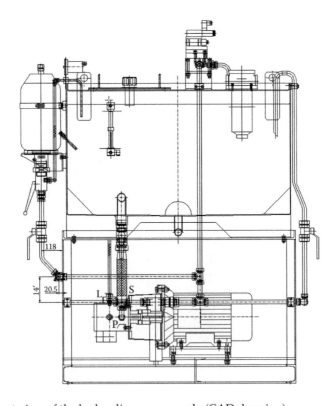

Figure 7.2.28 Front view of the hydraulic power supply (CAD drawing).

Figure 7.2.29 Hydraulic power supply of the test bench.

Figure 7.2.30 Lateral view of the servopump PV046R1K1T1NUPG.

Some servopump's responses to sudden closures of the load DFplus industrial servovalve, set up at different pressures are shown in Figures 7.2.31 through 7.2.33. The response time of the mechanical pressure compensator is very small: about 80 ms. The whole transients need less than 200 ms. The overall aspect of the pressure variation indicates a second-order element with a good damping [10–12]. These real results are in good agreement with the ones obtained by numerical simulations.

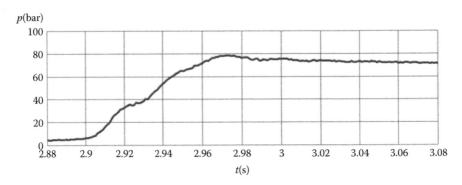

Figure 7.2.31 Servopump response to a sudden closure of the load servovave set up at 72 bar.

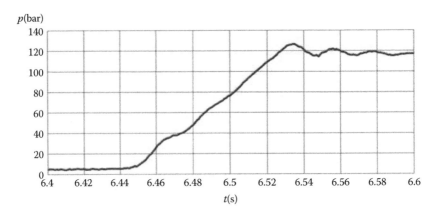

Figure 7.2.32 Servopump response to a sudden closure of the load servovave set up at 118 bar.

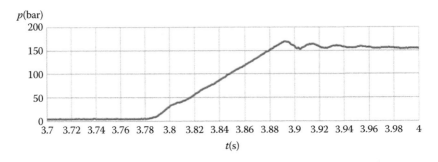

Figure 7.2.33 Servopump response to a sudden closure of the load servovave set up at 155 bar.

The high dynamic performance of the servopump pressure compensator, included in the pump control manifold shown in Figure 7.2.34, can be understood by a deeper analysis of the real structure of the geometrical details only. This study performed using the input signal from Figure 7.2.35 can explain the equalization of rectangular expensive slots by simple round holes manufactured in the sleeve, the replacement of

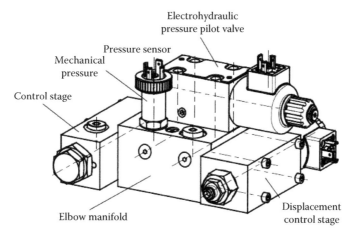

Figure 7.2.34 CAD view of the servopump control manifold.

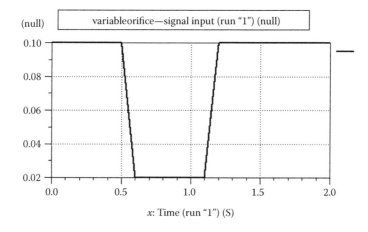

Figure 7.2.35 The relative variation of the load orifice area in a cycle of work.

the spool direct control by a conical poppet-pressure pilot valve, which changes the spool position according to the spring pretension, and so on. In order to use the load orifice area relative variation from Figure 7.2.35, the simulation run needs to specify the flow valve spool land parameters from Figures 7.2.36 and 7.2.37. The global parameters of the model are indicated in Figure 7.2.38. The positive effect of these options results from the good quality of the pressure and flow-rate variations is shown in Figures 7.2.39 and 7.2.40.

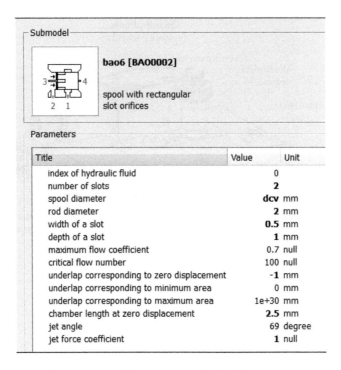

Figure 7.2.36 The optimal parameters of the input spool shoulder.

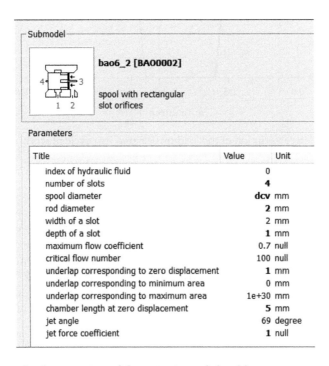

Figure 7.2.37 The optimal parameters of the output spool shoulder.

Set global parameters:

Name	Title	Type	Unit	Value	Minimum	Default	Maximu
Rpiston	radial position...	Real	m	0.03	0	0.03	1e+06
nbpist	number of pis...	Real	null	5	-1e+06	0	1e+06
pclear	clearance on ...	Real	mm	0.005	-1e+06	0	1e+06
dp	diameter of o...	Real	mm	10	-1e+06	0	1e+06
pcircuit	pressure in th...	Real	null	200	-1e+06	0	1e+06
dcv	diameter of th...	Real	mm	5	-1e+06	0	1e+06
beta_max	max swash pl...	Real	degree	12.5	-1e+06	0	1e+06

Figure 7.2.38 The general parameters of the simulation model.

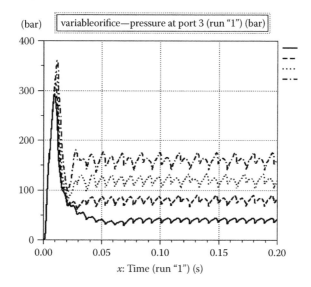

Figure 7.2.39 Pressure variations at the servopump output port for four compensator spring settings.

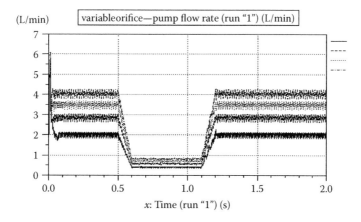

Figure 7.2.40 The pump flow rate variation during the transient produced by the elastic force.

7.2.4   Conclusion

The reaching of a high-quality dynamics by any hydraulic control system needs a complete stage of numerical simulations. The impact of the study may be the complete change of the structures or parameters with a valuable economy of time and money. However, the new generation of hydraulic control engineers gains more time by preliminary studies with theoretical tools when these are applied with simulator support. A good example in this field is the possibility of assessment of the stability of a system by the aid of the frequency response of nonlinear (common) systems using the dedicated block Frequency *Response Analyzer* available in the *Signal, Control* library of Amesim.

7.3   Open-circuits electrohydraulic servopumps dynamics

7.3.1   Structure of the electrohydraulic servopumps for open circuits

The modern servopumps are controlled by hydromechanic or electrohydraulic servosystems according to the application performance demands. Special applications such as military ones require hybrid systems, including electric control devices and mechanical feedback, rigid or elastic ones. Another structural matter regards the pressure supply of the control system. The classic servosystems are powered through auxiliary pumps with gears, put into motion by the controlled pumps, and are being protected against overpressure by piloted valves.

In the case of an open circuit, powering the servosystem through the controlled pump is possible, if the minimum backpressure of the latter is greater than the minimum pressure necessary for the control. A typical example for this principle is the PV Plus servopump family (Figures 7.3.1 and 7.3.2). The tilting angle of the swashplate is controlled by a three-way flow valve (Figure 7.3.3) with a small overlap actuated by a proportional force solenoid, which has a nominal force of less than 90 N. The spool-control edges have different geometry: one edge is very sharp and is controlling the displacement increase. The other control edge has a small chamber, increasing the displacement recovery time.

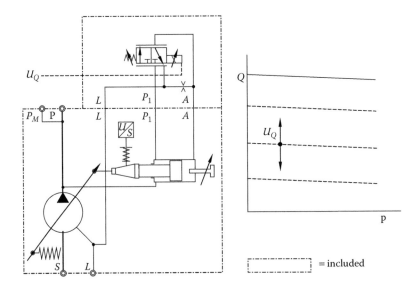

Figure 7.3.1 Hydraulic diagram of a PARKER PV Plus electrohydraulic servopump. (Source: Parker Hannifin Corporation)

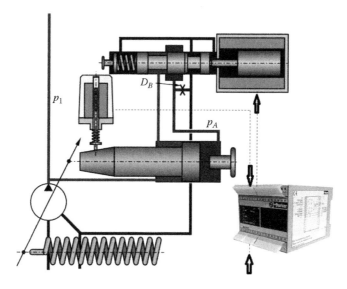

Figure 7.3.2 Schematic diagram of a PARKER PV Plus electrohydraulic servopump [1]. (Source: Parker Hannifin Corporation)

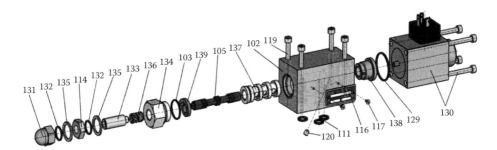

Figure 7.3.3 Expanded view of a proportional three-way flow valve. (Source: Parker Hannifin Corporation)

The position-control loop feedback at the hydraulic cylinder rod is given by an inductive transducer whose sensing shaft is permanently in contact with a conical area of the piston kept in position by a spring. The transducer is waterproofed (Figure 7.3.4).

The pump flow is set up by a PID compensator, which is followed by a PWM voltage interface of the proportional force solenoid.

This part presents the modeling, simulation, and experimental validation of the model of the servopump PV046R1K1T1NUPG [12] performed by the authors in the Electrohydraulic Amplifier Performance Certification Laboratory certified by Romanian Accreditation Association—RENAR. The experimental results are found in good agreement with the theoretical ones. The simulation model was included in the Amesim super components library.

The analog or digital servocontroller allows the voltage control (0–10 V), current control (0–20 mA), or by a PROFIBUS CAN. The reference of the transducer core can be adjusted with a potentiometer. The PID error amplifier generates a control signal for a PWM generator, which has a Dither signal with a frequency of 100 Hz, and an amplitude of 0.2 V. The servocontroller is DC powered (22–36 V) by the internal power supply (±15 V); it is galvanically isolated to prevent accidental overvoltage.

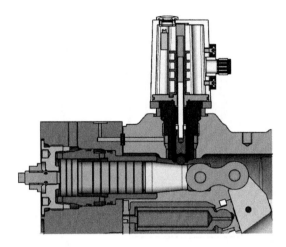

Figure 7.3.4 Measurement system for the tilting angle of the swashplate. (Source: Parker Hannifin Corporation)

In cyclic applications, the servomechanism is combined with a direct-action mechanical pressure compensator or with an electrohydraulic one controlled by the flow proportional servovalve using the information supplied by a pressure transducer, U_P. The pressure-limiting system is a proportional one, not needing a PID error amplifier, but only a time modulator of the command voltage of the flow proportional servovalve (Figure 7.3.5).

Other manufacturers prefer closed-loop pressure control after filtering the signal supplied by a piezoceramic pressure transducer and the precise control of the spool position of the proportional servovalve by a position feedback realized with an inductive LVDT. From the point of view of unitary information processing and pump capacity control, this solution is optimal, but its cost is bigger and the safety level is reduced. The mechanical–hydraulic pressure compensator offers a higher level of safety than the electronic one in the case of accidental working fluid degradation.

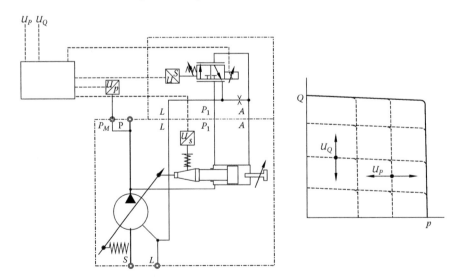

Figure 7.3.5 Open circuit electrohydraulic servopumps hydraulic scheme and steady-state characteristics. (Source: Parker Hannifin Corporation)

7.3.2 Numeric simulation of open-circuit servopump dynamics

The numeric simulation model from Figure 7.3.6 corresponds to the servopump, which is mounted on the hydraulic amplifier test bench from the laboratory designed by the authors. The modeling problems were solved [1,2,3] taking into account some technical

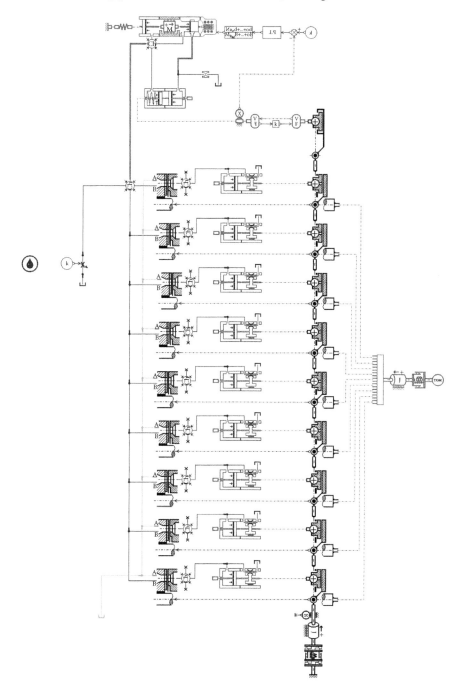

Figure 7.3.6 The Amesim simulation model of the swashplate flow servopump.

bulletins published by LMS Imagine [4–11]. The proposed simulation model was included in the Amesim Model library under the name *spehd.ame* [3].

Figures 7.3.7 through 7.3.12 show the evolution of the main parameters that define the dynamics of the servopump for a step voltage input signal. It is noted that for small

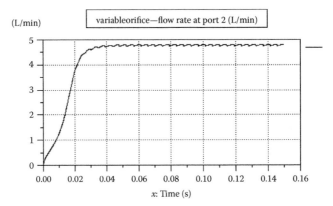

Figure 7.3.7 Servopump flow rate delivered to a small variable orifice for a small input voltage (0.5 V).

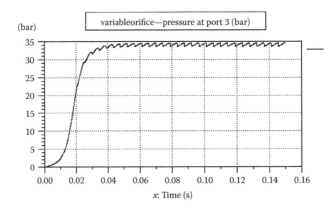

Figure 7.3.8 Servopump output pressure for a small input voltage (0.5 V) (the flow is supplied to a small variable orifice).

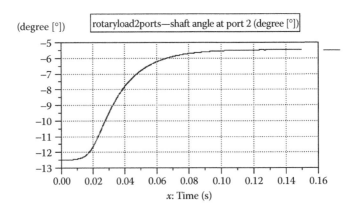

Figure 7.3.9 The swashplate angle variation during the transient.

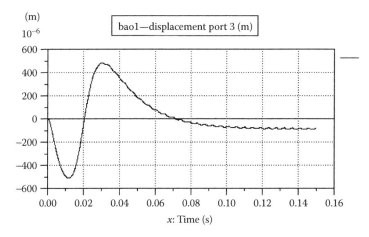

Figure 7.3.10 Control valve spool displacement during the transient.

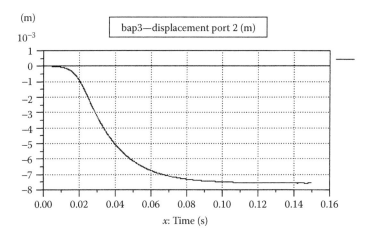

Figure 7.3.11 Piston displacement during the transient.

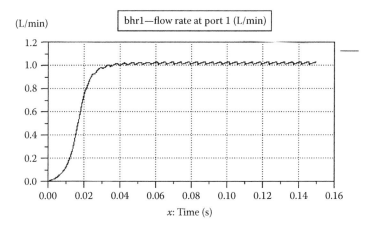

Figure 7.3.12 Flow rate passing through the flow control valve during the transient.

signals, the servomechanism responds similar to a second-order system, whereas for greater amplitudes the behavior is aperiodic. The time constant is less than 50 ms, quality of which allows the use of the servopump in fast positioning systems.

The next figures (7.3.13 through 7.3.22) present the effect of the solenoid force variation (Figure 7.3.13) on the control valve spool on different servopump parameters. The response of the solenoid for a step input current generated by the servocontroller is a first-order one with a constant time of about 0.035 s [10,11].

The pressure inside the servopump cylinders has a complex variation, shown in Figure 3.17.

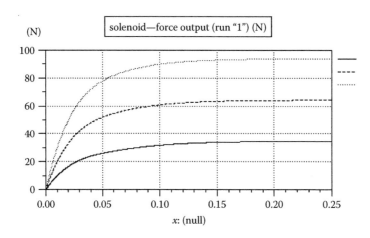

Figure 7.3.13 Response of the solenoid force for a step current input.

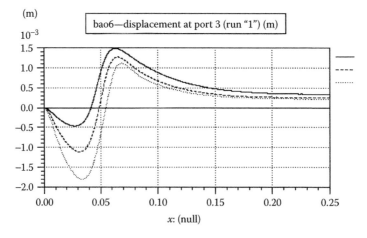

Figure 7.3.14 Control valve spool displacement during the transients generated by the solenoid step-force variation.

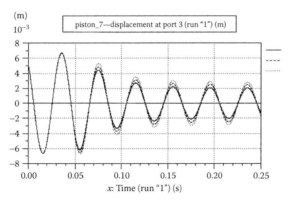

Figure 7.3.15 Stroking piston displacement during the transients.

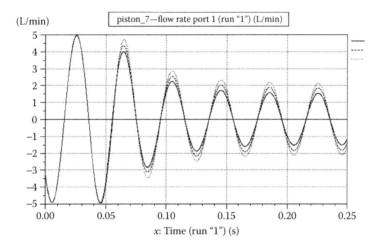

Figure 7.3.16 Flow rate at the input port of the stroking cylinder evolution.

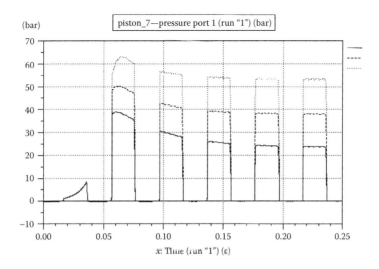

Figure 7.3.17 The pressure variation in a cylinder of the barrel during the transients.

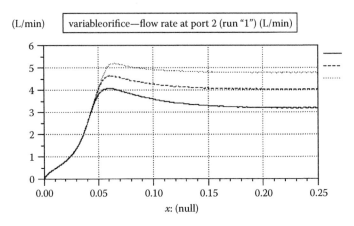

Figure 7.3.18 Flow rate delivered by the servopump to the variable load orifice.

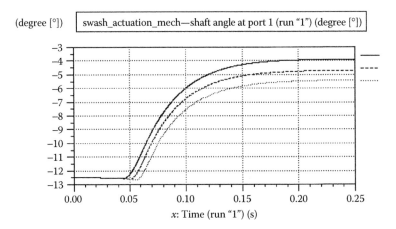

Figure 7.3.19 Variation of the angle of the swashplate during the transients.

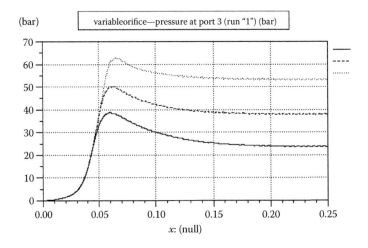

Figure 7.3.20 Pressure variation at the input of the load orifice.

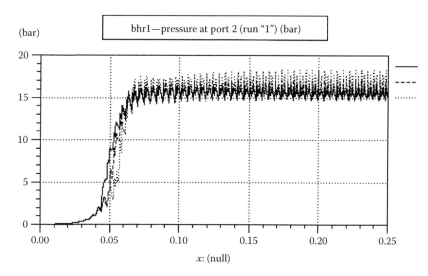

Figure 7.3.21 Pressure variation in the passive chamber of the swashplate.

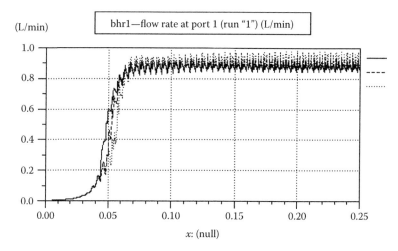

Figure 7.3.22 The flow rate consumed by the servopump through the hydraulic resistance of the passive actuator chamber during the transients.

Figures 7.3.23 through 7.3.25 show the simulated behavior of the servopump flow rate for periodic signals applied to the servocontroller, generating periodical solenoid forces. Three types of signals currently used in practice were applied: (1) sinusoidal, (2) triangular, and (3) rectangular ones. A very good tracking capacity is observed with a maximum delay of 100 ms.

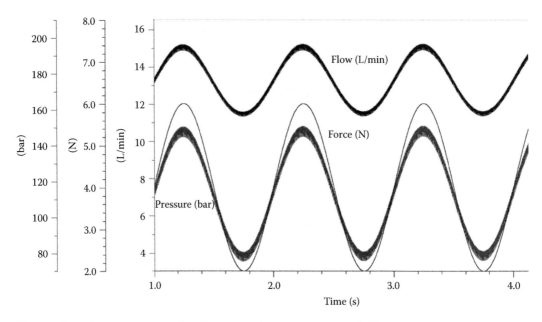

Figure 7.3.23 Servopump-simulated response for sine input signal (f = 1.0 Hz).

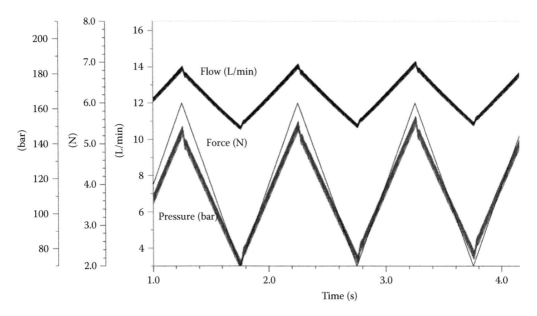

Figure 7.3.24 Servopump-simulated response for triangular signals (f = 1 Hz).

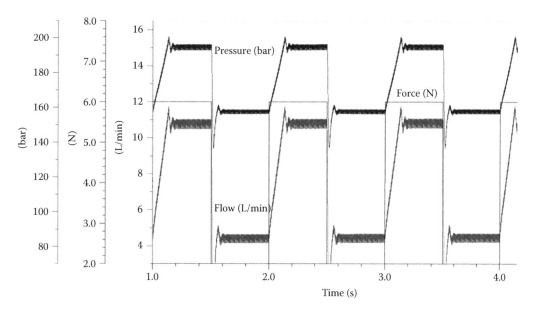

Figure 7.3.25 Servopump-response for rectangular signals ($f = 1$ Hz).

7.3.3 *Experimental validation of the simulations*

The flow-control module of the servopump was tested using sinusoidal and triangular waveforms supplied by Stanford Systems precision analog waveform generator. The medium servopump load pressure was generated using two types of resistances:

1. A high-dynamics two-stage servovalve from Parker (SE2E), which includes a digital servocontroller; this can be voltage controlled, so it is compatible with the signal generator.
2. A high-dynamics proportional industrial servovalve DFPlus (DDV) with connected load ports with a variable opening. In both cases, the control voltage was recorded by a PXI-based data acquisition system with LabVIEW software [14]. All experiments indicate a very good dynamic behavior of the flow servopump. The 100–200 ms delay that depends on the frequency of the excitation signal can be neglected in all industrial applications. Figures 7.3.26 through 7.3.28 show some of the representative dynamic tests.

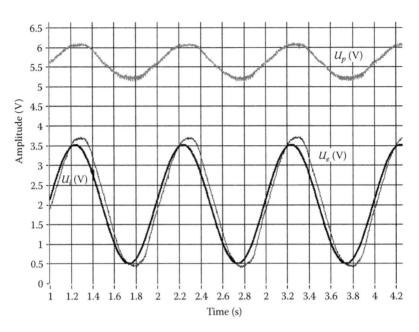

Figure 7.3.26 Experimental frequency response for sine input with $f = 1.0$ Hz and $p_{medium} = 196.0$ bar.

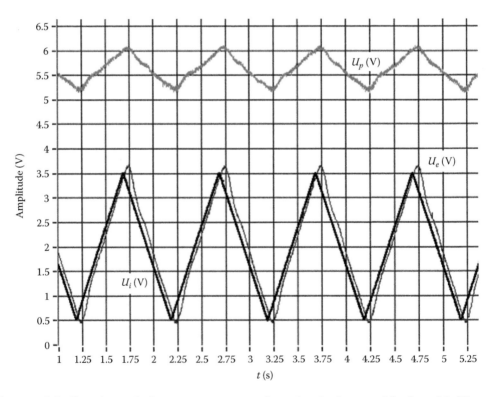

Figure 7.3.27 Experimental frequency response for triangle input with $f = 1.0$ Hz and $p_{medium} = 196.0$ bar.

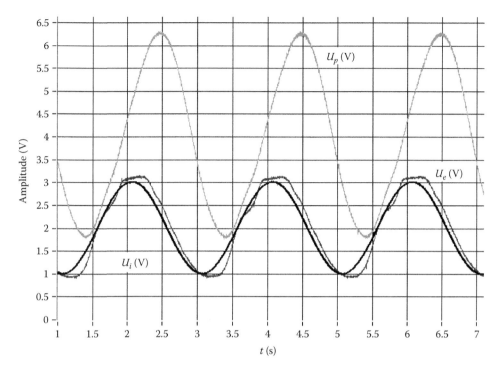

Figure 7.3.28 Experimental frequency response for $f = 0.5$ Hz and $p_{medium} = 140.0$ bar generated by a DFplus industrial servovalve with connected load ports.

7.3.4 Conclusion

Industrial automatic electrohydraulic systems development essentially depends on the progress in the field of the automatic digital systems, which include hybrid interfacing elements such as high-speed distribution valves or the fast-response flow servopumps with automatic pressure and power control.

The most promising research is being done for integrating the electrohydraulic automated systems with electrical ones, creating hybrid systems that are interconnected through CAN networks adapted to modern Real-Time systems [12,13].

In the modern high-complexity equipment field and in all the hydraulic energy consumers, the optimal control of primary energy sources by distributed and hierarchized computing systems is continuously researched. A good example of progress is the EtherCAT system, which is an Ethernet-based fieldbus system, invented by Beckhoff Automation [15]. The protocol is standardized in IEC 61158 and is suitable for both hard and soft Real-Time requirements in automation technology.

7.4 Numerical simulation of the mechanical feedback servopumps by Amesim

7.4.1 Mechanical feedback servopump structure

This section presents the mathematical modeling and the numerical simulation of the dynamics of a typical swashplate servopump with mechanical feedback produced by Eaton Corporation [1,2]. The servomechanism controlling the pump displacement is supplied by

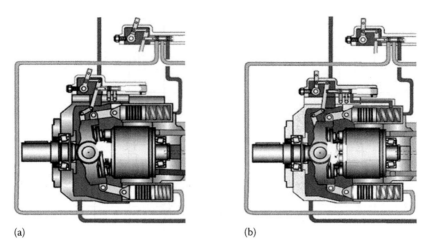

(a)   (b)

Figure 7.4.1 Mechanical feedback servopump: (a) in neutral swashplate position and (b) in a tilted swashplate position.

a low-pressure gear pump, and includes oversized pistons for rejecting the high-frequency tilting force. The response time is controlled by the diameter of a metering orifice sited on the supply port. The theoretical performance of the servopump is found in good agreement with the manufacturer's specifications. The mobile hydrostatic transmissions include hydromechanical servos with lever feedback and centering springs sited in the cylinders (Figure 7.4.1).

The security offered by these servos is one of the main quality needed by the mobile equipment operating in extreme environmental conditions. In the past decade, the mechanical feedback became a hybrid one: the input lever is controlled by two-position proportional solenoids for the two-flow direction. These devices are replacing the manual direct input by a joystick.

The last type of the displacement control system is a pure electrohydraulic servomechanism controlled by a *CAN Electronic Displacement Control*. The centering springs are acting directly on the hydraulic cylinder controlling the pump piston's stroke.

This chapter is devoted to the modeling and simulation of the mechanical part of such servopump by the aid of Amesim simulation language. Common components from the dedicated libraries are used. The servos controlling the pump displacement (Figure 7.4.2) has an input lever 4 actuating one end of the error bar 5. The other end of this bar is connected to the feedback lever 17. The positioning error is fed to the spool valve with a ratio of 1:2, improving the stability of the whole control system [3]. The spool underlap of about 1 mm introduces a dead band of about ±2.5° at the level of input lever. The neutral position of the swashplate is automatically obtained by releasing the input lever: a centering spring is sending the spool in the neutral position from any stroke.

The stability and the response time of the servomechanism are controlled by the aid of a sharp-edge metering orifice. The orifice size is established according to the customer demand. The usual diameter is about 1 mm, nondangerous from the obliteration point of view [4]. The response time of the stroking system directly depends on the orifice diameter

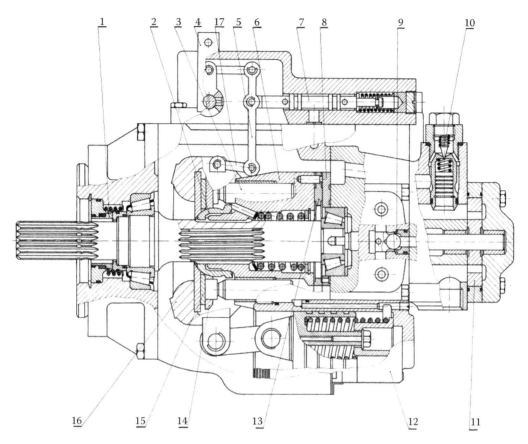

Figure 7.4.2 Mechanical feedback servopump with swashplate: 1—shaft mechanical sealing; 2—spherical bush; 3—piston; 4—input lever; 5—error bar; 6—barrel; 7—spool; 8—fixed valve plate; 9—check valve; 10—pressure relief valve; 11—auxiliary pump; 12—hydraulic cylinder spring centered; 13—mobile valve plate; 14—hydrostatic shoes retainer; 15—hydrostatic shoes; 16—swashplate; and 17—feedback lever.

between 0.71 and 2.59 mm, the time needed by pistons to accomplish that the whole stroke stays in the range 0.29 and 7.48 s. The big values correspond to big displacement pumps (250 cm³/rev). According to the numerical simulations and the experimental identifications [3], this is the simplest way of controlling the dynamic behavior of the hydraulic servomechanism.

7.4.2 Modeling the kinematics of the servomechanism

The mathematical model of the servomechanism contains a mechanical sequence, including all the moving components. This part can be generated using the module PLMASSEMBLY from Amesim (Figures 7.4.3 and 7.4.4).

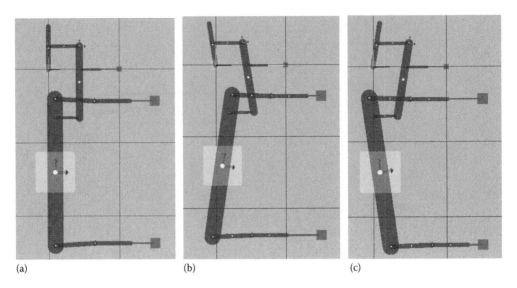

(a)   (b)   (c)

Figure 7.4.3 The servopump servomechanism mechanical components in different steady-state situations (PLMASSEMBLY model from Amesim): (a) neutral position (no input), (b) locked spool and constant force developed by the lower cylinder, and (c) locked spool and constant force developed by the upper cylinder.

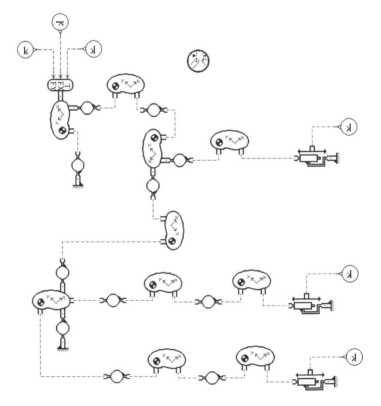

Figure 7.4.4 Servomechanism simulation network developed in Planar Mechanical module from Amesim.

7.4.3 Modeling and simulation of the servomechanism dynamics

The mathematical model of the servomechanism set up in Amesim language may be considered as a general one, but it reflects with accuracy the kinematic and the hydraulic structure of the studied system (Figure 7.4.5). This model was used for simulating three linear inputs applied to the input (control) lever from 0 to 5°, 10°, and 15° (Figure 7.4.6).

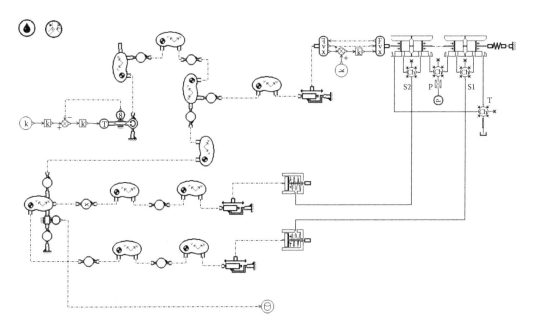

Figure 7.4.5 Amesim model of the servopump servomechanism with mechanical feedback.

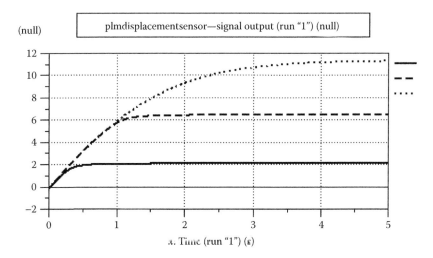

Figure 7.4.6 Swashplate angle variation during the transients.

The main variable evolution is shown in Figures 7.4.7 through 7.4.12. The servomechanism is supplied by the auxiliary pump with hydraulic fluid under low pressure (16 bar). Consequently, the pressure drop across the metering orifice becomes very important during the transients, increasing the control system response time. This one matches the requirements of the mobile equipments with high inertia components. The servovalve opening reaches 4.3 mm for a negative lap of about 1 mm and round metering holes sited in the valve body. The maximum speed of the stroking piston reaches small values (15 mm/s). The time constant of the first-order response can be adjusted between 0.8 and 2.2 s with damping orifices of 1.2–0.8 mm.

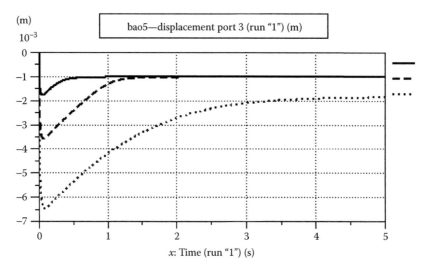

Figure 7.4.7 Servovalve spool-stroke variations during the transients.

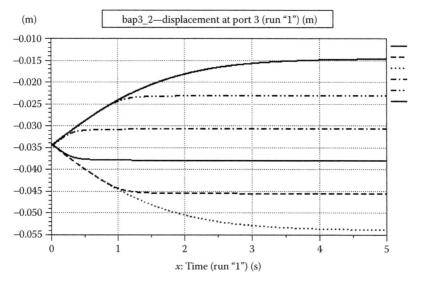

Figure 7.4.8 Low piston displacements during the transients.

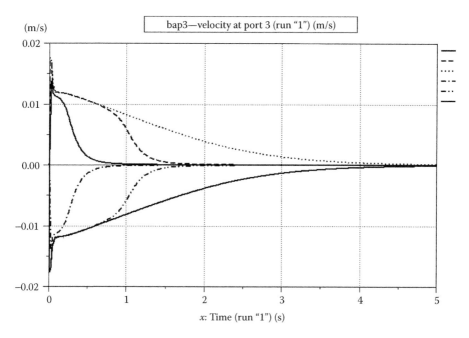

Figure 7.4.9 Hydraulic cylinder pistons velocity during the transients.

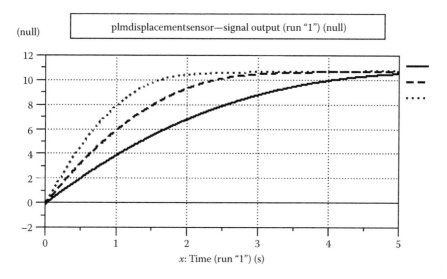

Figure 7.4.10 The influence of the damping orifices diameter on the system response time for a step input introduced by the control lever.

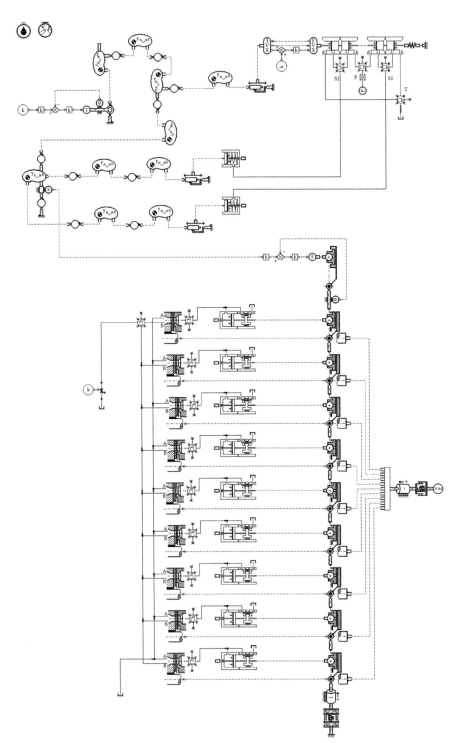

Figure 7.4.11 Amesim model of a swashplate servopump with 9 pistons supplying an orifice.

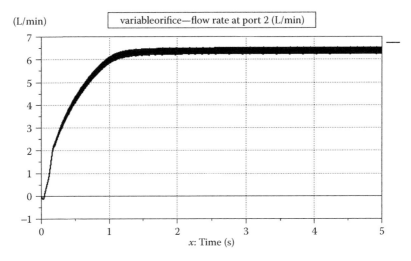

Figure 7.4.12 Servopump-flow variation generated by a step input.

7.4.4 Modeling and simulation of the servopump dynamics

The complete Amesim model of the studied axial piston servopump is shown in Figure 7.4.11. The output flow passes through an orifice. A step input applied on the input lever generates the flow variation as shown in Figure 7.4.12. The flow pulsations (Figure 7.4.13) correspond to the pistons number (9) and to the shaft speed (25 s^{-1}). The pressure variations are shown in Figures 7.4.14 and 7.4.15.

A typical piston displacement in the bore is shown in Figure 7.4.16. The motion of four adjacent pistons is shown in Figure 7.4.17.

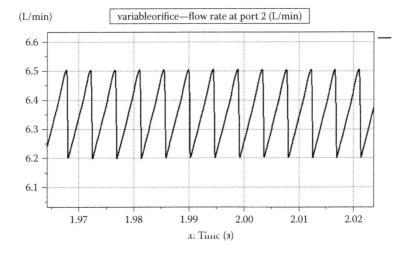

Figure 7.4.13 Zoom in the servopump flow variation generated by a step input.

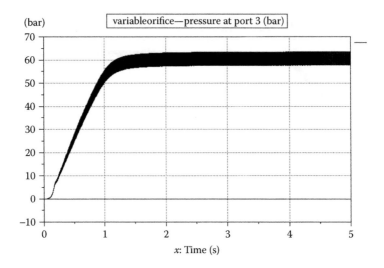

Figure 7.4.14 Pressure variation at the servopump output for a step input.

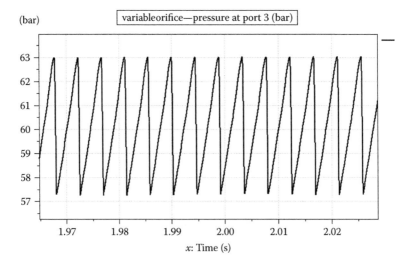

Figure 7.4.15 Zoom in the pressure variation at the servopump output for a step input.

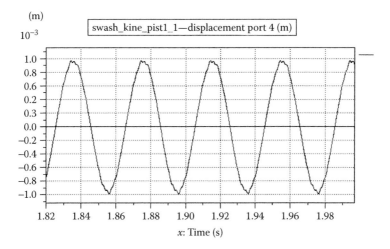

Figure 7.4.16 A pump piston displacement in the bore (small stroke of about 2 mm).

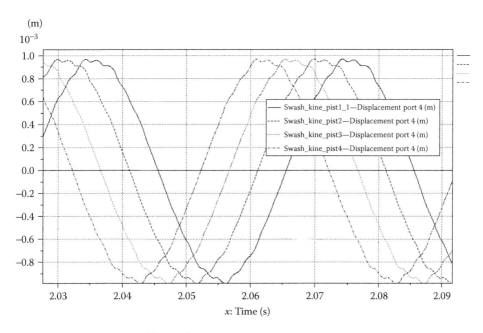

Figure 7.4.17 Displacements of four adjacent pistons in time.

The flow rate of pistons shown in Figures 7.4.18 and 7.4.19 shows a high-frequency component superposed on the kinematic flow.

The servopump control lever generates the control-valve spool displacement from Figure 7.4.20. Upper-stroking piston displacement and overall force developed by the servomechanism are shown in Figures 7.4.21 and 7.4.22.

The step input generates the pump shaft torque variation from Figure 7.4.23, using the flow supplied by the gerotor pump represented in Figure 7.4.24.

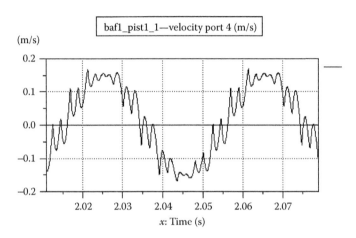

Figure 7.4.18 Piston velocity variation.

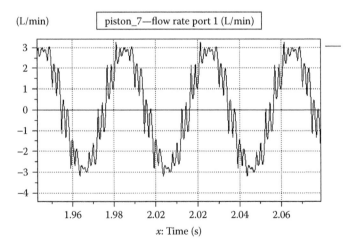

Figure 7.4.19 Flow rate of a piston.

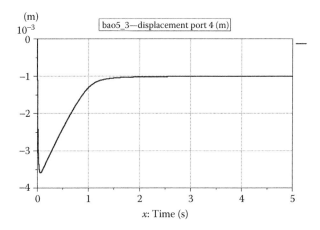

Figure 7.4.20 Spool displacement generated by an input step of the pump control lever.

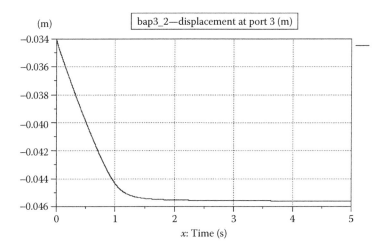

Figure 7.4.21 Upper stroking piston displacement for a step input of the pump control lever.

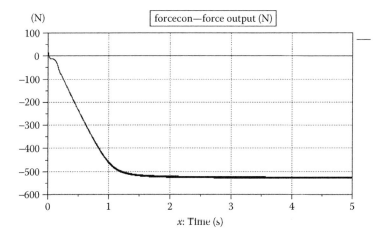

Figure 7.4.22 Overall force developed by the servomechanism.

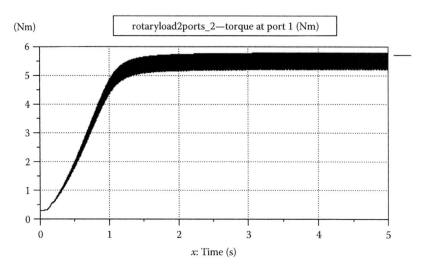

Figure 7.4.23 Pump shaft torque variation during the transient.

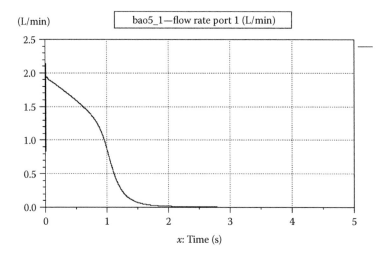

Figure 7.4.24 Control valve flow-rate variation during the transient.

7.4.5   Conclusion

The numerical simulations results obtained by Amesim environment for the global parameters of the servopump displacement control are in good agreement with the manufacturer technical specifications. Other series of numerical simulations performed by general purpose languages such as SIMULINK or LabVIEW are giving the same results. The new mathematical model can be used in a wide category of applications [4]. The servopump dynamic model can be included in the hydraulic library of the new release of Amesim language [5].

7.5 Numerical simulation of the dynamics of the electrohydraulic bent axis force feedback servomotors

7.5.1 Modern structures for the bent axis force feedback servomotors

This section presents the strategy of finding a good control algorithm for the bent axis electrohydraulic servomotors through numerical simulation [1,2]. Two simulation environments—SIMULINK and Amesim are used in order to point out the utility of the numerical simulation in any industrial design process. At the same time, the advantages and the drawbacks of each language are identified [3–5].

The theoretical results regarding the stability and the accuracy of the force feedback servomechanisms are found in good agreement with the experimental ones. A final numerical study on the ways to improve stability shows practical interesting results concerning the real effectiveness of the small metering orifices widely used as damping components in hydraulic control systems. These hydraulic resistances can simply improve the stability.

The modern hydrostatic transmissions requiring a continuously variable motor shaft speed are based on the electrohydraulic proportional control of a bent axis axial pistons motor displacement (Figure 7.5.1) [6,7]. The servovalve is governed by a DC solenoid attached to the control cover. When the solenoid current increases above the threshold current, the servopiston starts to move from the max toward the minimum displacement position. The shaft speed versus current relation is a nonlinear one. In order to minimize hysteresis, a pulse-width modulated control signal of 70–90 Hz has to be utilized [8]. This type of hydraulic motor has gained widespread acceptance in closed-loop industrial systems because they do not need an auxiliary pump: a three-way selector takes the small control oil flow from the motor power ports. The overall motor size is small enough to fit the driving wheels, winches, and so on. The swashplate variable displacement motors are heavier and expensive because they need an external servovalve, and an additional medium-pressure gear pump. They are promoted

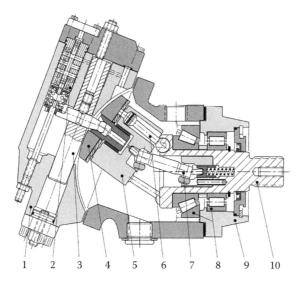

Figure 7.5.1 Bent axis hydraulic servomotor: 1—servomechanism body; 2—flow control valve; 3—stroking piston; and 4—sliding valve plate. (Source: Parker Hannifin)

especially in high-performance applications such as aerospace and military ones. The main problem of these servomotors dynamics is the control stability generated by the low-stiffness feedback spring.

Industrial hydraulic power systems need different types of control systems for servomotor displacement, but the hydromechanic and the electrohydraulic ones are widely used because of their full compatibility with the internal force feedback servomechanism. Figure 7.5.2 presents the hydraulic diagram for both types.

The use of a differential setting piston allows the use of a simple three-way flow control valve with round ports and slight positive overlap. The high-performance applications require a four-way flow control valve with critical center and a symmetric piston (Figure 7.5.3). The dynamic stability requires two metering orifices placed on the valve external ports.

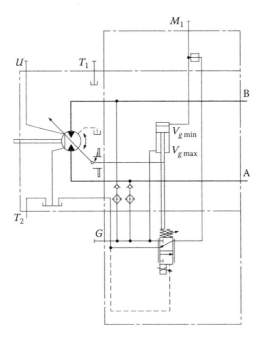

Figure 7.5.2 Structure of an electrohydraulic bent axis servomotor controlled by a proportional solenoid.

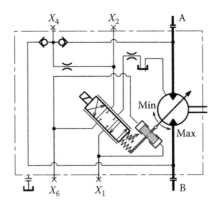

Figure 7.5.3 High dynamics electrohydraulic bent axis servomotor (V14-PARKER HANNIFIN).

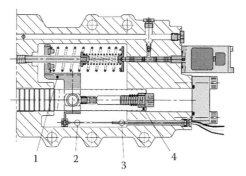

Figure 7.5.4 Electrohydraulic bent axis servomotor with additional electric feedback (V14): 1—feedback arm; 2—mobile armature; 3—position sensor; and 4—setting piston. (Source: Parker Hannifin)

The high-accuracy closed-loop control systems need an additional electric feedback supplied by an immersed LVDT attached to the setting piston (Figure 7.5.4). The electrical feedback is also useful for the Real-Time survey of the overall servomotor state.

The force feedback servomechanism included in a bent axis electrohydraulic servomotor has quasi-linear steady-state displacement–solenoid current characteristics (Figure 7.5.5) depending on the threshold current. The external steady-state characteristics of the electrohydraulic force feedback servomotor representing the relation between the shaft speed and the solenoid current have a parabolic shape (Figure 7.5.6), corresponding to the typical option of a fine speed control at the start of a power train hydrostatic transmission. The bent angle wide range field (6.5°–35°) allows a wide overall speed range (100–3000 rev/min) with a good efficiency in a large pressure range (210–420 bar); the average specific power reaches 3.0 KW/kg, and the specific torque overcomes 10 Nm/kg.

This section is devoted to the mathematical modeling and numerical simulation of the basic servomotor design from Rexroth. The system dynamics is studied by the aid of

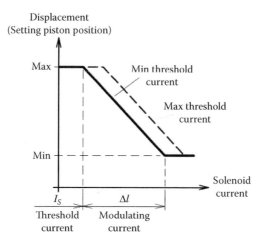

Figure 7.5.5 Steady-state characteristics V_g (I) of an electrohydraulic force feedback servomotor. (Source: Parker Hannifin)

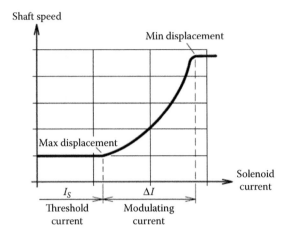

Figure 7.5.6 External steady-state characteristics of an electrohydraulic force feedback servomotor. (Source: Parker Hannifin)

both SIMULINK and Amesim simulation languages in order to point out their benefits and drawbacks.

7.5.2   *Mathematical modeling*

A typical force feedback servomechanism included in a bent axis electrohydraulic servomotor shown in Figure 7.5.7 includes a three-port valve shown in Figure 7.5.8.

The important advantage of single critical-axial dimension requires bias pressure acting on one side of an asymmetrical cylinder for direction reversal. Usually, the head-side piston area is twice the rod-side area and the supply pressure acts on the smaller area to provide the bias force for reversal. The simplest spool control system is a hydraulic one. It is used in heavy vehicle hydrostatic transmissions in close

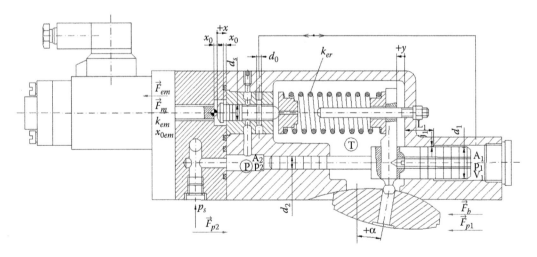

Figure 7.5.7 Main cut-view of a bent axis electrohydraulic servomotor with three-way control valve. (Source: Rexroth)

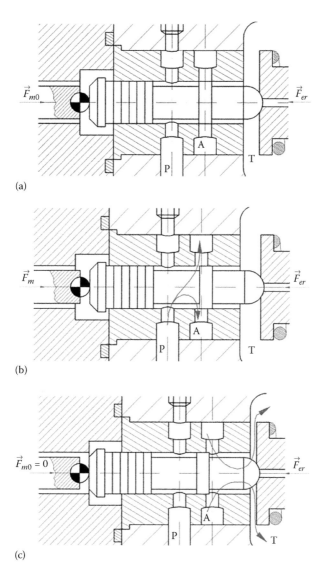

Figure 7.5.8 Control valve connections: (a) all the ports closed, (b) P→A and T closed, and (c) A→T and P closed.

connection with the main pump flow control pressure. The valve is supplied by the high-pressure motor line by a three-way selector. The same pressure is applied on the rod-side area.

The modern hydrostatic transmissions are controlled by industrial process computers, which need an electromagnetic solenoid to actuate the control valve. In this last case, the mathematical model of the studied system includes the following equations [1,2]:

1. Motion equation of the solenoid plunger and the spool:

$$m_x \ddot{x} = F_m - F_{em} - F_{er} - F_h - F_{fv} - F_{fs} \tag{7.5.1}$$

Here,

$$F_m = K_m \left(i - i_0 \right) \tag{7.5.2}$$

is the electromagnetic force corresponding to the difference between the input current and the threshold one denoted by

$$I = i - i_0 \tag{7.5.3}$$

In practical computation, the dynamics of the solenoid plunger can be regarded as a first-order lag with a time constant of about 35 ms [9]. The slope of the solenoid static characteristics is

$$K_m = \frac{F_m \left(I_{max} \right)}{I_{max}} \tag{7.5.4}$$

The solenoid plunger is retired inside the coil by the spring force:

$$F_{em} = K_{em} \left(x + x_{0em} \right) \tag{7.5.5}$$

which depends on the solenoid spring stiffness, K_{em}, the preload x_{0em}, and the spool travel from the neutral position, x is defined by the geometrical closure of all the three valve orifices: P, A, and T. The feedback force

$$F_{er} = K_{er} y \tag{7.5.6}$$

depends on the setting piston stroke y and the feedback spring stiffness, K_{er}. The maximum setting piston stroke can be approximated by the relation

$$y_{max} \cong R_d \left(\alpha_{max} - \alpha_{min} \right) \tag{7.5.7}$$

Here, R_d is the radius of the cylindrical surface of the servomechanism body in contact with the motor sliding valve plate. The flow force depends on the spool travel direction. For $x > 0$,

$$F_h^+ = \rho Q^+ v^+ \cos \theta \tag{7.5.8}$$

Here,

$$Q^+ = 2 c_d A^+ (x) \sqrt{\frac{2 \left(p_s - p_1 \right)}{\rho}} \tag{7.5.9}$$

and

$$v^+ = c_v \sqrt{\frac{2 \left(p_s - p_1 \right)}{\rho}} \tag{7.5.10}$$

Here, θ is the jet angle regarding the spool axis. The valve sleeve has two round orifices with a diameter of d_0. Finally,

$$F_h^+ = K_h \cdot A(x) \cdot (p_s - p_1) \tag{7.5.11}$$

where:
   p_s is the input motor line pressure
   p_1 is the pressure in the hydraulic cylinder *passive* chamber

The flow force constant,

$$K_h = 2c_d c_v \cos\theta \tag{7.5.12}$$

has no dimensions. This force trends to close the valve.
   If $x < 0$,

$$F_h^- = \rho Q^- v^- \cos\theta \tag{7.5.13}$$

Here,

$$Q^- = 2c_d A^- (|x|) \sqrt{\frac{2p_1}{\rho}} \tag{7.5.14}$$

and

$$v^- = c_v \sqrt{\frac{2p_1}{\rho}} \tag{7.5.15}$$

The flow force F_h^- also trends to close the valve:

$$F_h^- = K_h A^- (|x|) p_1 \tag{7.5.16}$$

The viscous friction force,

$$F_{fv} = K_{fv} \cdot \dot{x} \tag{7.5.17}$$

includes a coefficient that has to obtained through experiments, K_{fv}. The same problem arises by the Coulomb static friction force:

$$F_{fs} = K_{fs} \cdot \text{sign}\,\dot{x} \tag{7.5.18}$$

The moving valve assembly includes the equivalent plunger mass, m_m; the spool mass, m_s; and a part of the feedback spring mass m_{er}:

$$m_x = m_m + m_s + 0.35\,m_{er} \tag{7.5.19}$$

2. The setting piston motion equation is

$$m_y \ddot{y} = F_{p1} - F_{p2} - F_{er} + F_b - F_{fvp} - F_{fsp} \tag{7.5.20}$$

where:

$$m_y = m_p + m_t + m_{bc} \qquad (7.5.21)$$

is the stroking assembly equivalent mass composed by the stroking piston mass m_p, feedback arm mass m_t, and the barrel equivalent mass m_{bc}. The other forces can be computed as follows:

$$F_{p1} = p_1 A_1 \qquad (7.5.22)$$

$$F_{p2} = p_2 A_2 \qquad (7.5.23)$$

$$F_{er} = K_{er} y \qquad (7.5.24)$$

$$F_{fvp} = K_{fvp} \dot{y} \qquad (7.5.25)$$

$$F_{fsp} = K_{fsp} \operatorname{sign} \dot{y} \qquad (7.5.26)$$

$$F_b = p_s K_b - y K_y \qquad (7.5.27)$$

The barrel tilting force coefficients K_b and K_y have to be found by experiments. Systematic experimental research carried out in RWTH from Aachen [9] has shown that the component depending on the stroking piston trends to increase the motor displacement. This means a positive feedback in the displacement control loop!

However, the value of the coefficient K_b is usually much more than the value of the coefficient K_y. For example, for the hydraulic motor HYDROMATIK A6V, $K_b = 2.6$ N/bar and $K_y = 1.35$ N/mm.
3. The flow equation corresponding to the *passive* chamber of the stroking piston depends on the spool motion direction from the neutral position.
   If $0 \le x \le x_{\max}$,

$$Q_d^+ - \dot{y} A_1 - Q_{l1} = \frac{V_1}{\varepsilon} \cdot \frac{dp_1}{dt} \qquad (7.5.28)$$

The control flow valve can be computed by the relation:

$$Q_d^+ = 2 c_d A \left(|x| \right) \sqrt{\frac{2 \left(p_s - p_1 \right)}{\rho}} \ge 0 \qquad (7.5.29)$$

Usually, the leakage flow has a basic laminar component:

$$Q_{l1} = K_{l1}p_1 > 0 \tag{7.5.30}$$

in which

$$K_{l1} \cong 2.5\frac{\pi d_1 j_1^3}{12\eta l_1} \tag{7.5.31}$$

and

$$V_1 = V_{10} + A_1 y \tag{7.5.32}$$

is the *passive* chamber volume. For $y = 0$, $V_1 = V_{10}$ (the initial volume). The flow equation becomes

$$\dot{p}_1 = \frac{\varepsilon}{V_1}\left(Q_d^+ - \dot{y}A_1 - Q_{l1}\right) \tag{7.5.33}$$

If $x_{\min} \le x < 0$, the flow equation becomes

$$\dot{p}_1 = -\frac{\varepsilon}{v_1}\left(Q_d^- + \dot{y}A_1 + Q_{l1}\right) \tag{7.5.34}$$

where:

$$Q_d^- = 2c_d A\left(|x|\right)\sqrt{\frac{2p_1}{\rho}} \ge 0 \tag{7.5.35}$$

The metering orifices can be computed by the following relation:

$$A(x_r) = \frac{A_0}{\pi}\left[\arccos\left(1-2x_r\right) - 2\left(1-2x_r\right)\sqrt{x_r - x_r^2}\right] \tag{7.5.36}$$

where:
   x_r is the relative opening of the valve
   $x_r = |x|/d_0$
   $A_0 = \pi d_0^2/4$ is the maximum orifices area

7.5.3   *Numerical simulation by SIMULINK©*

The above-mentioned mathematical model was integrated by SIMULINK–MATLAB© using the *compressed* network as shown in Figure 5.9. The basic model parameters are

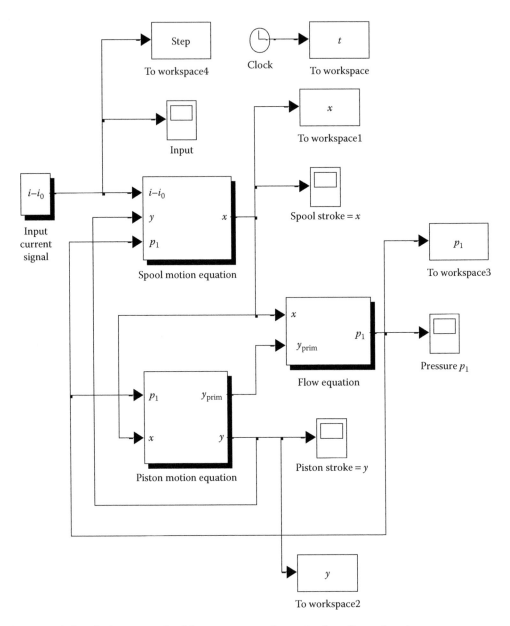

Figure 7.5.9 Simulation network of the servomotor dynamics for a linear input.

the following: $i_0 = 0.09934$ A; $K_m = 129$ N/m; $1/m_x = 7.0922$ kg^{-1}; $x_{0em} + x_0 = 1 \times 10^{-3}$ m; $K_{er} = 1620$ N/m; $K_h = 0.428$; $A_0/\pi = 1.5625 \times 10^{-6}$ m^2; $K_{fv} = 200$ N/(m/s); $K_{fs} = 100$ N/m; $A_1 = 3.14 \times 10^{-4}$ m^2; $A_2 = 1.57 \times 10^{-4}$ m^2; $K_b = 2.6$ N/bar; $K_{fvp} = 4000$ N/m; $K_{fsp} = 2000$ N/m; $1/m_y = 0.2717$ kg^{-1}; $c_d = 0.61$; $\rho = 900$ kg/m^3; $K_{11} = 1.924 \times 10^{-13}$ N/bar; $A_d = 0.196 \times 10^{-6}$ m^2; $\varepsilon/V_1 = 4 \times 10^{13}$ N/m^2; and $K_y = 1350$ N/m.

Figures 7.5.10 through 7.5.12 show the variation of the main servomotor parameters for a linear input signal of 2 s and $p_s = 100$ bar.

Some significant parameters of the mathematical model were optimized step by step, in order to facilitate the wide use of this type of servomotor in practical applications.

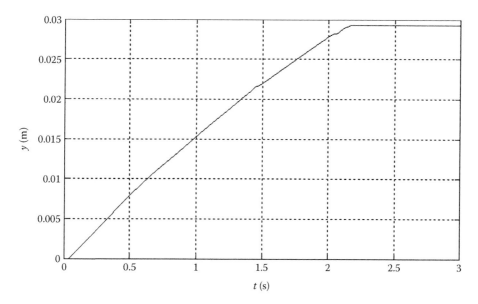

Figure 7.5.10 The servomechanism response for a 75% ramp input signal.

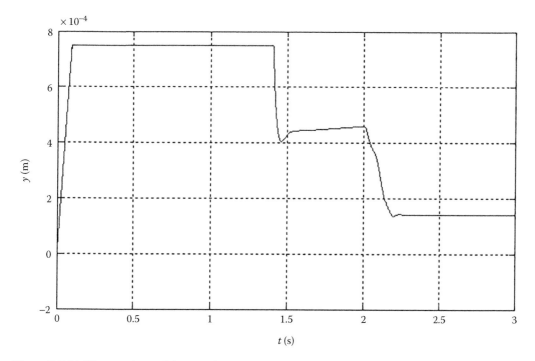

Figure 7.5.11 The spool travel for a 75% ramp input signal.

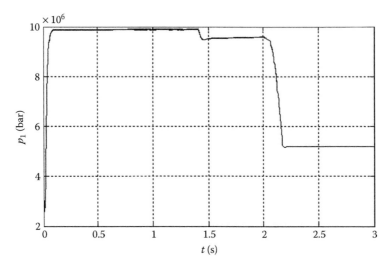

Figure 7.5.12 Pressure variation in the passive chamber of the stroking piston for a typical ramp input signal.

7.5.4   Numerical simulation by Amesim

The above-mentioned simulations performed by the aid of SIMULINK have shown a poor dumping factor of the displacement control system. A significant improvement can be achieved by introducing a small metering orifice between the three-way nonlinear flow control valve and the pressure control chamber of the piston.

The study of the influence of the metering orifice size, d_a, was performed by the mathematical model of the same servomotor built by the aid of general-purpose symbols from the libraries of Amesim simulation environment (Figure 7.5.13).

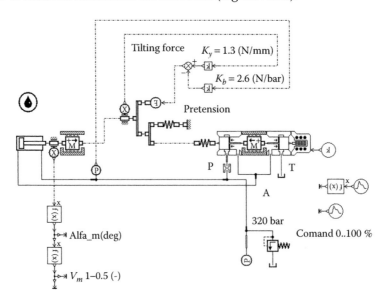

Figure 7.5.13 Amesim simulation network for a bent axis electrohydraulic servomotor with three-way control valve.

Figures 7.5.14 through 7.5.22 show the variations of different internal servomechanism parameters for a step force of 50% from the nominal one (100 N).

The small degree of damping of the servomechanism can be improved using a ramp input signal instead of a step one (Figure 7.5.23). Some relevant parameter variations as functions of the motor supply pressure are shown in Figures 7.5.24 through 7.5.26.

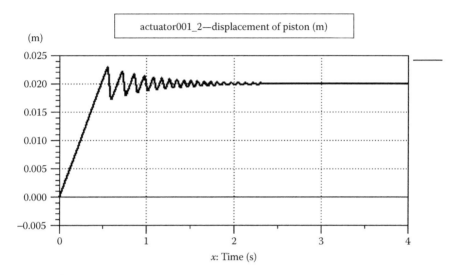

Figure 7.5.14 The servomechanism response for a step input signal ($p_s = 200$ bar, $d_a = 0.7$ mm).

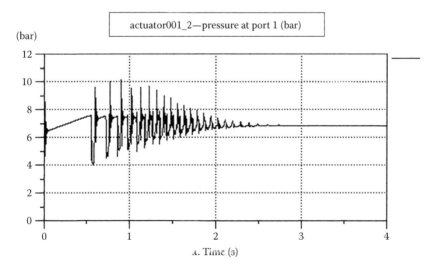

Figure 7.5.15 Typical variations of the pressure in the control chamber of the hydraulic cylinder.

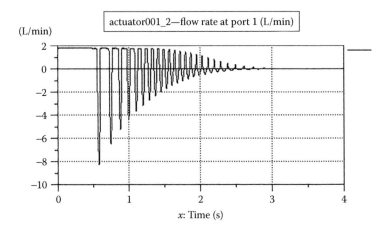

Figure 7.5.16 Flow rate to the control chamber of the stroking piston.

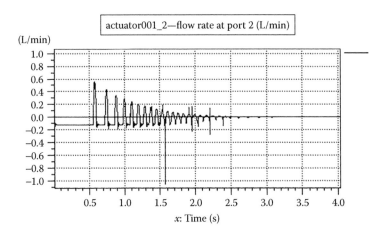

Figure 7.5.17 The flow rate to the tank through the three-way valve.

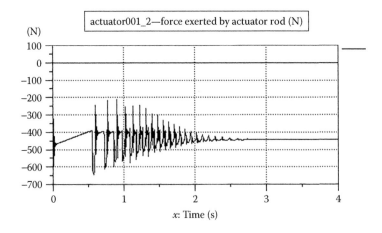

Figure 7.5.18 Force exerted by actuator rod on the moving assembly of the motor.

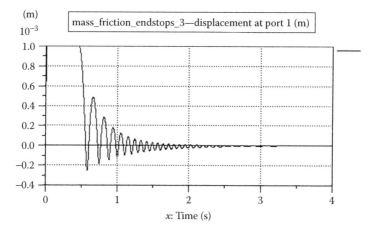

Figure 7.5.19 Stroke of the control valve spool.

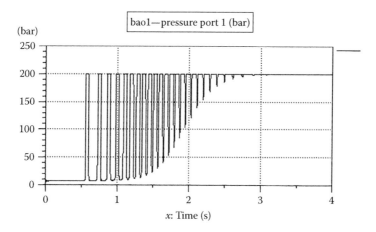

Figure 7.5.20 Pressure variation in the small area of the actuator.

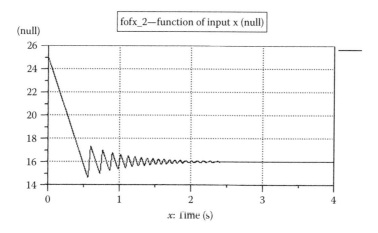

Figure 7.5.21 Tilting angle variation during the transient.

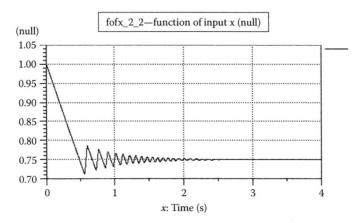

Figure 7.5.22 Relative displacement of the motor variation during the transient under 150 bar supply pressure.

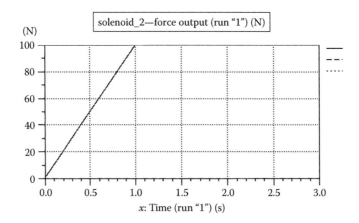

Figure 7.5.23 A ramp of nominal force input signal supplied by the solenoid.

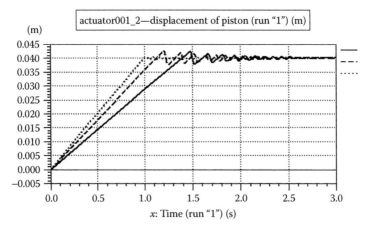

Figure 7.5.24 Displacement of the stroking piston for different supply pressure (100, 150, and 200 bar).

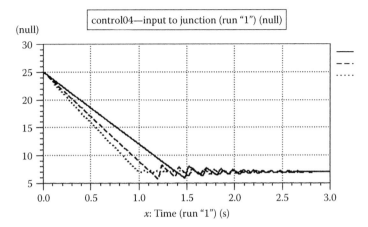

Figure 7.5.25 Tilting angle of the motor barrel for different supply pressure (100, 150, and 200 bar).

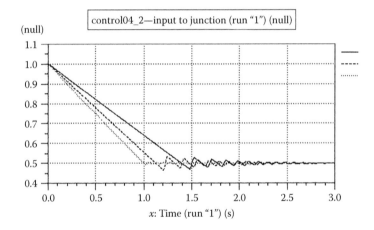

Figure 7.5.26 The relative motor displacement for different supply pressure (100, 150, and 200 bar).

The use of a slightly overlap three-way control valve instead of round holes' control valves leads to a very small damping factor of the force feedback system with relatively small stiffness feedback spring. The following two attempts to eliminate the great amplitude oscillations generated by any step input were performed:

1. Introducing a metering orifice on the input port of the control valve
2. Introducing a metering orifice on the input port of the passive chamber of the stroking piston

Both of them are increasing the response time of the step input response of the servo-mechanism, but they reduce the actuator piston oscillations. A feasible explanation of this situation regards the combination between the small fluid volumes involved in the control process, the small forces demanded by the displacement control, and especially the high level of the pressure in the input port of the servomechanism. The above-mentioned remarks are sustaining the idea of avoiding a low stiffness force feedback spring, and the utility of an additional electrical position feedback. The dynamic behavior of the motor displacement control system is good enough *for industrial purposes.*

7.5.5   Conclusion

The numerical results supplied by the two simulation environments are nearly identical. The big difference between them comes from the time needed to build the simulation networks. Both languages allow the full control of the integration process, but some variable step algorithms developed for Amesim are more accurate in the high-speed phenomenon description. However, the variable step integration is difficult to implement on a general-purpose Real-Time simulation platform such as dSPACE [10] without the conversion in SIMULINK.

A new generation of Amesim simulation models is now prepared for running on PXI and CompactRIO platforms [11] in different fields such as automotive engineering, speed governors for power units, mobile hydraulics, and so on. Other analysis and synthesis software such as AUTOMATION STUDIO are developed now in the same way [12]. It seems also that the high-speed development of the modern industry is sustained by Hardware-in-the Loop (HiL) simulation experiments using different hardware platforms.

Section 1: Bibliography

1. MOOG, *Electro Hydraulic Valves–Applications, Selection Guide, Technology, Terminology, Characteristics,* East Aurora, MA, 2010.
2. ISO 10770-1-2009, Hydraulic Fluid Power-electrically modulated Hydraulic Control Valves. Part 1. Test Methods for four-way Directional Control Valves, 2009.
3. Vasiliu D. Researches on the servopumps and the servomotors of the hydrostatic transmissions, PhD Thesis, University POLITEHNICA of Bucharest, Romania, 1997.
4. The Math Works Inc., *Simulink R5,* Natick, MA, 2007.
5. Lebrun M., and Richards C. How to create good models without writing a single line of code, *Sixth Scandinavian International Conference on Fluid Power,* Linköping, Sweden, 1997.
6. LMS International, *AMESim R8.2b,* Roanne, France, 2009.
7. Parker Hannifin GmbH, Catalogue HY11-3243/UK–Axial Piston Pump, Series PV, Variable Displacement, Hydraulic Control Division, Kaarst, Germany, 2004.
8. https://www.hk-hydraulik.com/en/hydraulic-pumps/brueninghaus-hydraulic-pumps/brueninghaus-a2v.

Section 2: Bibliography

1. Parker Hannifin GmbH, Catalogue HY11-3243/UK–Axial Piston Pump, Series PV, Variable Displacement, Hydraulic Control Division, Kaarst, Germany, 2004.
2. Parker Hannifin GmbH, Catalogue HY02-8001/UK–Axial Piston Pump, Series PV, Variable Displacement, Hydraulic Control Division, Kaarst, Germany, 2004.
3. Parker Hannifin GmbH, Axial Piston Pumps, Variable Displacement, Dynamic Performance, Hydraulic Control Division, Kaarst, Germany, 2004.
4. Vasiliu D. Researches on the Servo Pumps and the Servo Motors of the Hydrostatic Transmissions, PhD Thesis, University POLITEHNICA of Bucharest, 1997.
5. Călinoiu, C., Vasiliu, D., Vasiliu, N., Catană, I. *Modelling, Simulation, and Experimental Identification of the Hydraulic Servo Mechanisms,* Technical Press House, Bucharest, 1999 (in Romanian).
6. Lebrun, M., and Richards, C. How to create good models without writing a single line of code, *Sixth Scandinavian International Conference on Fluid Power,* Linköping, Sweden, 1997.
7. LMS International, *AMESim R8.2b,* Roanne, France, 2009.

8. IMAGINE S.A. Modeling an axial piston pump, Technical Bulletin no. 207, Roanne, France, 2003.
9. Ivantysyn, J., Ivantysynova M. *Hydrostatic Pumps and Motors.* ABI, New Delhi, 2001.
10. National Instruments Corporation. LabVIEW Course, Manu, Texas, 2010.
11. Nasser K. *Digital Signal Processing System Design: LabVIEW Based Hybrid Programming*, Elsevier, San Diego, CA, 2008.
12. Sonderstrom L., Stoica P. *Systems Identification*, Prentice Hall, New York, 1989.

Section 3: Bibliography

1. Vasiliu D. Researches on the transients from the servopumps and servomotors of the hydrostatic transmissions, PhD Thesis, University POLITEHNICA of Bucharest, Romania, 1997.
2. Negoiţă C.G., Vasiliu D., Vasiliu N., Călinoiu C. Modeling, simulation and identification of the servo pumps. *25th IAHR Symposium on Hydraulic Machinery and Systems*, Timisoara, September 20–24, 2010 (paper no. IAHRXXV205DG published by Professional Engineering Publishing), 2010.
3. Negoiţă G.C. Researches on the dynamics of the hydrostatic transmissions, PhD Thesis, University POLITEHNICA of Bucharest, Romania, 2011.
4. Numerical Challenges Posed by Modeling Hydraulic Systems. Technical Bulletin n 114, IMAGINE, 2001.
5. A Brief Technical Overview. Technical Bulletin n 100, IMAGINE, 2001.
6. The Hydraulic Component Design Library. Technical Bulletin n108, IMAGINE, 2001.
7. AMESim: Interfaces with other Software. Technical Bulletin n109, IMAGINE, 2001.
8. LMS Imagine. *Lab Amesim Rev.13*. Leuven, 2013.
9. Lebrun M., Vasiliu D., Vasiliu N. Numerical simulation of the Fluid Control Systems by AMESim. Studies in Informatics and Control with Emphasis on Useful Applications of Advanced Technology, 18(2): 111–118, 2009.
10. Vasiliu N., Călinoiu C., Vasiliu D. Electrohydraulic servomechanisms with two stages DDV for heavy load simulators controlled by ADwin. *Recent Advances in Aerospace Actuation Systems and Components*, Toulouse, France, November 24–26, 2004.
11. Puhalschi R., Feher S., Vasiliu D., Vasiliu N., Irimia C. Simulation languages facilities for innovation, *EUROSIS 2013 SIMEX Conference*, Bruxelles, 2013.
12. www.parker.com
13. www.boschrexroth.com
14. www.ni.com
15. www.beckhoff.com

Section 4: Bibliography

1. www.eaton.com
2. Negoiţă G.C. Researches on the dynamics of the hydrostatic transmissions, PhD Thesis, University POLITEHNICA of Bucharest, Romania, 2011.
3. Vasiliu N., Vasiliu D. *Fluid Power Systems, Vol. I.* Technical Press House, Bucharest, Romania, 2005.
4. Puhalschi R., Feher S., Vasiliu D., Vasiliu N., Irimia P.C. *Simulation Languages Facilities for Innovation.* EUROSIS 2013 SIMEX Conference, Bruxelles, Romania, 2013.
5. www.siemens.com

Section 5: Bibliography

1. Vasiliu D. Researches on the servopumps and the servomotors of the hydrostatic transmissions, PhD Thesis, University POLITEHNICA of Bucharest, Romania, 1997.
2. Negoiţă G.C. Researches on the dynamics of the hydrostatic transmission, PhD Thesis, University POLITEHNICA of Bucharest, Romania, 2011.
3. Lebrun M., Vasiliu D., Vasiliu N. Numerical simulation of the fluid control systems by AMESim. Studies in Informatics and Control with Emphasis on Useful Applications of Advanced Technology, 18(2): 111–118, 2009.

4. Popescu T.C. et al. *Numerical Simulation–A Design Tool for Electro Hydraulic Servo Systems*, in "Numerical Simulations, Applications, Examples and Theory," INTECH PRESS, Zieglergasse 14, 1070 Vienna, Austria, 2011.

5. Vasiliu N., Vasiliu G.C., Vasiliu D., Irimia P.C. *Modern Developments of the Hydraulic Transmissions in Automotive Technology*, Romanian Academy of the Technical Sciences and AGIR PRESS HOUSE, ISSN 2066-6586, Bucharest, Romania, 2012.

6. PARKER HANNIFIN. Catalogue HY30-8223/UK.1.5, 08/2008 PC. *Hydraulic Motors Series V12, V14 and T12*, 2008.

7. REXROTH – BOSCH GROUP. Catalog RE-E 91604. *Axial Piston Variable Motor A6VM Series 63.* Edition: 05.2016.

8. Muraru V., Vasiliu N., Vasiliu D. Structural analysis of the proportional solenoids for control systems using ANSYS Software. *World ANSYS Conference*, Pittsburgh, 2000.

9. Haas H.J. Sekundargeregelte Hydrostatische Antriebe im Drehzah und Drehwinkelregelkreis. PhD Thesis, RWTH, Aachen, 1989.

10. www.dspace.com

11. Vasiliu C. Real-time simulation of the electric powertrains, PhD Thesis, University POLITEHNICA of Bucharest, Romania, 2011.

12. Vasiliu N., Vasiliu G.C., Vasiliu D., Irimia P.C. *Modern Developments of the Hydraulic Transmissions in Automotive Technology*, Romanian Academy of the Technical Sciences and AGIR PRESS HOUSE, Bucharest, Romania, 2012.

chapter eight

Numerical simulation of the hydrostatic transmissions

8.1 Design problems of the hydrostatic transmissions

8.1.1 Structure and applications of hydrostatic transmissions

The mechanical characteristics of the engines are adapted to the needs of the working machines by transmissions. The mechanical characteristic of an engine represents the relation between the torque delivered by the shaft M and the shaft speed ω. The relation may be a curve or a surface, depending on the engine control facilities. For example, a diesel engine steady-state behavior may be described by a family of curves (Figure 8.1.1) obtained by cutting the characteristic surface, $M = f(\omega,\varphi)$ with planes defined by a constant relative fuel mass injected during each active stroke in the cylinders, $\varphi = ct$ [1]. The dimensionless traction characteristics (Figure 8.1.2) show the variation of the relative traction force Z/Z_0 and running resistance F/F_0 as functions of the car relative speed, v/v_0, for different stages selected in the gear box.

The modern technical systems are using mechanic, electric, hydraulic, pneumatic, and hybrid transmissions. A hydraulic transmission includes a pump and a motor connected in a closed loop. The pump turns the mechanical energy supplied by a prime mover into pressure energy, which is turned back into mechanical energy by the hydraulic motor and supplied to the driven machine. Different devices, according to the needs of the working machine, can control the torque and the speed of the hydraulic motors continuously. The flexibility is the main advantage of the hydraulic transmission regarding the mechanical ones, although their overall efficiency is affected by a double energy conversion. Three categories of hydraulic transmissions can be defined according to the types of pumps and motors used in the power loop: hydrostatic, hydrodynamic, and sonic. A hydrostatic (or volumetric) transmission includes mainly two reversible positive displacement hydraulic machines connected in a closed power loop. A typical example is shown in Figure 8.1.3 from Eaton Corporation [2]. Practically, the two machines are changing pressure energy only, but the fluid velocity can reach 10 m/s in the main connections and a few 100 m/s in some safety and control devices.

A global image of the complex systems studied in this book can be obtained from Figure 8.1.4, which contains a view of a wide-spread industrial hydrostatic transmission produced by EATON CORPORATION from the United States.

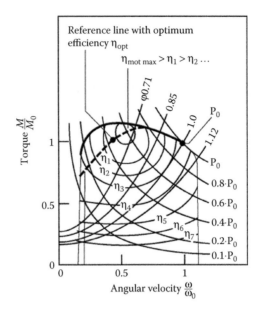

Figure 8.1.1 Universal dimensionless characteristic of a thermal engine.

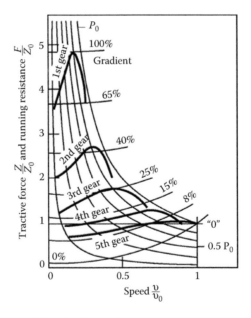

Figure 8.1.2 Traction characteristic of a standard passenger car.

The *hydrodynamic* transmissions include mainly a centrifugal pump and a centripetal turbine connected by pipes in a closed loop. The hydraulic machines are changing both pressure and kinetic energy. The overall efficiency is poor, but the lack of the mechanical connection between the two machines gives a high flexibility to this type of transmission. The *turbo clutch* is still widely used in many automatic car transmissions. The energy can

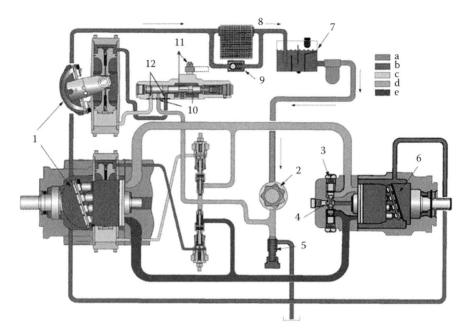

Figure 8.1.3 Hydraulic diagram of a typical hydrostatic transmission (EATON): 1—swash plate axial pistons servopump; 2—auxiliary pump; 3—shuttle valve; 4—directly controlled pressure relief valve (anticavitation); 5—directly controlled pressure relief valve (replenishing pressure); 6—fixed displacement swash plate axial piston hydraulic motor; 7—oil tank with suction filter; 8—cooler; 9—by-pass valve; 10—inline sharp edge orifice; 11—hydromechanical displacement control system; and 12—hydraulic cylinder line.

also be transmitted through a liquid by pressure waves generated by a *sonic* pump and received by a *sonic* motor in the frame of a *sonic* transmission.

Hydrostatic transmissions can be classified by the aid of system theory in the following: hydraulic power transmissions, open-loop hydraulic control systems, and closed-loop hydraulic control systems. The hydrostatic power transmissions have no feedback. Their accuracy depends on the operator's skills. Usually, these systems are transmitting medium mechanical powers with good overall efficiency. The excavator shown in Figure 8.1.5 is entirely actuated by rotary and linear hydraulic motors controlled by electrohydraulic proportional valves. The hydraulic diagram for a typical application of the hydrostatic power systems is presented in Figure 8.1.6 for a forklift [3].

The open-loop hydraulic control systems are transmitting small powers and their motors drive the control devices of much higher power mechanical transmissions. The most important applications can be found in the field of the automated mechanical transmissions of the modern car, trucks, mobile platforms, and so on.

The closed-loop hydraulic control systems use a feedback to reduce the difference between the output of a system and the reference input (the control error) in the presence of different disturbances. These systems may achieve a high accuracy but they have a major problem: stability. The tendency to overcorrect the control error can cause oscillations.

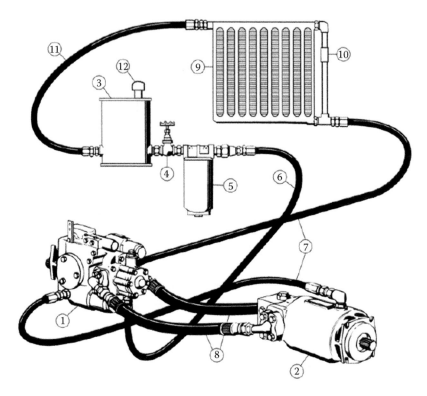

Figure 8.1.4 General-purpose hydrostatic transmission (EATON): 1—swash plate mechanically controlled servopump; 2—swash plate-fixed displacement hydraulic motor; 3—oil tank; 4—two-way flow valve; 5—suction filter; 6—auxiliary pump suction line; 7—drain hoses; 8—power hoses; 9—cooler; 10—by-pass pressure valve; 11—return hose; and 12—fill filter.

Figure 8.1.5 Overview of an earth-moving machine with electrohydraulic control. (Source: HITACHI.)

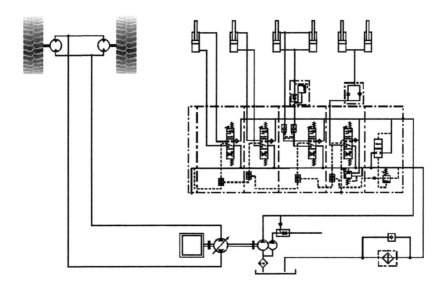

Figure 8.1.6 The hydraulic diagram of a hydrostatic forklift.

The closed-loop hydraulic control systems are widely used for controlling the position, speed, force, torque, power, and so on in different complex systems. The overall digital control by industrial process computers facilitates the use of the hydraulic control systems in any technical field.

8.1.2   *Electrohydraulic control systems in automotive powertrains*

The electrohydraulic control systems are widely used in the field of the modern automotive applications. In addition to the main transmission, they are involved in the fuel injection (Common rail) systems, antilocking braking systems, hydraulic steering systems, hydropneumatic adaptive suspensions, electrohydraulic stability systems, and so on. The electrohydraulic control systems are involved in the automated transmissions that are using turbo clutches, the continuous variation ratio-automated transmissions (Figures 8.1.7 and 8.1.8), and especially the new generation of the double-clutch sequential automated gearboxes.

A synthetic image of the electrohydraulic-automated sequential transmissions for automotive applications is presented in Figure 8.1.9 [5]. Electrohydraulic constant pressure supply rotary and linear servomechanisms are used both for clutch drive and stages changes.

8.1.3   *Hydrostatic transmission performances*

Some feature of the hydrostatic transmissions explains both their wide usage and common restrictions [6].

1. The hydraulic motors can be placed in any position inside a working machine for fulfilling any task. This feature gives the designers a possibility to use any option regarding the machine architecture. Instead of complex mechanisms, only pipes and hoses are used. The security level of hoses increased continuously, thereby allowing the use of mobile ports hydraulic motors.

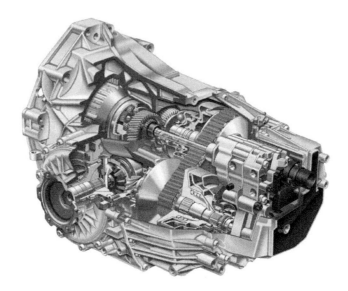

Figure 8.1.7 Partial section of an electrohydraulic continuous-automated transmission Multitronic 01J. (Source: AUDI.)

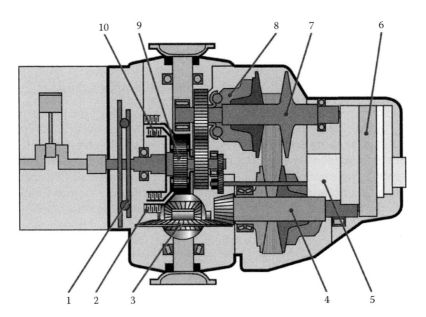

Figure 8.1.8 Principle scheme of the electrohydraulic continuous-automated transmission MULTITRONIC 01J (AUDI): 1—clutch damper; 2—reverse clutch; 3—differential crown; 4—secondary conical pulley; 5—gear pump; 6—electrohydraulic manifold; 7—primary conical pulley; 8—torque transducer; 9—planetary gear; 10—*forward* run clutch.

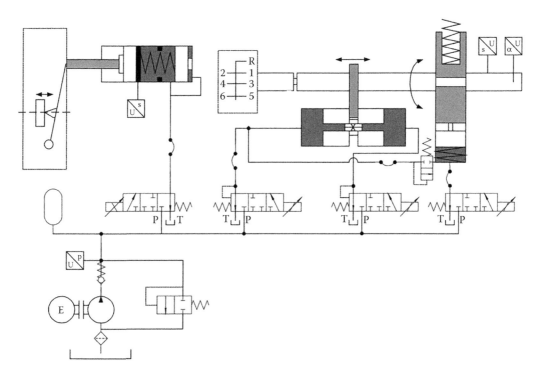

Figure 8.1.9 Hydraulic diagram of the electrohydraulic control system of the automated transmission SMG Drivelogic for BMW Series 3 cars.

2. The control devices of the hydrostatic transmissions require small forces or torques from the operators, and they can be placed in optimal positions. This disposal facility gives special ergonomic qualities to any kind of machine.

3. The torque developed by the electric actuators is proportional to the current; the magnetic field saturation and the insulation overheating limit the torque. The torque developed by the hydraulic motors and cylinders depends directly on the pressure drop across the power ports and it is limited by the mechanical strength of their components only.

4. The heat generated by the internal losses of the hydraulic pumps and motors, which limits the power of any machine, is transferred to the hydraulic fluid; an optional cooler that is sited into a convenient place transfers the heat into the surrounding environment. This process explains the high value of the specific power of the positive displacement hydraulic machines (usually—more than 1 kW/kg).

5. The usual hydraulic fluids are lubricant with special properties regarding the wear, temperature range, and so on. The lifetime of the hydrostatic transmissions is long enough for any industrial application.

6. The hydraulic motors are running stable in a wide range of speeds. The minimum stable speed depends mainly on three factors: (1) the type of mechanism used to obtain variable displacement chambers, (2) the valving system type, and (3) the manufacturing

quality. Both volumetric efficiency and mechanical one have high values. Consequently, the static stiffness of these motors is relatively high. This quality is useful in position control systems, which need a high accuracy achieved by a good disturbance rejection.

7. The electric motors offer a linear connection between speed and voltage; the ratio between the active torque and the inertia has relatively low values. Hydraulic motors give a linear connection between the flow and the angular speed, and the ratio between the active torque and the inertia is very high. Controlled by high-speed ser-vovalves, they can start, stop, and reverse the rotation direction in a very short time. The power gain and the frequency response of the hydrostatic transmission are good enough for industrial applications.

8. Linear hydraulic cylinders can develop huge forces with small overall dimensions, according to the high level of the supply pressure. The high ratio between the pressure force and the moving parts' mass allows a high-speed dynamic response, which is very useful for the position servosystems. The internal leakages of these actuators are very low. This means a high static stiffness and a very low steady-state speed.

9. The operating parameters of the hydraulic motors can be precisely controlled by the aid of variable displacement pumps or variable hydraulic resistances. Industrial process computers or microcontrollers can drive these actuators by the aid of the electrohydraulic servovalves. Such a way, the control signals are electric, and the hydraulic actuators are controlling the motors displacements.

10. Hydraulic energy can be easily stored in oleo-pneumatic accumulators. This feature is very important in high-security applications such as aerospace ones.

The earlier advantages point out the following conclusion: The hydraulic motors are the first option for all the mobile equipment's drives, which need minimum size and weight components. Linear motors are used in all applications that need important forces obtained with compact components.

The use of hydrostatic transmissions is limited by some disadvantages, which are described as follows:

1. Hydrostatic transmissions are expensive; besides the hydraulic pump and motors, they include many additional components for operational parameters' control, security, fluid filtering, transport, and storing. Most of these components require high technologies, special materials, and high manufacturing precision to obtain the required efficiency, security, and long enough lifetimes.

2. The power losses occurring during the energy changes inside the hydraulic machines, and the other components reduce the overall efficiency of the working machines, which include hydrostatic transmissions. Nevertheless, the overall power efficiency increases by eliminating the complex mechanical transmissions.

3. Hydrostatic transmissions can affect the environment quality by fluid leakages. The hydraulic fluid can be entirely lost by the damage of a single elastic sealing element.

4. The liquid fog exiting from the hydraulic elements under pressure by thin cracks or fissures is very flammable due to the volatile components of the hydrocarbons included in most of the common hydraulic fluids.

5. Hydraulic fluids' self-ignition or the loss of the lubricant properties limits up the operating temperature of any hydrostatic transmissions. This danger can be avoided by the aid of the synthetic high-temperature or nonflammable hydraulic fluids, which are now available for any field of application such as aerospace or heavy industry.

6. The contamination of the hydraulic fluids is responsible for the premature fail of the hydrostatic transmissions. Abrasive particles are increasing the clearances between the sliding surfaces in contact such as pistons and barrels. A continuous increase in the leakages lowers the overall transmissions' performances in time. The small control orifices' clogging supplies false control signals, which can generate accidents or even disasters.
7. Air penetration in a liquid system drastically reduces the hydraulic fluid bulk modulus. The increase in the liquid compressibility can cause severe limitations of the transmissions static and dynamic performances by altering the overall response time.
8. The maintenance, fault detection, and reparation of the hydrostatic transmissions need specific technical staff having a multidisciplinary qualification.
9. The complexity of the analysis and synthesis methods requires high-level scientific and technical staff. The design phase needs high-level mathematical knowledge and tools.

From the above-mentioned performance analysis, the following practical conclusion results: The choice of an optimum transmission type for the given application conditions is not only a technical problem but also an economic one. The proper solution can be found by a deep knowledge of all the possible solutions only. All the strong R&D companies are using interdisciplinary teams to find a competitive solution for designing or implementing a hydrostatic transmission in any technical system.

8.1.4 Innovations in the field of the hydrostatic transmissions

In the last decade, the technical innovation generated new principles and designs in the field of the hydrostatic transmissions. The most important innovation can be considered as the joint of a linear ballistic compression ignition engine and a linear positive displacement pump (Figure 8.1.10). The new assembly was called by the innovative company INNAS BV—*CHIRON powertrain* [7], and it was experimentally set up on a forklift (Figure 8.1.11). The thermal engine allows a high flexibility granted by the variable compression degree

Figure 8.1.10 The CHIRON powertrain. (Source: INNAS BV.)

Figure 8.1.11 Forklift with CHIRON powertrain.

controlled by a set of high-speed proportional valves and by the Real-Time control of the fuel injection with a piezoceramic injector.

The powertrain includes two bladder-type accumulators, which allow the constant pressure supply of the electrohydraulic proportional valves that control all the hydraulic motors. The overall performances of the above-mentioned powertrain are improved by the high dynamic qualities of a new type of hydraulic motors called *floating cup motors* (Figure 8.1.12). The pistons (22) are working on two rows against two-inclined fixed disk that play the role of valve plates too. The pistons are sliding into individual mobile cylinders

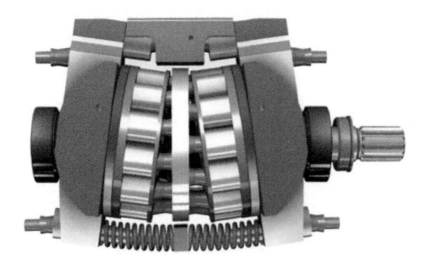

Figure 8.1.12 Floating cup hydraulic motor (pump). (Source: INNAS BV.)

that play the role of hydrostatic bearings too and include the connection openings. By these features, the floating cup motors (pumps) are running without noise and shocks, much quietly than the bent axis ones [8,9].

The most important scientific problem regarding the hydrostatic transmissions is the optimal implementation in complex control systems, using mathematical modeling, numerical simulation, identification, and Real-Time simulation. The first practical objective of the researches is the significant reduction of both fuel consumption and the pollutant emission level by the common digital control of the thermal engine and the hydrostatic transmissions with distributed microcontrollers, which is interconnected by CAN or other digital bus. This target can be reached by detailed dynamic analysis of all the components only, using modern theoretical and experimental procedures [10–13].

8.1.5   Approached problems and solving methods

The term *synthesis* has a broad meaning. From a practical point of view, it is associated with the creative activities that are aiming to generate the technical documentation required to produce a system in the given technical, economic, and social conditions. The complex phenomena associated with the fluids flow through power systems require some iteration always to achieve the given static and dynamic performances.

In the common design activity, the term *synthesis* is associated with the engineering activities for selection and sizing of hardware to form a good operating system. This objective cannot be reached with reasonable efforts without mathematical modeling and numerical simulation.

An ideal synthesis methodology should mean the direct or iterative solving of a set of equations and inequality, which are having the optimal system parameters as unknown quantities (geometric, hydraulic, mechanic, electric, etc.). The performance optimization needs the integration of a set of differential equations. A realistic dynamic approach involves nonlinear differential equations that are solved quickly by computer aid. The original design options can be improved systematically until the required performances (reserve of stability, response time, accuracy, etc.) are fulfilled. The real behavior is frequently far from the theoretical one. Discrepancies have two important sources: (1) multivalued nonlinearities such as backlash, which causes limit cycle oscillations and (2) the types of quantity involved in hydraulic analysis: hard or soft, as they were clearly defined by H. Merritt in 1967. The procedure *cut and try* is sometimes applied to reach the desired system behavior by iterations, which include computation, sizing, design, manufacturing, and test.

Thirty years ago, a *paper and pencil* analysis was always done before numerical simulation needed for final refinements. *The sound engineering judgments was the key of the success.* In the computers era, the facilities of the simulation languages allow more emphasis on the physics and mathematical formulation of problems and less emphasis on the technical solution. The *dual* approach is still applied in a complementary manner, but the mouse replaced the pencil.

This chapter presents a systematic research of the hydrostatic transmissions that are aiming to develop some embedded mathematical models, which can be integrated in the hydraulic library of Simcenter Amesim, developed by LMS International Corporation. The designers of complex fluid power systems will have the possibility to promote the hydrostatic transmissions easily without considering the mechanical design details.

The new embedded models are now used in the Real-Time Simulation Laboratory For Mechatronic Systems, which is developed in the frame of a long-term cooperation by LMS

Test Division—CAE Division—Engineering Services from Romania and the Fluid Power Laboratory from the University Politehnica of Bucharest.

8.2   Dynamics of the hydrostatic transmissions for mobile equipments

8.2.1   Design criteria for the hydraulic scheme

The design of a hydrostatic transmission can be gradually performed from the simple choice of the whole assembly, offered by the dedicated manufacturers on the basis of some preliminary data obtained from similar vehicles to a complete hydraulic, thermal, mechanical, electrical, and control computation that is passing through the following stages:

1. Design of the main components of the transmission
2. Sizing the main components needed for specified performances
3. Choice of the common components from the catalogs of the available suppliers
4. Optimization of the operation parameters of all the components by systematic numerical simulations

This chapter treats mainly the first and the last stages with the target of being able to point out the capabilities of the numerical simulation for optimization of the whole system's behavior without taking into account the concrete details of the components.

The overall option for the design stages of such a complex system depends on the application type. The simplest solution is the choice of a compact transmission composed by a servopump and a servomotor with fixed or variable displacement as shown in Figure 8.2.1a [1]. In this case, the dialog between the designer and the manufacturer is the only way to obtain the required static, dynamic, and maintenance performances. The typical example of this type is offered by the heavy-duty Parker GOLD CUP® series of pumps and motors, which are included in hydrostatic transmissions (Figure 8.2.1b) with a wide range of applications in marine, drilling, shredding, military equipment, and so on. The tried and true design of the GOLD CUP® product line incorporates features such as integral servo and replenishing pump, hot oil shuttle, and a unique servocontrol system, all of which combine to provide a rugged self-contained package that can withstand the harshest conditions and can continue to perform with trouble-free long life.

In common applications, the equipment's designer is obliged to place the servopump in the same place with the thermal engine and to specify the position of the servomotor (or servomotors) in different places and positions such as inside the wheel or outside the wheel, with long hydraulic hybrid connections (pipes and hoses). From the customers' point of view, a distributed positioning system creates the chance of direct access to the servomotors and other components for maintenance operations. The best dynamic performance can be obtained in the first case only. The structure of a hydrostatic transmission essentially depends on the following specifications of the automotive equipment [2,3]:

1. Application field; usually, this is the main transmission of a mobile equipment: forklifts, front loaders, mobile cranes, concrete mixer trucks, agricultural machines, articulated steering forestry tractors, civil work machines such as land levelers, and so on.
2. Range of steady-state velocity control with good enough overall efficiency.
3. Type of positive displacement hydraulic machines available on the market (usually—with axial or radial pistons).

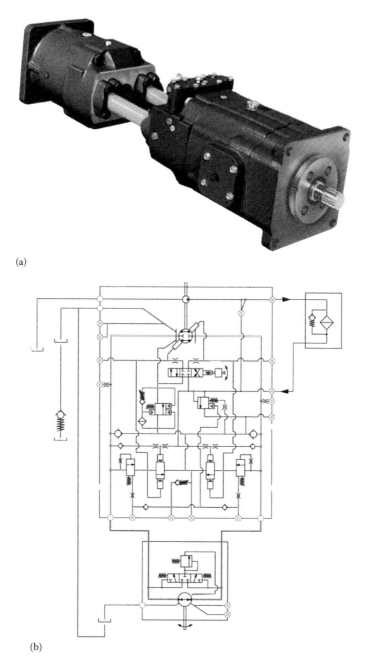

(a)

(b)

Figure 8.2.1 (a) Typical high-performance compact hydrostatic transmission. (Source: PARKER DENISON Hydraulics - GOLD CUP.) (b) Hydraulic scheme of a Parker compact hydrostatic transmission.

4. Type of the displacement control system: mechanical, hydromechanical, electrohydraulic, or hybrid one.
5. Type of the servopump driving engine; usually, diesel engines are used with mechanical or electrohydraulic fuel injection systems.
6. Maximum moving vehicle mass, depending on the destination and the size.

7. Maximum slope of the accessible roads.
8. Maximum velocity on the maximum slope road.
9. System nominal pressure.
10. Maximum velocity on horizontal road.
11. Dynamic radius of the motor wheels.
12. Mechanical transmission ratio of the gears placed between the hydraulic motors and the wheels.
13. Universal characteristics of the servopump driving engine (diesel).

A typical hydraulic scheme of the main hydrostatic transmission of a mobile equipment is shown in Figure 8.2.2.

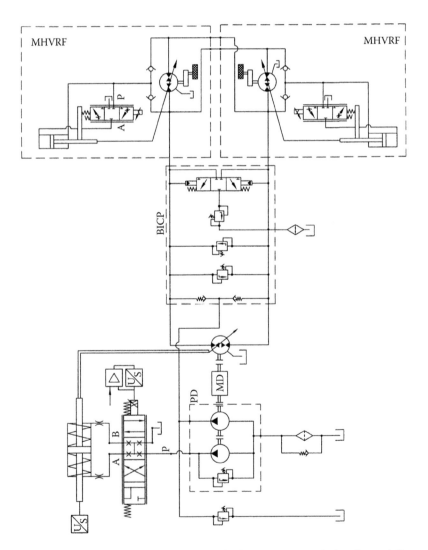

Figure 8.2.2 Typical hydraulic scheme of the main hydrostatic transmission for mobile equipment.

8.2.2   *Optimization of the hydrostatic transmissions by numerical simulation*

In order to illustrate the capabilities of the numerical simulation to help the designer to obtain the specified performance, we have studied a complete dynamic model of a hydrostatic transmission composed mainly by an electrohydraulic swash plate axial pistons servopump and one electrohydraulic bent axis servomotor, which drives a vehicle through a planetary gear and a classical differential. The four-wheel vehicle mathematical model was chosen from the Amesim library $AME/demo/Libraries/TR/ ManualGearbox.ame.

In a first stage, using the model from Figure 8.2.3 we have studied the variation of the car body longitudinal velocity during the start generated by a linear increase in the servopump displacement (0%–100%) in 4 s followed by the decrease in the servomotor displacement from 100% to 50% (Figure 8.2.4). The following parameters were inspected during such a transient:

- The relative variation of the pump displacement (from 0% to 100%), servomotor displacement (from 100% to 50%), and the variable sine wave source relative output corresponding to a sine road slope variation between 0% and 15% (Figure 8.2.5)
- Variations of the car body longitudinal acceleration and longitudinal velocity (Figure 8.2.6) during the motion cycle as shown in Figure 8.2.5
- Variation of the pressure in the servopump main ports (Figure 8.2.7) during the motion cycle from Figure 8.2.5
- Variation of servopump shaft torque and the servomotor shaft torque (Figure 8.2.8) during the motion cycle from Figure 8.2.5
- The influence of the equivalent bulk modulus of the main connection hoses of the transmission (1200 bar, 2000 bar, and 4000 bar) on the servopump output port pressure during the first stage of the vehicle start when the servopump displacement is continuously increasing (Figure 8.2.9)

Figure 8.2.10 shows the variation of the pressure in the servopump delivery port during the complete acceleration of the car in the first 10 s of the transient, for different equivalent bulk modulus of the main connection hoses of the transmission (1200 bar, 2000 bar, and 4000 bar). The influence of the stiffness and the length of the main connections are very important, especially in the case of a large amount of micron air bubbles in the hydraulic fluid.

The most dangerous phenomenon that can occur in the operation of a hydrostatic transmission is the cavitation. The setting of the cracking pressure of the replenishing pump at a low value generates a sudden pressure drop in the servopump suction line in the first second of the increase in the flow rate in the main loop of the transmission (Figure 8.2.11). This is the reason for setting the lowest pressure in the main loop at minimum 16 bar. The heavy-duty transmissions, working at 700 bar, are pressurized at 45 bar! Another important information obtained by simulation is regarding the minimum flow rate of the replenishing pump. The use of the swash plate axial pistons reversible machines offers an overall good mechanical efficiency and a good start torque applied to the load, but the size of the replenishing pumps is greater than that for other types of machines. The main leakages are generated by the sleepers when the oil lubricant properties are too poor at high operating temperature. All the suppliers of compact hydrostatic transmissions are turning any step input signal for the servopump displacement into a ramp (Figure 8.2.12) to reduce the cavitation danger. The simulations revealed a lot of other important criteria for accepting the chosen parameters for different components.

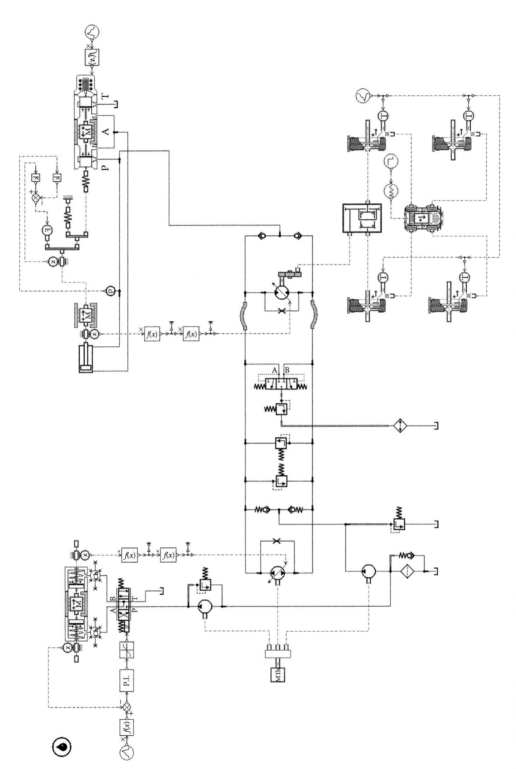

Figure 8.2.3 Simulation network for the hydrostatic transmission with primary and secondary controls.

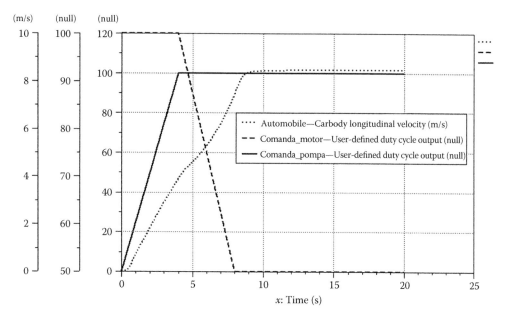

Figure 8.2.4 Variation of the car body longitudinal velocity during the start generated by a linear increase in the servopump displacement from 0% to 100% followed by the decrease in the servomotor displacement from 100% to 50%.

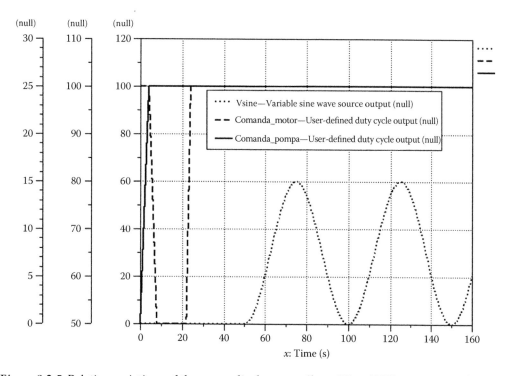

Figure 8.2.5 Relative variations of the pump displacement (from 0% to 100%), servomotor displacement (from 100% to 50%), and the variable sine wave source relative output corresponding to a sine load slope variation between 0% and 15%.

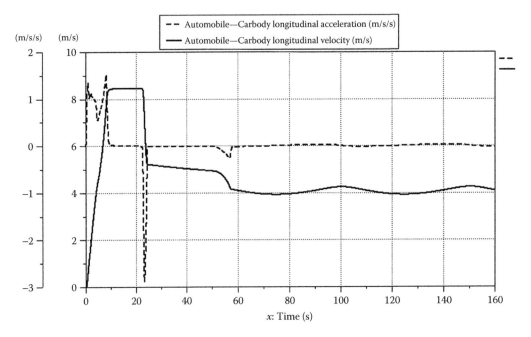

Figure 8.2.6 Variations of the car body longitudinal acceleration and longitudinal velocity during the motion cycle from Figure 8.2.5.

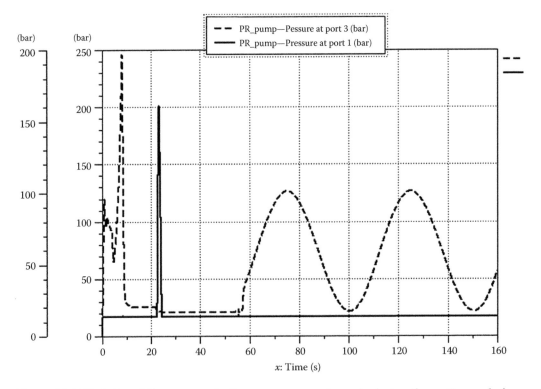

Figure 8.2.7 Variation of the pressure in the servopump main ports during the motion cycle from Figure 8.2.5.

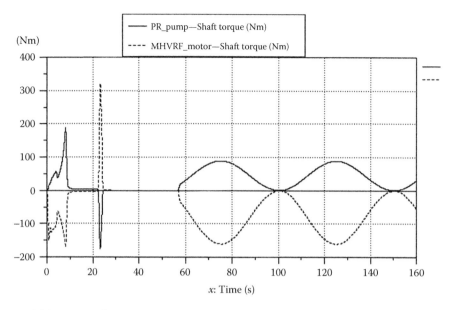

Figure 8.2.8 Variation of servopump shaft torque and the servomotor shaft torque during the motion cycle from Figure 8.2.5.

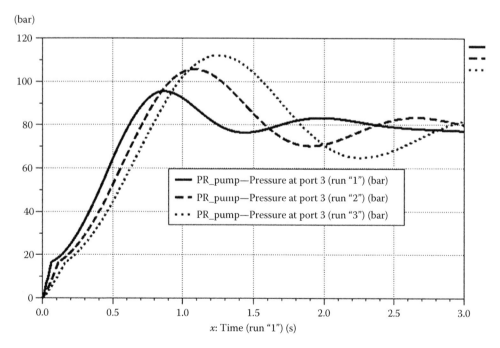

Figure 8.2.9 The influence of the equivalent bulk modulus of the main connection hoses of the transmission (1200, 2000, and 4000 bar) on the servopump output port pressure during the vehicle start.

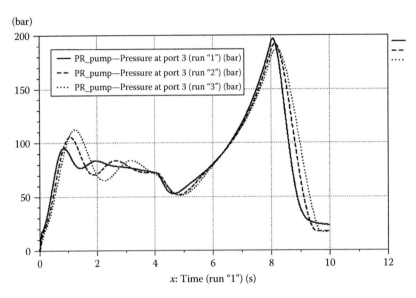

Figure 8.2.10 Variation of the pressure in the servopump delivery port during the complete acceleration of the car in the first 10 s of the transient for different equivalent bulk moduli of the main connection hoses of the transmission (1200, 2000, and 4000 bar).

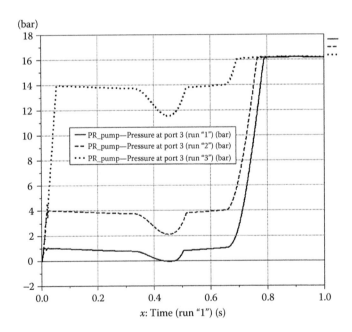

Figure 8.2.11 Pressure variation in the suction line of the servopump during the first second of increase in the displacement for different setting pressures of the replenishing pump (2, 5, and 16 bar).

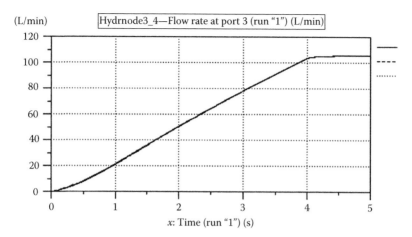

Figure 8.2.12 Servopump flow rate variation during a normal start of the vehicle on a horizontal road.

For example, the maximum value of the pressure in the main loop occurs in the moment of the sudden increase in the servomotor displacement (at 22 s from the beginning of the motion), but it is not overcoming the cracking pressure of the two-stage pressure relief valves that protect the main loop against the burst (350 bar).

8.2.3 Conclusion

The numerical simulation using realistic models based on adequate super components such as hydrostatic slippers is a very useful tool to choose the structure and the parameters of the hydrostatic transmissions properly [3]. The continuous world progress in the use of the fluid power systems needs the development of new tools more and more for the correct description of the border phenomena that occurs in the complex technical systems controlled by digital devices. The compatibility of the electronic, electromechanic, and fluid power systems remains a real challenge for all the innovative companies that are sustained by the modern universities [4–8]. The study by finite element method (FEM) and other modern digital analysis methods of the new materials behavior in extreme mechanical conditions [9–11] will increase the performances and the lifetime of the hydrostatic transmissions. The implementation of the new generation of control algorithms as fuzzy ones [12], together with the adequate Real-Time simulation environments [13], can increase the implementation time of the hybrid control systems.

Section 1: Bibliography

1. BOSCH. *Bosch Automotive Handbook*, 9th ed., John Wiley & Sons Ltd, London, UK, 2014.
2. www.eaton.com/Eaton/ProductsServices/Hydraulics/index.htm
3. www.hitachi-automotive.co.jp/en/products/dcs/index.html
4. www.audi.com/en/innovation/design.html
5. www.bmw.com/en/all-models/3-series/sedan/2015/at-a-glance.html
6. Vasiliu N., Vasiliu D. *Fluid Power Systems*, Vol. I, Technical Press Bucharest (in Romanian), 2005.

7. www.innas.com/floating-cup.html
8. Achten P. *Improving Pump Control*, 10.IFK, Dresden, Germany, March 9, 2016.
9. https://www.innas.com/assets/barrel-tipping-in-axial-piston-pumps-and-motors_2017.pdf
10. www.boschrexroth.com
11. www.poclain-hydraulics.com
12. www.sauer-danfoss.com
13. www.parker.com

Section 2: Bibliography

1. PARKER Denison GOLD CUP® Product Catalog Piston Pumps & Motors for Open & Closed Circuits HY28-2667-01/GC/NA, EU. Effective: October 21, 2014.
2. Brossard J.P. *Dynamique du véhicule: Modélisation des systèmes complexes*, Sciences Appliques, INSA Lyon - PPUR, 2006.
3. Vasiliu D. *Researches on the Hydrostatic Servopump and Servomotors of the Hydrostatic Transmissions*. PhD Thesis, University POLITEHNICA of Bucharest, 1997.
4. Johansson A., Blackman L.D. *Predicting pump dynamics: The dynamic behavior of a Parker Hannifin/ Abex NWL hydraulic-axial-piston pump was simulated to determine information on pump flow ripple for longer life and reduced noise*. SAE Off-Highway Engineering, June 2001.
5. Ivantysyn J., Ivantysynova M. *Hydrostatic Pumps and Motors*. ABI, New Delhi, 2001.
6. Grabbel J., Ivantysynova M. An investigation of swash plate control concepts for displacement controlled actuators. *International Journal of Fluid Power*, 6(2): 19–36, 2005.
7. Ivantysynova M., Baker J. Power loss in the lubricating gap between cylinder block and valve plate of swash plate type axial piston machines. *International Journal of Fluid Power*, 10(2): 29–43, 2009.
8. Pelosi M., Ivantysynova M. A new fluid-structure interaction model for the slipper-swashplate interface. *Proceedings of the 5th FPNI PhD Symposium*, Cracow, Poland, pp. 219–236, 2008.
9. Milella A., Di Paola D., Cicirelli G. *Mechatronics Systems, Simulation, Modelling and Control*. In-Tech, Croatia, 2010.
10. Roşu C., Sorohan Şt., Vasiliu N. FEM analysis of a hydrostatic pump. *1st FPNI—PhD Symposium*, Hamburg, Germany, 2000.
11. Vasiliu N., Roşu C.A., Sorohan Şt., Marin S. Theoretical and experimental researches on the axial piston pumps yoke. *ANSYS Conference*, Pittsburgh, PA, 2002.
12. Zadeh L.A., Tufis D., Filip F.G., Dzitac I. (Eds.), *From Natural Language to Soft Computing: New Paradigms in Artificial Intelligence*, Editing House of the Romanian Academy, Bucharest, 2008.
13. Ion Guta D.D. Researches on the real-time simulation of the fluid power systems. PhD thesis, University POLITEHNICA of Bucharest, 2008.

chapter nine

Design of the speed governors for hydraulic turbines by Amesim

9.1 Modeling and simulation of the high-head Francis turbines

9.1.1 Problem formulation

From the control theory point of view, the hydraulic turbine together with the power generator represents the controlled system that determines, depending on its functions, the performance requirements for the frequency and power control systems. Among the most important operating modes of a hydropower unit are start-up until the synchronizing speed is achieved with the generator islander from the power grid, normal operation within the power grid, and unload with the generator from the power grid. Regarding the control system, the most difficult operating modes are the ones where the generator is not coupled to the power grid.

When operating as part of a large power grid, the generator develops a very large synchronization torque as soon as a tendency to move away from synchronous speed exists. Since the generator is mechanically connected to the turbine shaft, this synchronization torque provides the hydropower unit with a very good stability, almost independent of the control system. As the power of the hydropower units being built increases, so does the difficulty of designing adequate control systems. Everincreasing water flows, and the increase in water supply duct length have as a result an increase in the time constant of the water supply duct and a decrease in the turbine time constant, which negatively affects the stability of the system [2–8]. Moreover, Union for Coordination of the Transmission of Electricity (UTCE) has adopted a strict regulation concerning the parallel operation of Europe's power plants. The quality of the electricity supplied to more than 400 million inhabitants of Europe by interconnected hydropower plants, which produces more than 2100 TWh per year, requires control systems with very high level of performance. By developing new control system architectures, radical improvements can be achieved in the following fields:

- Increasing operational safety
- Increasing reliability for hydropower units and interface components
- Increasing flexibility regarding possible operating modes
- Increasing yearly production of green energy

The control system of the hydropower unit must achieve a required level of performance, defined in connection to the following elements:

- The major hydropower unit components that are considered part of the speed governor, such as the electrohydraulic servomechanism of the wicket gate, and the speed transducer

- The design goal of the control system, which is directly connected to the control laws developed, the tuning of controller parameters in order to insure the desired response of the system, and therefore good steady-state and transient performance [11,13]

In order to fully utilize the performance of modern digital fully programmable control hardware, the first step in the design of a control system for hydropower units is establishing an accurate mathematical model of the hydropower unit, which allows the problem to be solved using established control theory algorithms and methods, which results in the design of an optimal control system structure.

Control theory works with models, usually standard representations (transfer functions or state space), which incorporates the main characteristics of the real-life system, but do not reproduce it with good enough accuracy in any condition. The design of control system starts from the premise that models are a limited representation, and are only to be used as a reference, relative to which the real phenomenon takes place in a vicinity that needs to be precisely estimated and controlled. Therefore, the quality of the resulting control system depends directly on the precision of the mathematical model developed for the controlled system or process. Unfortunately, the more complex a system is, the harder will be the development of an accurate model.

An accurate model of a system or process might prove difficult to develop, and too complex for the performance of modern hardware, especially when Real-Time and/or Hardware-in-the-Loop (HiL) simulation is needed.

As a result, one must seek a model that achieves the optimum compromise between complexity and accuracy. Robust control systems should be designed on the basis of such a model, which still achieve the stability of the real system despite the simplifications of the modeling process, and any common variation of functional parameters.

9.1.2 Mathematical model of a high-head Francis turbine

On a concept level, hydraulic turbines are devices that convert the potential energy of a water flow into mechanical work. Reaction hydraulic turbines are a type of turbines that rely on converting the potential energy of the water stored at a higher level into kinetic energy by accelerating a water column in the supply duct, and converting it into work inside the turbine due to the pressure difference between the two faces of each rotor blade. Because all the energy conversion takes place in a closed duct system, this has to be resistant enough to withstand pressure variations. High-head reaction turbines (Figure 9.1.1) are relatively sensitive to water hammer, which strongly depends on the supply duct length.

In order to reduce this effect, pressure governors with bypass valves are used (Figure 9.1.2), whose role is to allow the passing of water downstream without passing through the turbine when the wicket gate is closing quickly. Water hammer depends on the flow-speed variation in the duct, and a bypass valve reduces this variation speed [16].

Modern solutions use conical valves placed at the end of the supply duct. One of the main advantages of these systems is the possibility of mounting the actuators (hydraulic cylinders outside the duct), where they do not come into contact with water. A typical high-head Francis turbine ($H = 522$ m), whose runner was designed by the authors, and combined with a bypass valve is shown in Figures 9.1.3 through 9.1.5 [1].

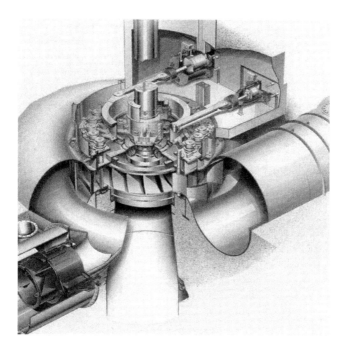

Figure 9.1.1 High-head Francis turbine with bypass valve. (Source: VOITH)

Figure 9.1.2 Bypass conical valve of a Francis turbine. (Source: FLOW CONTROL APPLICATIONS)

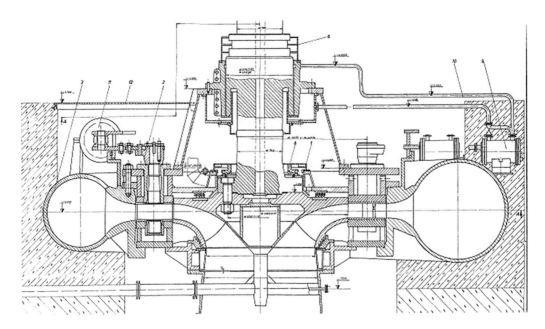

Figure 9.1.3 Francis turbine from Râul Mare Retezat Hydropower Plant. (Source: U.C.M. REŞIŢA-ROMANIA)

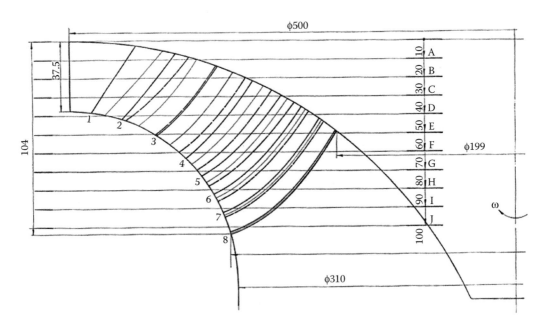

Figure 9.1.4 Meridian plane sections of the rotor blades of the Francis turbine from Hydropower Plant Râul Mare Retezat (Romania) [1].

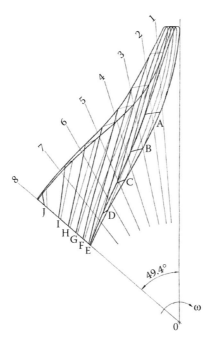

Figure 9.1.5 Topographic curves of the rotor blades from the Francis turbine from Hydropower Plant Râul Mare Retezat [1].

The power is adjusted by controlling the water flow rate through the turbine, which, in case of Francis turbines is done by the wicket gates. The nominal power comes from the following equation:

$$P = \rho g Q H \eta \tag{9.1.1}$$

where:
    P is the power at the turbine shaft
    ρ is the water density
    g is the gravitational acceleration
    Q is the flow rate entering the turbine
    H is the head
    η is the efficiency

The turbine efficiency η represents how much of the system's potential energy is actually used. The efficiency depends on turbine characteristics and flow. Theoretically, the efficiency is defined as the ratio between the mechanical power at the turbine shaft and the consumed hydraulic power.

The relationships that explain the dependency between various turbine characteristics (flow rate, head, speed, wicket gate opening, efficiency, etc.) are difficult to calculate theoretically and are usually provided as graphs, called *efficiency hill chart* (Figure 9.1.6) [2].

Although an efficiency hill is quite difficult to use in a mathematical model due to the problems encountered in actually determining a relationship between the variables, the following general conclusion can be easily drawn: The shaft torque strongly depends on the turbine speed for a large number of speeds and wicket gate positions. Generally, the

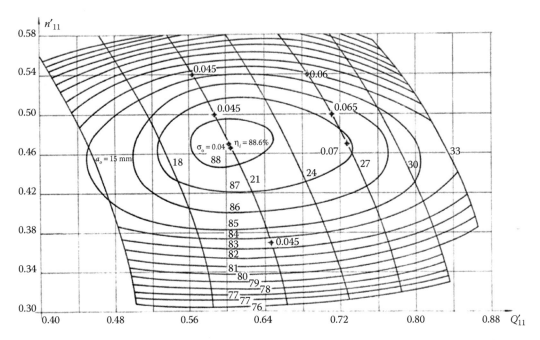

Figure 9.1.6 The efficiency hill chart of the Francis turbines from Hydropower Plant Râul Mare Retezat [1].

efficiency hills are built based on results of tests performed by the turbine's manufacturer in steady flows, but it can be assumed that the results hold true to an extent in transient flows as well. An efficiency hill chart allows the efficiency of the turbine to be determined in a given point, corresponding to a combination of speed, head, and wicket gate position, specific energy, flow, speed, and wicket gate position. In order to graphically represent the efficiency hill chart, one of the above-mentioned needs to be eliminated using the similitude laws for hydraulic machinery. For analyzing transient flow, the most convenient is to use a representation that eliminates specific energy [3].

The mathematical model of a turbine's characteristic can be written as follows:

$$Q_{11} = f\left(n_{11}, y\right) \tag{9.1.2}$$

$$M_{11} = f\left(n_{11}, y\right) \tag{9.1.3}$$

where:

Q_{11} is the unitary flow rate
M_{11} is the unitary torque
n_{11} is the unitary speed
y is the wicket gate position (%) [2]

Currently, there are no exact mathematical models describing Q_{11} and M_{11}. Their values are usually obtained by interpolation. The expressions for n_{11} and Q_{11} can be written as

$$n_{11} = \frac{nD_{\text{ref}}}{\sqrt{H}} \tag{9.1.4}$$

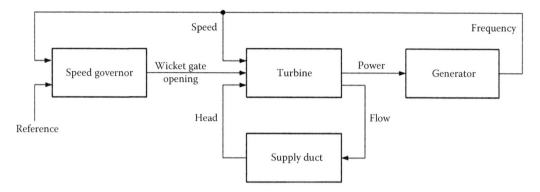

Figure 9.1.7 Theoretical scheme of the speed and pressure governor system.

$$Q_{11} = \frac{Q}{D_{ref}^2 \sqrt{H_n}} \tag{9.1.5}$$

$$M_{11} = \frac{M}{D_{ref}^3 H_n} \tag{9.1.6}$$

The characteristic curves are extrapolated for small openings of the wicket gate. Therefore, it is necessary to know the flow value until which the turbine speed is 0. The wicket gate opening for which the turbine achieves synchronous speed must also be known.

The dynamic characteristics of the turbine are the main elements influencing the system's stability. Figure 9.1.7 presents the schematic of the model for the whole system.

The mechanical power at the turbine shaft is a function of head and flow. The flow is a function of head, speed, and wicket gate opening. We can write as follows:

$$p = f_p(H, n, y) \tag{9.1.7}$$

$$q = f_q(H, n, y) \tag{9.1.8}$$

The turbine model is connected to the generator model through the mechanical power at the turbine shaft P_m. The relationship between power p_m and torque m_m is expressed unitary as follows:

$$p_m = \frac{n}{\omega_s} m_m \tag{9.1.9}$$

Here, ω_s is the synchronous angular speed ($2\pi f$). Normally, we can consider that $n = \omega_s$ at synchronous operation. This simplification does not mean that the speed is constant, but rather that the variations are small with little effect on the model. Equation 9.1.9 becomes

$$p_m = m_m \tag{9.1.10}$$

The nominal power of a hydraulic turbine depends on the energy the fluid loses in the supply duct. Since the turbine is not 100% efficient, the no load flow q_{NL} is subtracted from the effective flow as follows:

$$p_m = A_t H(q - q_{NL})\eta \tag{9.1.11}$$

Turbine manufacturers offer detailed information about the variations in turbine efficiency η for different operational points, but the use of such information in a mathematical model requires the use of a database and interpolation procedures. If variations in efficiency are neglected, Equation 9.1.11 becomes

$$p_m = A_t H \left(q - q_{NL} \right) \tag{9.1.12}$$

A_t is a proportional constant for calculating the nominal turbine power (in MW) and nominal generator power (in MVA).

The pressure drop in the turbine depends on the unitary flow as follows [4]:

$$q = y\sqrt{H} \tag{9.1.13}$$

To finalize the mathematical model of a high-head Francis turbine, the influence of the supply duct must be taken into account, mathematically represented as:

$$q = f(H) \tag{9.1.14}$$

Water hammer models can calculate this value as a function of wicket gate position, but we can obtain a simplified relationship as follows [8]: Consider a rigid duct of length L and section A, which connects the turbine to the lake, resulting in a head H. Any change in the turbine's operating conditions requires the acceleration or deceleration of the water column from the supply duct determined by the pressure force, according to Newton's second law:

$$F = m\frac{dv}{dt} \tag{9.1.15}$$

It is also known that

$$F = -\rho g A \Delta H \tag{9.1.16}$$

$$m = \rho AL \tag{9.1.17}$$

Replacing Equations 1.16 and 1.17, Equation 1.15 becomes

$$\rho g A \Delta H = -\rho AL \frac{dQ}{A dt} \tag{9.1.18}$$

or

$$\Delta H = -\frac{L}{Ag}\frac{dQ}{dt} \tag{9.1.19}$$

Expressing Equation 1.19 in relative units, we obtain the following:

$$\Delta H^* = -T_w \frac{d\Delta Q^*}{dt} \tag{9.1.20}$$

where

$$\Delta H^* = \frac{\Delta H}{H_n}$$

$$\Delta Q^* = \frac{\Delta Q}{Q_n}$$

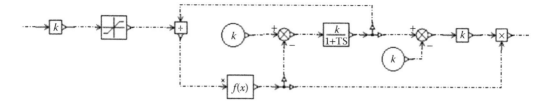

Figure 9.1.8 Amesim model of the turbine.

Here, $T_w = LQ_n/gAH_n$ represents the time constant of the supply duct and indicates the time needed to reach flow rate Q_0 at a head H. Applying the Laplace transform to Equation 9.1.20, we get the following:

$$\frac{h(s)}{q(s)} = -T_w s \tag{9.1.21}$$

Figure 9.1.8 presents a simplified model of a Francis turbine based on Equations 9.1.12, 9.1.13, and 9.1.21 built in Amesim. The model is not complete, as its input needs to be attached to the position feedback of the electrohydraulic servomechanism model, and its output to the input of the generator model, as described in the mathematical model below [6].

9.1.3 Mathematical modeling of the synchronous generator

As previously stated, when the turbine is operating with the generator insulated from the power grid, the synchronization torque is almost completely lacking. In this case, the generator's influence on the system is almost completely mechanical, mainly rotor inertia, so a first-order response. On the basis of these considerations and experimental identification, a simplified model can be built [2,9].

When there is an imbalance between the torques acting on the rotor, the resulting torque (which causes the rotor to accelerate or decelerate) can be expressed as the following:

$$T_a = T_m - T_e \tag{9.1.22}$$

where:
T_m is the mechanical torque
T_e is the electromagnetic torque
T_a is the resulting torque

The amount by which the rotor's speed increases or decreases depends on its inertia, therefore

$$J\frac{d\omega_m}{dt} = T_a = T_m - T_e \tag{9.1.23}$$

where:
J is the turbine-generator assembly moment of inertia
ω_m is the rotor angular speed

The inertial constant H is defined as

$$H = \frac{1}{2}\frac{J\omega^2_{0m}}{VA_{base}} \tag{9.1.24}$$

where VA_{base} is the apparent nominal power of the generator.

Therefore:

$$J = \frac{2H}{\omega^2_{0m}}VA_{base} \tag{9.1.25}$$

Replacing Equation 9.1.25 in 9.1.23 we get

$$\frac{2H}{\omega^2_{0m}}VA_{base}\frac{d\omega_m}{dt} = T_m - T_e \tag{9.1.26}$$

or

$$2H\frac{d}{dt}\left(\frac{\omega_m}{\omega_{0m}}\right) = \frac{(T_m - T_e)\omega_{0m}}{VA_{base}} \tag{9.1.27}$$

Knowing that

$$T_{base} = \frac{VA_{base}}{\omega_{0m}} \tag{9.1.28}$$

Equation 9.1.27 becomes

$$2H\frac{d\overline{\omega_r}}{dt} = \overline{T_m} - \overline{T}_e \tag{9.1.29}$$

$$\overline{\omega_r} = \frac{\omega_m}{\omega_{0m}} = \frac{\omega_r/p_f}{\omega_0/p_f} = \frac{\omega_r}{\omega_0} \tag{9.1.30}$$

where:

ω_r is the angular speed
ω_0 is the nominal angular speed
p_f is the generator number of poles

Considering δ being the rotor's position (in radians) relative to a synchronous reference position δ_0, the value at $t = 0$ is

$$\delta = \omega_r t - \omega_0 t + \delta_0 \tag{9.1.31}$$

By derivation two times in relation to time, we obtain

$$\frac{d\delta}{dt} = \omega_r - \omega_0 = \Delta\omega_r \tag{9.1.32}$$

and

$$\frac{d^2\delta}{dt^2} = \frac{d\omega_r}{dt} = \frac{d\Delta\omega_r}{dt} = \omega_0\frac{d\overline{\omega_r}}{dt} = \omega_0\frac{d\overline{\Delta\omega_r}}{dt} \tag{9.1.33}$$

Replacing in Equation 9.1.29 yields the following:

$$\frac{2H}{\omega_0} \frac{d^2\delta}{dt^2} = \overline{T_m} - \overline{T_e} \tag{9.1.34}$$

Generally, taking the damping torque into account is needed for calculating T_e. This is achieved by adding to Equation 9.1.34 a term proportional to $\Delta\omega_r$

$$\frac{2H}{\omega_0} \frac{d^2\delta}{dt^2} = \overline{T_m} - \overline{T_e} - K_D\Delta\overline{\omega_r} \tag{9.1.35}$$

This represents the motion equation of a synchronous electrical machine. From Equation 9.1.32 we get:

$$\Delta\overline{\omega_r} = \frac{\Delta\omega_r}{\omega_0} = \frac{1}{\omega_0} \frac{d\delta}{dt} \tag{9.1.36}$$

Replacing in Equation 9.1.35 yields the following:

$$\frac{2H}{\omega_0} \frac{d^2\delta}{dt^2} = \overline{T_m} - \overline{T_e} - \frac{K_D}{\omega_0} \frac{d\delta}{dt} \tag{9.1.37}$$

From Equation 9.1.29, we get the following:

$$\frac{d\overline{\omega_r}}{dt} = \frac{1}{2H} \overline{T_a} \tag{9.1.38}$$

Integrating in relation to time we find

$$\overline{\omega_r} = \frac{1}{2H} \int_0^t \overline{T_a}dt \tag{9.1.39}$$

T_M is considered as the mechanical time constant of the generator or the time needed for the nominal torque to accelerate the generator from standstill to nominal speed. Considering $\overline{\omega_r} = 1$ and $\overline{T_a} = 1$, from Equation 9.1.39 we get the following:

$$1 = \frac{1}{2H} \int_0^{T_M} \overline{T_a}dt = \frac{T_{mM}}{2H} \Rightarrow T_M = 2H \tag{9.1.40}$$

In order to make an Amesim implementation possible, the motion equations need to be expressed as a set of differential equations:

$$\frac{d\overline{\omega_r}}{dt} = \frac{1}{T_M} \left(\overline{T_m} - \overline{T_e} - K_D\Delta\overline{\omega_r} \right) \tag{9.1.41}$$

$$\frac{d\delta}{dt} = \omega_0\Delta\overline{\omega_r} \tag{9.1.42}$$

Figure 9.1.9 presents a simplified generator model built in Amesim. The model is not complete, its input has been attached to the output of the turbine model developed previously, and its output needs to be attached to the input of the speed governor [8].

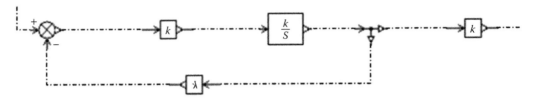

Figure 9.1.9 Amesim model of the generator.

9.1.4   Main characteristics of the high power servovalves

Modern high-power electrohydraulic amplifiers are double-stage electrohydraulic ser-vovalves with built-in interface electronics [9]. The pilot stage is a low-flow proportional valve with very good dynamic performance. The power stage of the amplifier is a hydrau-lic command valve, whose spool is positioned with a closed-loop control system. The valve spool position is measured by a displacement transducer whose output represents the feedback of the control loop. The flow rate of the amplifier can be calculated as

$$Q(x) = c_d A(x) \sqrt{\frac{2\Delta p}{\rho}} \tag{9.1.43}$$

where:
   $Q(x)$ is the flow rate passing through the valve (m³/s)
   $A(x)$ is the distribution window area (m²)
   Δp is the pressure drop on the edges of the distribution window (N/m²)
   c_d is the flow coefficient of the ports
   ρ is the oil density (kg/m³)
   x is the spool displacement (m)

From Equation 9.1.43 results that the shape and area of the distribution windows directly affect the flow-rate characteristics and the maximum flow-rate value passing through the valve for a given pressure drop. If the area of the distribution windows is a nonlinear func-tion of spool displacement, so is the flow characteristic. For example,

$$A(x) = \begin{cases} b_1 x & \text{for } 0 \le x \le x_1 \\ A_1 + b_2(x - x_1) & \text{for } x_1 < x \le x_{\max} \end{cases} \tag{9.1.44}$$

Here, $A_1 = b_1 \cdot x_1$. Figure 9.1.10 shows a typical electrohydraulic amplifier with built-in elec-tronics. The spool of the main stage can be seen in Figure 9.1.11. In the shape of the dis-tribution windows, two areas can be observed: (1) one purely linear, covering about 40% of the maximum spool displacement and (2) a circular one, utilized for high-amplitude flow rates.

   Figure 9.1.12 shows the theoretical flow characteristics of this amplifier with two distinct slopes; the slope change point occurs around 40% of the maximum spool dis-placement. The shape of the static characteristics of an electrohydraulic amplifier varies according to its intended usage.

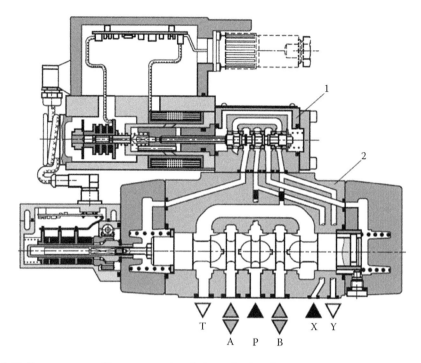

T ▽   △ ▲ △   ▲ ▽ X Y
        ▽ △ ▽
        A P B

Figure 9.1.10 Two-stage nonlinear proprtional electrohydraulic servovalve DN 25. (Source: BOSCH)

Figure 9.1.11 Main spool of the servovalve with the core of the LVDT.

From a control theory standpoint, hydraulic valves are mechanical hydraulic amplifiers that turn spool displacement into flow rate. The steady-state behavior of such a device can be described by the following equation:

$$F\left(p_s, x, Q, P\right) = 0 \qquad (9.1.45)$$

where:

p_s is the supply pressure

x is the spool displacement

Q is the flow rate through the hydraulic motor

P is the pressure drop on the hydraulic motor

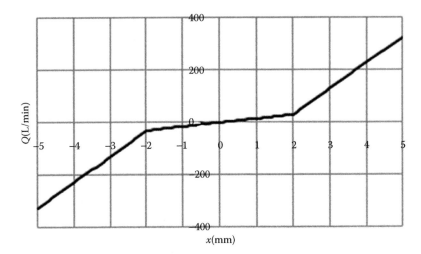

Figure 9.1.12 Theoretical flow characteristics of the amplifier.

The steady-state characteristics of the valve, ignoring cavitation, are

$$Q = c_d A(x) \sqrt{\frac{p_s}{\rho} \left(1 - \frac{x}{|x|} \frac{P}{p_s} \right)}$$ (9.1.46)

For rectangular distribution windows, with two increasing width stages ($b_1 < b_2$), the flow gain around the null point can be determined with the following relations:

$$K_{Qx} = \frac{\partial Q}{\partial x} = \begin{cases} c_d b_1 \sqrt{\dfrac{p_s - P}{\rho}} & x \in [0, x_1] \\ c_d b_2 \sqrt{\dfrac{p_s - P}{\rho}} & x \in (x_1, x_{max}] \end{cases}$$ (9.1.47)

The flow-pressure coefficient is calculated by

$$K_{QP} = -\frac{\partial Q}{\partial P} = \frac{c_d A(x)}{2\sqrt{\rho(p_s - P)}}$$ (9.1.48)

But, in practice, this relation is replaced by another one obtained by experiments.

The values of the flow-gain and the flow-pressure coefficient depend on the supply pressure, spool displacement, and pressure drop on the hydraulic motor, and can be found by experiments only. For closed-loop hydraulic actuators, these values need to be determined around the valve's neutral position ($x = 0$).

For an ideal valve, the K_{QP} coefficient is theoretically 0 around the neutral position, but this is not the case in reality. In literature, various methods and procedures for determining this coefficient can be found. For a real valve, it can be found from the following relation:

$$K_{QP} = -K_{Qx} \cdot \frac{1}{K_{Px}}$$ (9.1.49)

Where K_{Px} represents the pressure-sensitivity coefficient. This value can be taken from the pressure-gain characteristics supplied by the servovalve manufacturer.

9.1.5 Numerical simulation of the dynamic behavior of nonlinear electrohydraulic servovalves

Amesim was used for the numerical simulation of a nonlinear electrohydraulic amplifier with built-in electronics (onboard electronics—OBE). A predefined valve was used for the pilot stage, whereas the power stage was built from independent mechanical components (more detailed approach). The simulation architecture is shown in Figure 9.1.13 [13].

Figures 9.1.14 and 9.1.15 show the system response (flow rate) to a step signal with two-step inputs (1.0 and 10.0 V). The system's response to a step signal depends on the value of the command voltage, but the time constant is around 0.05 s for a supply pressure of $p_s = 10$ bar. This value is much smaller when the valve is supplied at the nominal pressure. The response of the amplifier (displacement and flow rate) to a sine wave with frequency $f = 0.5$ Hz and amplitude $U = 10$ V is shown in Figures 9.1.16 and 9.1.17.

It can be seen that the spool displacement curve follows the sine wave, and the flow curve has discontinuities corresponding to the discontinuities in the function describing the area of the distribution windows.

The frequency characteristics of the amplifier have been determined for the supply pressure $p_s = 100$ bar and two values of the command signal: (1) 0.5 and (2) 10 V. The results are shown in Figures 9.1.18 and 9.1.19. The cutoff frequency of the amplifier depends on the amplitude of the command signal and is cited in the 15–20 Hz interval.

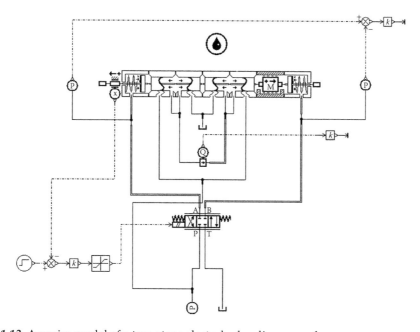

Figure 9.1.13 Amesim model of a two-stage electrohydraulic servovalve.

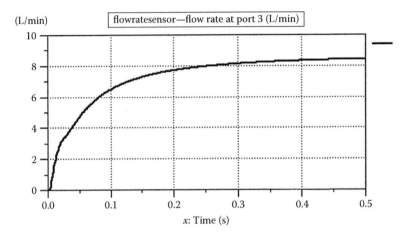

Figure 9.1.14 Flow rate for a small step signal (1.0 V).

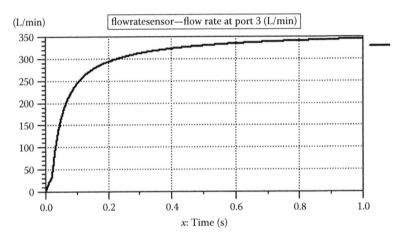

Figure 9.1.15 Flow-rate variation for the nominal step signal (10 V).

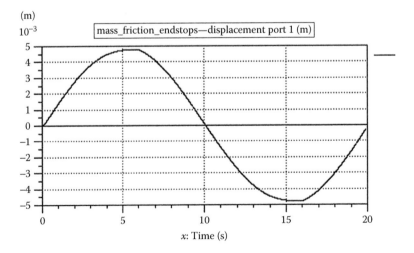

Figure 9.1.16 Displacement for a sine wave signal ($f = 0.05$ Hz and $U = 10$ V).

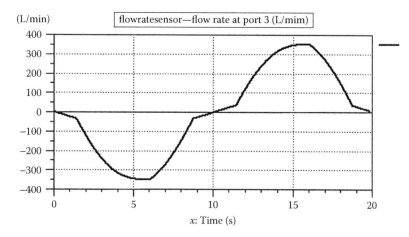

Figure 9.1.17 Flow rate for a sine wave signal ($f = 0.05$ Hz and $U = 10$ V).

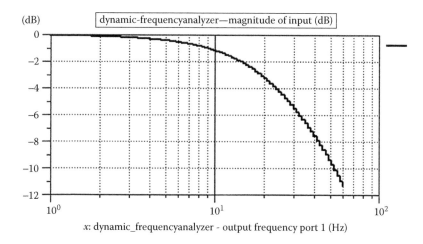

Figure 9.1.18 Frequency characteristic of the amplifier (magnitude).

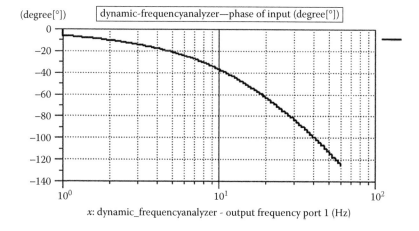

Figure 9.1.19 Frequency characteristic of the amplifier (phase).

9.1.6   Modeling and simulation of a speed governor for high-head turbines with Amesim

Starting from the simplified turbine and generator models presented earlier, the Amesim model from Figure 9.1.20 has been built in order to study the performances of the electrohydraulic servomechanisms as a solution for wicket gate actuators (and later bypass valve) of high-head turbines. The simulations have been focused on the turbine start-up with the generator disconnected from the power grid until the turbine is brought to the synchronous speed. The detailed modeling facilities provided by Amesim have allowed the authors to obtain detailed diagrams regarding the evolution of numerous operation parameters.

In the operation of medium- and high-power hydropower units connected to power grids, there are three main problems:

1. Bringing the generator to the synchronous speed
2. Normal operation, connected to the power grid
3. Identification of malfunctions

When bringing the generator to the synchronous speed, the turbine speed must be increased from zero to the synchronous one in a limited time (60 s). This duration is

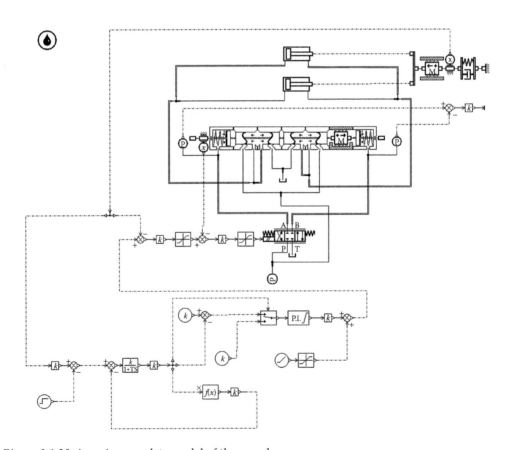

Figure 9.1.20 Amesim complete model of the speed governor.

defined in internal regulations and is one of the most important functional characteristics of the hydropower units. In order to achieve optimum start-up time, the turbine speed is first increased to a speed close to the synchronous speed, and then controlled around the synchronous speed. In order to accelerate the turbine rapidly, the wicket gate is quickly opened to 18%–20% of maximum opening, because the precise control is not needed in this phase. This operation requires a high oil flow rate to the servomotors moving to the wicket gate and, consequently, the valves must be opened enough to provide this flow. Around the synchronous speed, frequency deviations higher than 5–10 MHz are not accepted. This means that the flow coefficient of the valve must have two different values for the above-mentioned two stages, because the amplification factor (gain) of the electronic compensator is constant. Changing the amplification factor of the electronic compensator by software has proven, in practice, to be less reliable than changing the flow coefficient of the valve.

When connected to the power grid, the hydropower unit must maintain a constant speed and must follow the power reference transmitted from the process operator console. The speed is kept constant by the speed governor by rejecting disturbances. If these are small enough (10–20 mHz) then the valve operates in the small flow coefficient area. If the disturbances are large (100–200 mHz), the valve operates in the large flow coefficient area. In addition, changing the power reference implies changing the opening of the wicket gate with a speed imposed by specific regulations. This is also a situation when the valve operates in the maximum flow rate area. Malfunctions (unload) require the closing of the wicket gate with a speed determined by the characteristics of the hydropower unit (closing time is about 6 s–10 s). Consequently, the flow through the servomotors is very large, which means that the valve operates in the maximum flow rate region.

The electrohydraulic servomechanism, as the main object of this simulation, has been modeled with the greatest amount of details at the physical phenomenon level (spool motion, including inertia, flow through the valve, viscous friction etc.), for the other elements (PI controller, turbine, and generator) using a transfer function model only. As a high-level modeling and simulation language, Amesim easily allowed the building of a very detailed model of the electrohydraulic servomechanism without requiring an explicit mathematical model for the system. From the author's experience, the full realistic mathematical models are too complex, hence difficult to implement in simulation software.

9.1.7   Synthesis of the speed governor

The previously presented Amesim model of the complete system (Figure 9.1.20) can serve for the tuning of a PI controller (Figure 9.1.21) according to the following functional principle. From start-up until reaching a speed corresponding to a frequency of 40 Hz, the wicket gate is kept 25% open. Beyond 40 Hz, the PI controller is started that ensures that

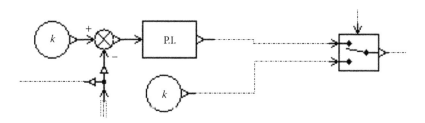

Figure 9.1.21 Amesim PID controller.

the 50 Hz frequency is reached and maintained, as this is needed in order to connect the generator to the power grid.

The tuning of the PI controller has been done through experimental methods. First, the integral coefficient K_I has been set to 0, and simulations have been performed to determine the K_P value for which the system starts to enter self-oscillation (Figure 9.1.22).

Following the simulations for K_P values between 1 and 6, we can observe (Figure 9.1.23) that the system enters self-oscillation for none of the verified values but starts to present significant oscillations when $K_P = 6$.

According to the recommendations from the literature, for optimal stability, the value of K_P should be half of the value where this phenomenon appears, therefore $K_P = 3$ [3,8].

For the chosen value, simulations have been performed in order to determine a value of the integral coefficient K_I that insures an optimal compromise between quickly eliminating the steady-state errors and the overshoot (Figures 9.1.24 and 9.1.25). The analyzed values have been in the interval 0.1–0.7.

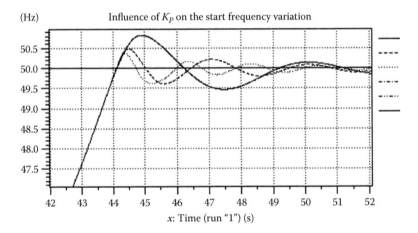

Figure 9.1.22 Batch simulation for various K_P values (1, 3, and 6).

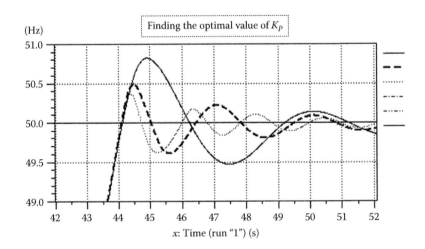

Figure 9.1.23 Batch simulation for various K_P values (detail).

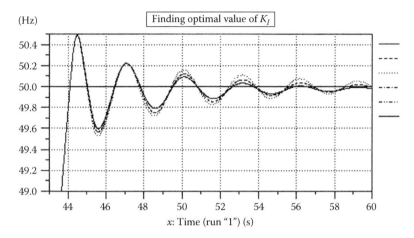

Figure 9.1.24 Batch simulation for various K_I values.

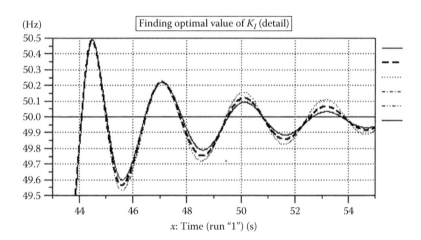

Figure 9.1.25 Batch simulation for various K_I values (detail).

As it can be seen in Figure 9.1.25, the value $K_I = 0.4$ offers second-best overshoot among the simulations, and quickest elimination of steady-state error. The final simulation results for the system with the chosen parameters can be seen in Figure 9.1.26.

9.1.8 Real-Time simulation with MATLAB©/SIMULINK© of a speed governor for high-head turbines

In order to study the Real-Time simulation capabilities of Windows platforms, a laboratory stand for HiL Real-Time simulation of a complete speed governor model has been designed. On the basis of the mathematical models presented previously (Figure 9.1.27), the model was set up in the Fluid Power Laboratory of the Politehnica University of Bucharest, Romania [10–13].

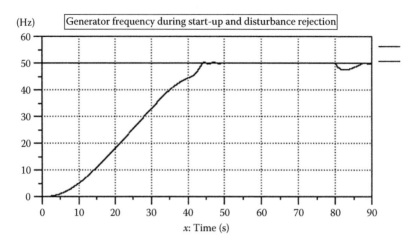

Figure 9.1.26 Generator frequency variation during start-up.

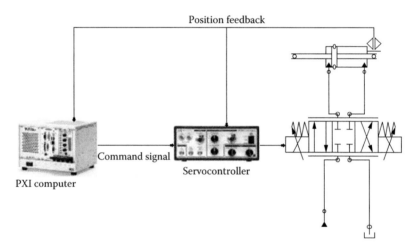

Figure 9.1.27 Principle schematic of the HiL test bench.

The laboratory stand (Figures 9.1.28 and 9.1.29) consists of a National Instruments PXI-1031 machine-running Microsoft Windows XP and MATLAB/SIMULINK with Real-time Windows Target for Real-Time simulation of the model. The data-acquisition board transmits the command signal for the wicket gate generated by the speed governor to an analogic servocontroller connected to an electrohydraulic servomechanism (modeling the wicket gate blade actuators). From the position transducer of the servomechanism, the data-acquisition board receives the wicket gate opening information, which is then provided to the turbine model.

For the speed governor, turbine, and generator, the mathematical models developed previously in Amesim have been used via co-simulation.

The first step is to replace the servomechanism in the Amesim model from Figure 9.1.20 with a SIMULINK interface block (Figure 9.1.30).

This block was created from the Modeling menu > Interface block > Create interface icon. Selected interface type must be SimuCosim and the number of inputs and outputs must be defined accordingly (in this case, one input and one output (Figure 9.1.31).

Figure 9.1.28 Side view of the test stand.

Figure 9.1.29 Front view of the test stand.

Then, a SIMULINK model was created incorporating the Amesim model and the inter-faces to the hardware part of the test stand. In order to do so, MATLAB must be run from within Amesim using the Tools > Start MATLAB command. This way, SIMULINK can be launched normally from MATLAB. This procedure allows access to the Amesim Interfaces library in SIMULINK (Figure 9.1.32). For co-simulation, the block AME2SLCoSim must be used. The complete SIMULINK model can be seen in Figure 9.1.33.

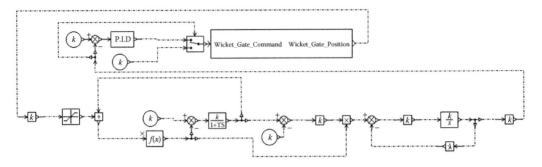

Figure 9.1.30 Co-simulation Amesim model.

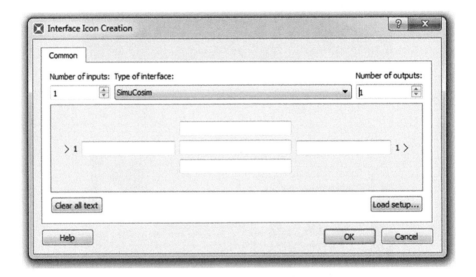

Figure 9.1.31 SIMULINK co-simulation interface block creation.

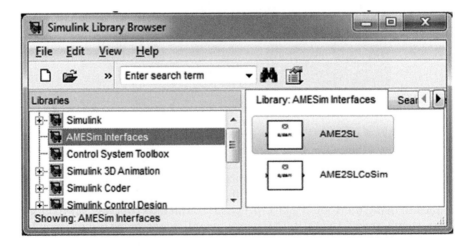

Figure 9.1.32 Amesim Interfaces Library in SIMULINK.

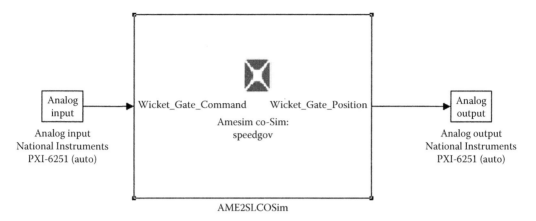

Figure 9.1.33 Co-simulation SIMULINK–Amesim model.

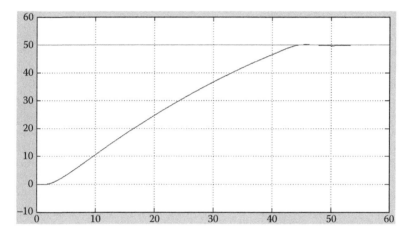

Figure 9.1.34 Generator frequency during start-up: frequency (Hz) versus time (s) from 0 to 50 Hz.

Because the servomechanism that was available for the hardware component of the stand has a very high performance and low inertia used as mechanical input generator, the simulation is not very relevant for determining the performances of the speed governor as the servomechanisms used in the wicket gates are significantly largely opened and have more inertia.

However, multiple tests were performed analyzing the system's behavior from start-up until the synchronous speed is reached. Figures 9.1.34 and 9.1.35 present the variation of the generator frequency during start-up.

9.1.9   Simulation of a redundant position control system with Amesim

Electrohydraulic actuators and control systems are often used in the structure of critical systems in fields such as aerospace, military, or power engineering. In such fields, ensuring the normal operation of the actuating and control systems has a very high priority. Malfunctions in these systems can lead to serious damage and/or loss of life. In the field of aerospace, malfunctions in the actuators of wing control surfaces lead almost invariably to crashes. In addition, in military fields, the ability of aircraft control systems to keep

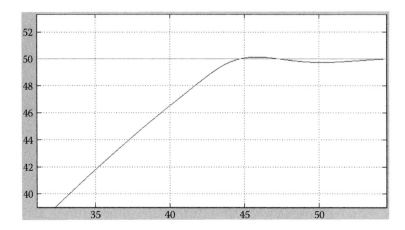

Figure 9.1.35 Generator frequency during start-up (detail): frequency (Hz) versus time (s) from 38 to 50 Hz.

operating even after damage from enemy fire is one of the design requirements. In order to meet such demands, the hydromechanic components and the control logic units are often doubled or even tripled.

The main redundant designs used in the field of hydraulic and pneumatic systems can be split into active redundancy and passive redundancy. In the case of passive redundancy, only one system is working at a given time. When the main system malfunctions, the backup system takes over. The malfunctions are detected by constant supervision of the active system. For this type of redundancy, it is important that the backup system has a greater operational safety than the main system and that the supervision system does not malfunction. These types of systems have several disadvantages, such as the time needed to activate the backup system, the backup system needs to be as large as the main system to accomplish the operating requirements (force, torque, etc.) are met. Active redundancy has two or more systems operating at the same time. The tasks are split between them, the output of the actuators being a sum of all forces. When a malfunction occurs, the damaged system is disconnected and the operation is taken over by the functional system. An example of a passively redundant control system will be built for a simple electrohydraulic servomechanism (Figure 9.1.36). The system's response at a step input can be seen in Figure 9.1.37.

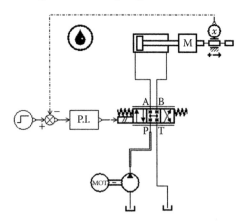

Figure 9.1.36 Basic electrohydraulic servomechanism.

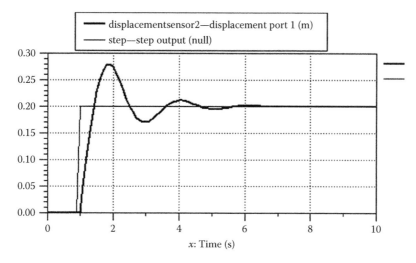

Figure 9.1.37 Typical system response for a step input.

In order to build a passive redundant system, two identical PI controllers will be used, together with a relay that switches to the second PI controller if the first one malfunctions. Malfunctions will be detected by comparing the outputs of the two controllers. The second system will be assumed to have a higher operating safety, a key requirement of passive redundancy, as stated earlier. Therefore, if the outputs of the two PI controllers are different, the main controller is assumed to be malfunctioning.

The redundant system can be seen in Figure 9.1.38. In order to test this redundant control system, several malfunctions will be simulated that are as follows: main controller taken offline (planned or malfunction), position transducer signal to the main controller missing, and incorrect main controller output.

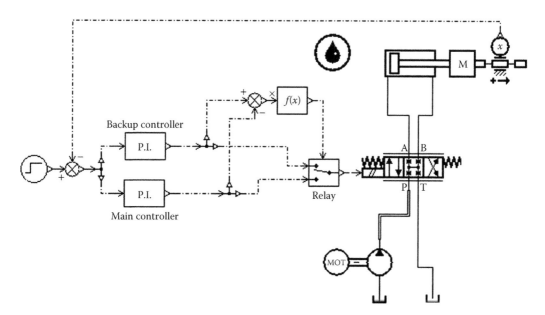

Figure 9.1.38 Redundant control system.

The Amesim model that simulates the main controller going offline at $t = 1.5$ s can be seen in Figure 9.1.39. Figure 9.1.40 shows a comparison of the control signals for the main and backup PI controllers, and Figure 9.1.41 shows the system response.

The Amesim model that simulates the main controller losing the signal for the position transducer at $t = 2$ s can be seen in Figure 9.1.42. Figure 9.1.43 shows a comparison of the control signals for the main and backup PI controllers, and Figure 9.1.44 shows the system response.

The Amesim model that simulates errors in the main controller output starting from $t = 2$ s can be seen in Figure 9.1.45. The output errors are simulated by switching the main

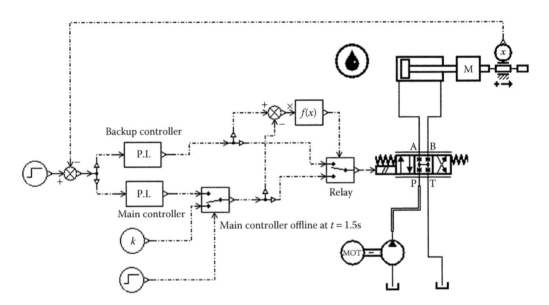

Figure 9.1.39 Main controller going off-line at $t = 1.5$ s.

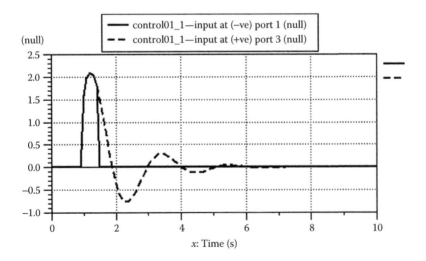

Figure 9.1.40 The control signals of the two PI controllers.

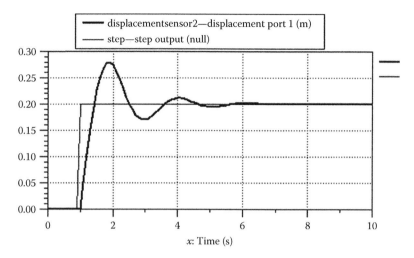

Figure 9.1.41 System response for a step input, including a PID malfunction.

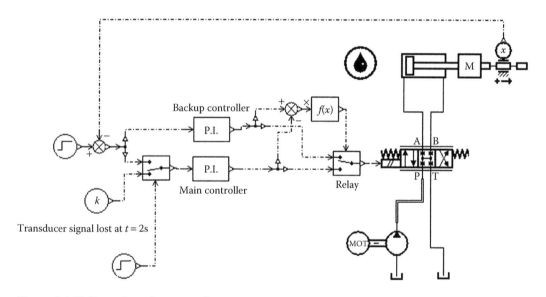

Figure 9.1.42 Transducer lost at $t = 2$ s.

controller output to a third PI controller with different values for the proportional and integral constants after $t = 2$ s. Figure 9.1.46 shows a comparison of the control signals for the main and backup PI controllers, and Figure 9.1.47 shows the system response.

The theoretical results obtained by numerical simulations were validated by systematic experiments performed by the authors in the Fluid Power Laboratory of U.P.B. on a special test bench (Figure 9.1.48), including three PXI controllers from National Instruments, and a controller COMPACT RIO from the same corporation for generating the switching signals. The above-mentioned results were tested in different conditions that can occur in the operation of a speed governor.

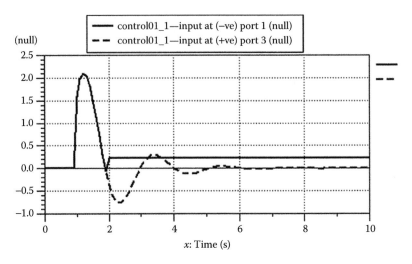

Figure 9.1.43 The control signals generated by the two PI controllers.

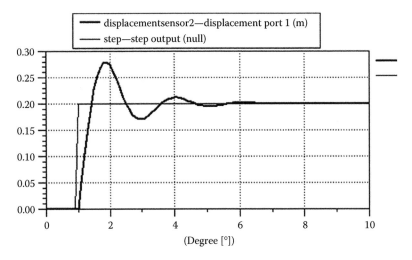

Figure 9.1.44 System response for a step input, including a transducer malfunction.

9.1.10   Conclusion

This section contains a small part of the author's works targeting the change of the old speed governors for the different hydropower stations. The speed governors patented by the authors are running from 2003 without any troubleshot (hydraulic or electronic ones). The use of the highest quality controller from ADWin family together with a low-speed tachogenerator built by the authors around an ABB heavy duty controller solved all the exploitation problems for the SCADA system of the two hydropower units restructured in the frame of Ramnicu Valcea Hydropower Station. The synthesis of the two speed

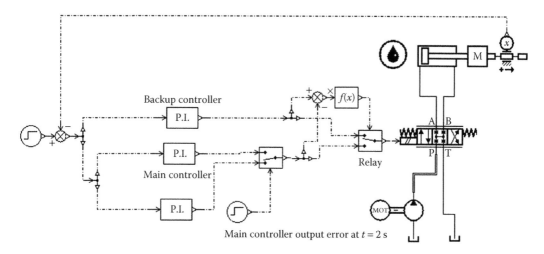

Figure 9.1.45 Main controller output errors after t = 2s.

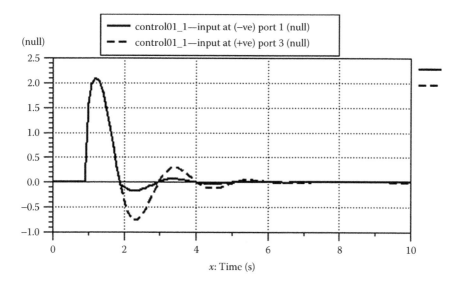

Figure 9.1.46 The control signals of the two PI controllers.

governors, approved by the national authority in the field, was very much aided by the numerical simulation and the Real-Time simulations performed by Amesim.

This section could not be closed without paying a special homage to the memory of a great Swiss power engineer and professor Lucien Borel, head of Laboratory of Thermodynamics and Energetics at Ecole Polytechnique of Lausanne, Lausanne, Switzerland. Writing the first complete book in the world about the control stability of the hydropower systems [15], he claimed the utility of a simulator needed for replacing the huge amount of graphical works that he was obliged to perform for tuning a speed governor for a hydraulic turbine.

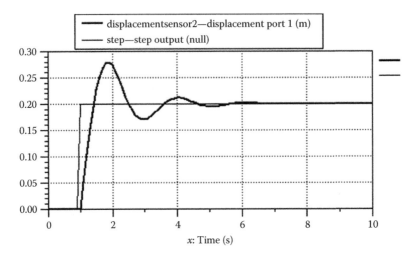

Figure 9.1.47 Real input step system response, including one PID malfunction.

Figure 9.1.48 Test bench for the study of the redundant position control systems in the Fluid Power Laboratory of U.P.B.

9.2 Example of sizing and tuning the speed governors for Kaplan turbines by Amesim

9.2.1 General design options

This part contains a report on a long series of theoretical and experimental activities aiming to create a new type of electrohydraulic digital speed governor for hydraulic KAPLAN turbines (Figure 9.2.1). Computational methods, control software, design problems, and experimental validation are shortly presented [1–4].

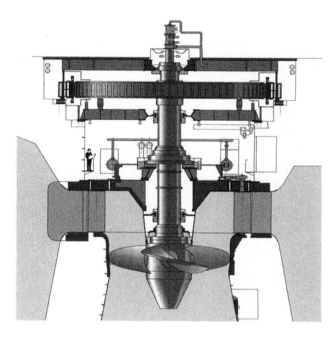

Figure 9.2.1 Medium-sized Kaplan turbine. (Source: VOITH)

From an industrial point of view, the main idea of the new concept is the use of only high-quality industrial electrohydraulic and electronic components, in order to avoid manufacturing difficulties, high prices, and the need of the manufacturer's permanent technical assistance. This target generated a new approach of the design and tuning complex nonlinear speed governors by numerical simulation.

A detailed model of a high-power servomechanism containing a two-stage nonlinear electrohydraulic proportional valve was designed in a simulation environment taking into account the real geometry of the metering spool windows. This way the authors created and patented a new generation of modern speed governors for hydraulic turbines. The whole hydropower unit dynamic behavior was simulated with Amesim, and finally the results were compared to field measurements made by the authors at the hydropower units of Râmnicu–Vâlcea hydropower plant. The simulated and the real responses for different input were nearly identical. The actual performance of the prototype was found in good agreement with the CEI demands.

The variation of speed (by increasing the power) during the operation of a hydropower unit is achieved by adjusting the water flow that passes through the unit with the wicket gates or the injectors (depending on hydropower unit type). High-accuracy electrohydraulic servomechanisms actuate these control elements.

The control power depends directly on the hydropower unit size, type, and the required dynamic performance, which depends on the quality requirements of the electrical power produced by the unit. The static and the dynamic forces that appear during the hydropower unit operation are rather large. The pistons of the hydraulic cylinders have rather large diameters, whereas the oil pressure is usually in the range of 40–63 bars.

The closing time in case of damage of a hydropower unit is usually lower than 10 s, in order to avoid dangerous overspeeds. The requirements regarding the quality of the electrical power provided by hydropower units are extremely strict, and meeting them requires speed governors to react at less than 2 mHz. This performance condition requires very precise positioning of the hydraulic cylinders' rods, keeping a reasonable stability reserve for the speed governor. These contradictory requirements can be satisfied by the aid of a two-stage proportional flow valve (Figure 9.2.2), which has two-slope flow characteristics (Figure 9.2.3) [4].

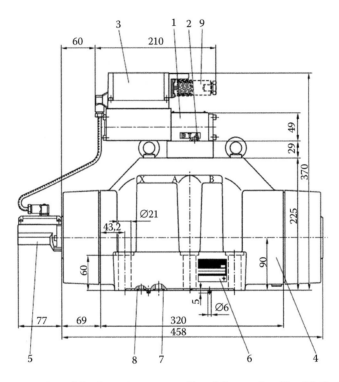

Figure 9.2.2 Last generation of the two-stage proportional flow valve Size 32: 1—hydraulic control valve (pilot stage); 2—adapter plate; 3—OBE industrial servovalve; 4—power stage; 5—LVDT. (Source: BOSCH REXROTH)

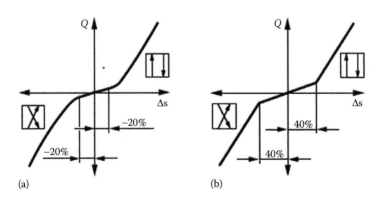

Figure 9.2.3 Flow characteristics of the valve power stage: (a) M type: progressive with fine metering and (b) P type: nonlinear, linear 40%.

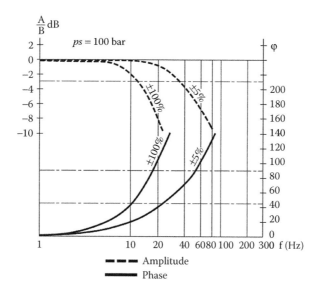

Figure 9.2.4 Proportional valve dynamics for the two-stage proportional flow valve size Size 32. (Source: BOSCH REXROTH)

The valve dynamics is suited for high-speed control process (Figure 9.2.4). However, to optimize such complex governor architecture, any designer always needs a proper mathematical model and a detailed simulation.

9.2.2   Tuning the speed governor

A careful comparative analysis of similar simulation software environments, including MATLAB/SIMULINK from MathWorks [13], Automation Studio from Famic Technologies [14], FluidSIM from Festo [15], and LabVIEW from National Instruments [16], was performed in order to choose a complete simulation environment. Finally, Amesim software produced by LMS IMAGINE Company, now a member of Siemens group, was selected as a current tool [5–8]. The AD Win PRO controller [17] and AD Basic software [18] offered the best combination for implementing the control algorithm studied in Amesim.

This complex software offers numerous advantages: (1) rich library of hydraulic symbols and components, which allows the authors to use existing, proven models for well-known components (valves, cylinders) and (2) ability to simulate different parts of the system at different levels of complexity, which allows the authors to model different parts of the system at different levels of detail, as required. Amesim models are fully compatible with LabVIEW for Real-Time and HiL simulations, that can be imported in LabVIEW and connected to a Real-Time or HiL simulation system. The power stage of the flow valve, which has the greatest influence on the dynamic behavior of the system, was modeled in deep detail, at the physical process level, using the Hydraulic Component library of Amesim in order to obtain access to all the internal variables. The pilot stages, the cylinders, the pressure sources, and so on have been modeled at a more concise level, using predesigned blocks from the Amesim Hydraulic and Mechanical libraries. Ultimately, the turbine-generator assembly has been modeled by a transfer function, because the internal variables are not particularly important for the current simulation. The simplest Amesim model developed by the authors for the speed governor can be seen in Figure 9.2.5.

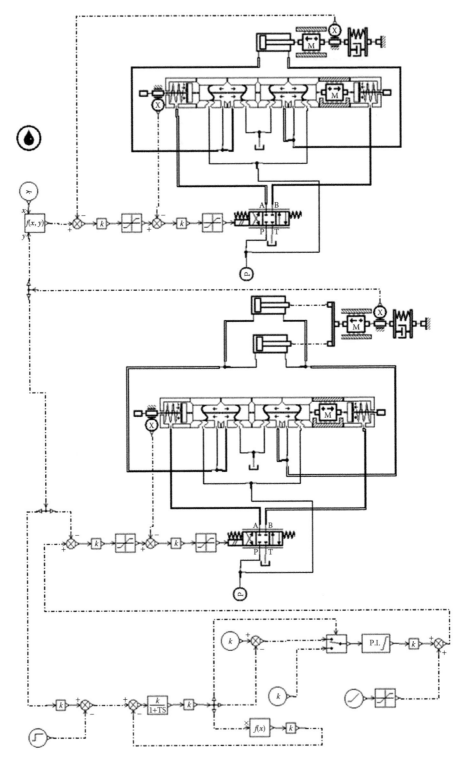

Figure 9.2.5 Amesim simulation network of the governor, including the second servomechanism that controls the runner blades position.

Two types of simulation have been performed using this model. First, the authors have simulated a normal start-up procedure, starting from zero speed, and closed wicket gates, until nominal frequency (50 Hz) is reached at the generator output (Figure 9.2.6). The second simulation takes the start-up procedure from the previous step and introduces a 10% power disturbance at 80 s from the simulation start, which is approximately 30 s after the start-up sequence has completed and the generator frequency has stabilized at 50 Hz (Figure 9.2.7). The evolution of some important parameters as the flow through the power stage of the valve and the main valve stage spool displacement are shown in Figures 9.2.8 and 9.2.9.

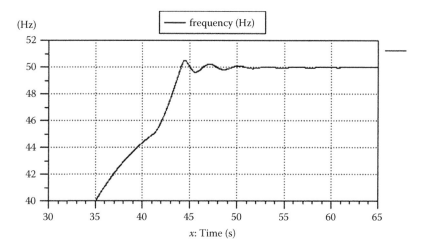

Figure 9.2.6 Frequency variation at hydropower unit start (detail).

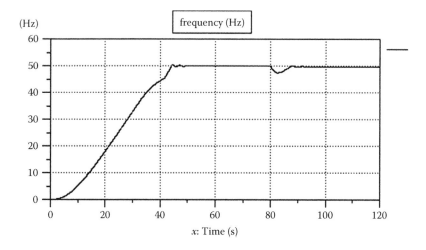

Figure 9.2.7 Frequency variation during hydropower unit start.

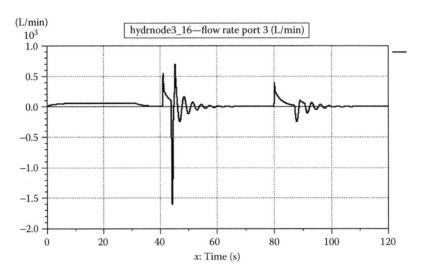

Figure 9.2.8 Flow through the power stage of the valve.

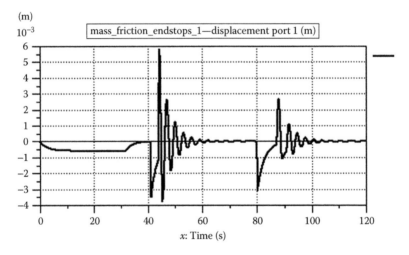

Figure 9.2.9 Main valve stage spool displacement.

9.2.3   *Experimental validation of the design*

The results obtained in the previous chapter by numerical simulation have been compared with experimental data collected by the authors while working on a similar speed governor for the hydropower units of Ramnicu–Valcea hydropower plant. The research team of the Fluid Power Laboratory from the Power Faculty of the Politehnica University of Bucharest, Romania has designed a digital speed governor patented in Romania by N. Vasiliu and C. Călinoiu in 2003, and set up on both hydropower units of Ramnicu–Valcea power plant [1,2]. The functional structure of the new concept is shown in Figure 9.2.10.

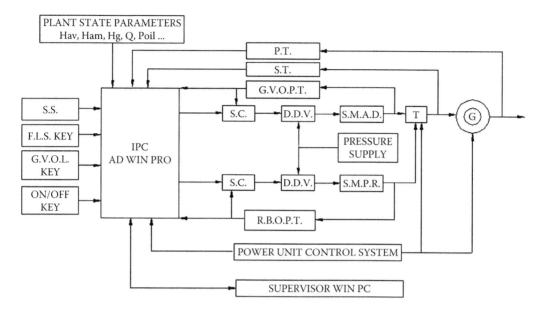

Figure 9.2.10 Speed governor functional structure.

The functional diagram of the speed governor includes the following components and systems: SS—synchronizing system; FLS KEY—frequency/load setting key; GVOL KEY—guide vanes opening limiter key; PT—power transducer; ST—speed transducer; GVOPT—guide vanes opening position transducer; SC—analog servocontroller; DDV—direct drive valve; SMAD—guide vanes servomotor; SMPR—runner blades servomotor; RBOPT—runner blades opening position transducer; and T—hydraulic turbine; G—electric generator.

The hydraulic diagram of the speed governor is shown in Figure 9.2.11.

The hydraulic control interface of the speed governor is shown in Figure 9.2.12.

The governor electronic digital control panel view (outside and inside) is shown in Figures 9.2.13 and 9.2.14.

The governed hydropower units are 23.5 MW Kaplan turbines. The control systems have the following characteristics: wicket gate servomechanism stroke $Y_{AD} = 657$ mm and rotor blade servomotor displacement $Y_R = 190$ mm. The governor can control automatically hydropower unit start-up and stop in normal or damage condition, keeping the hydropower unit generator to 50 Hz frequency; the generator can operate insulated or connected to the national power grid.

The hydraulic panel of the speed governors uses high-speed two-stage proportional flow valves manufactured by Bosch Group. The turbine control system acts through a pair of servomechanisms: (1) one for the wicket gate (AD) and (2) another for the rotor blades (PR); each servo contains an analog PT1 servocontroller, very reliable in heavy environments. The main control loop of the speed governor includes an original low-speed transducer, also designed by the authors for interacting with the hydropower unit SCADA system.

The following measurements have been performed after long time of operation under normal working condition with no events using first the Test Point and then the LabVIEW software for data acquisition. The following parameters have been measured: (1) frequency, (2) generator active power, (3) wicket gate opening, (4) wicket gate servomotor command signal, (5) rotor blade position, and (6) rotor servomotor stroke command

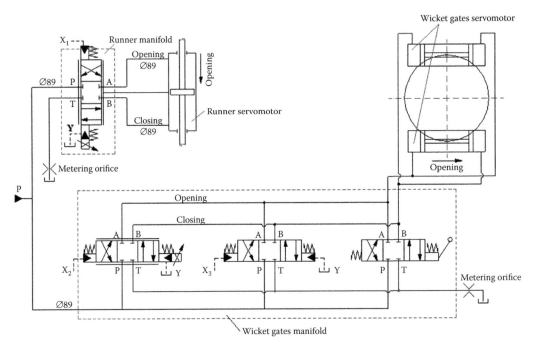

Figure 9.2.11 Speed governor hydraulic diagram.

Figure 9.2.12 The hydraulic control interface of the speed governor.

signal (Figure 9.2.15). If the power-grid frequency is constant, according to CEI 61362 [9], the parameters characterizing hydropower unit start-up are: the synchronization band—the interval between 0.995—1.010 of grid frequency; $t_{0.8}$—the time in which the turbine reaches 80% of the nominal speed; t_{SR}—time in which the system is ready for synchronization; t_S—time after which the generator is coupled to the grid (the system is ready for

Figure 9.2.13 Front panel of the electronic digital system of the speed governor.

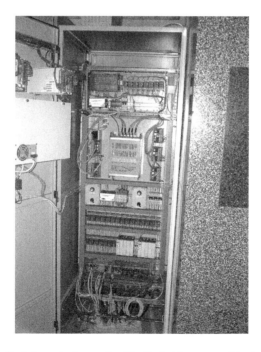

Figure 9.2.14 Governor digital control panel with digital low speed tachogenerator (ιιρ), AD Win PRO controller, manual control blocks, relays block, 220Vcc/24Vcc intelligent sources, and automatic safety line.

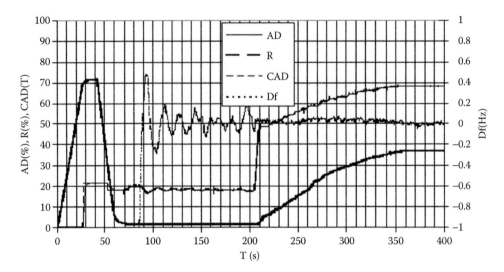

Figure 9.2.15 Evolution of parameters during the unit start: AD—wicket gates opening; R—runner opening; CAD—order for wicket gates opening; and Df—frequency error.

synchronization when the frequency variation speed in the synchronization band is less than 0.003 s^{-1}); and the value of $t_S/t_{0.8} = 1.5$–5.0 (larger values are normal for smaller power hydropower units).

After reaching 80% of nominal speed, the speed variation is controlled mainly by the governor, with the goal of preparing the unit for synchronization in an acceptable time. Figure 9.2.16 shows the evolution of the parameters mentioned earlier during unit start-up, running in the synchronization band and connecting to power grid. After data analysis, the following values have been obtained: frequency variation speed $\Delta f/f_0/\Delta t = 0.0006$ s^{-1} < 0.003 s^{-1}; $t_{0.8} = 35$ s; $t_{SR} = 63$ s; $t_S = 175$ s; $t_S/t_{0.8} = 5.0$.

The next two tests have been performed to study the behavior of the system around the minimum power (10 MW) and nominal power (22 MW) as follows: 200 mHz step

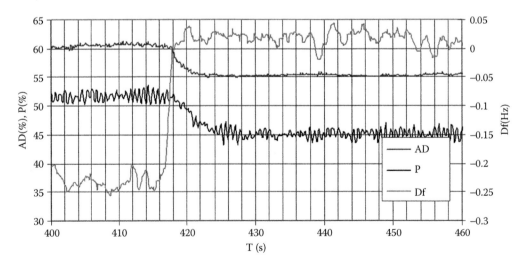

Figure 9.2.16 Experimental results for static droop around the power $P = 10$ MW.

signals have been applied to the frequency reference, both up and down. This is the so-called *static droop test*.

For the minimum power level, a 200 mHz step signal is applied up, followed by returning to the initial frequency (frequency variation 50.00–50.20–50.00Hz), and for the maximum power level, a 200 mHz step signal is applied down, followed by returning to the initial frequency (frequency variation 50.00–49.80–50.00 Hz).

For $P = 10$ MW, the following values have been obtained: wicket gate opening changes for the up frequency step from 56% to 60%; for the down frequency step—from 60% to 56%; wicket gate stabilization time for the up frequency step: 10 s < 30 s according to UCTE standards; wicket gate stabilization time for the down frequency step: 10 s < 30 s according to UCTE standards; and time to reach 75% of the stable value, for the up frequency step: 3 s < 15 s according to UCTE standards; time to reach 75% of the stable value, for the down frequency step: 3 s < 15 s according to UCTE standards. The above-mentioned results are shown in Figure 9.2.16.

For $P = 22$ MW, the following values have been obtained: wicket gate opening changes for the down frequency step: from 91% to 86%, for the up frequency step from 86% to 91%; wicket gate stabilization time for the up frequency step: 10 s < 30 s according to UCTE standards; wicket gate stabilization time for the down frequency step: 10 s < 30 s according to UCTE standards; and time to reach 75% of the stable value, for the up frequency step 5 s < 15 s according to UCTE standards; time to reach 75% of the stable value, for the down frequency step: 5 s (Figure 9.2.17).

The last relevant test has identified the speed governor's behavior in case of unload, and has been performed in the following conditions: initial power—$P = 15$ MW; initial wicket gate opening—68%. Figure 9.2.18 shows the evolution of the measured variables during this test <15 s according to UCTE standards.

After data analysis, the following values have been obtained: maximum speed after unload—$n_{max} = 121.2$ rev/min (60.6 Hz); minimum speed after unload—$n_{min} = 92$ rev/min (46 Hz); stabilization time (time after the speed variation compared to the idle speed is less than 1%)—$t_E = 37$ s; time to reach maximum speed—$t_M = 5.2$ s; ratio $t_E/t_M = 37/5.2 = 7.1$; and $n_{min}/n_0 = 46/50 = 0.92$. CEI 61362 standard allows for n_{min}/n_0 values between 0.85–0.95 and for t_E/t_M—values between 2.5 and 10. The fulfillment of all these quality criteria proves that the new governor ensures normal hydropower unit behavior in any case of sudden unload.

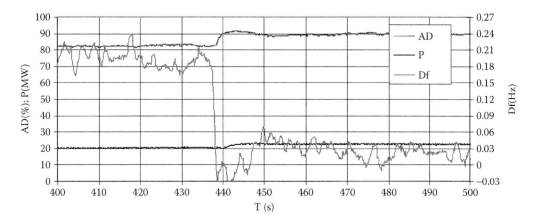

Figure 9.2.17 Experimental results for static droop around $P = 22$ MW.

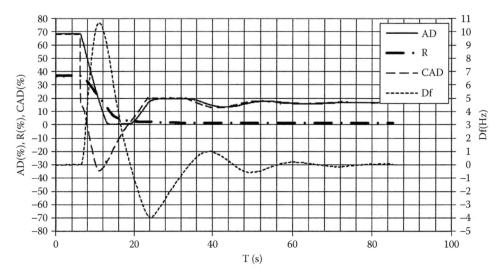

Figure 9.2.18 Dynamic response of the speed governor for a partial shutdown while remaining connected to the system.

9.2.4   Conclusion

The governor prototype was first tested in the Fluid Power Laboratory of the Politehnica University of Bucharest, Romania (Figure 9.2.19) using an original HiL test bench [2], which includes hardware and software components similar to the industrial system. The loads applied to the hydraulic cylinders were included in the RTS models.

The first set of experiments was devoted to the calibration of all the components. The second step was used for testing the PID controller for different operation regimes. A long series of researches were carried out on the algorithm needed by the low-speed tacho-generator to cover all the turbine exploitation situations. The compatibility between this important device and the hydropower unit SCADA system is a very complex problem to solve in only laboratory conditions.

An important chapter of the design was devoted to the electromagnetic compatibility of all the electronic components with the power lines. After this long lab test period, the first governor was set up in the hydropower plant. The first governor was fully tested for a long period according to the international standards [3]. All the official quality indices were accomplished in normal conditions. After many years of normal operation, the main conclusion is a positive one: The new type of electrohydraulic speed governor can be implemented in new or refurbished hydropower plants.

All the design, test, and identification stages of this project pointed out that Amesim provided a strong solver and numerical core for transient simulation. As modeling a complex multiphysics system is not the main objective of engineers, it is important to have tools and interfaces that accelerate and optimize the design. From this point of view, Amesim is a complete software perfectly adapted for model creation and deployment. The wide field of application is continuously extended by the users or serves as a strong inspiration source [10–16].

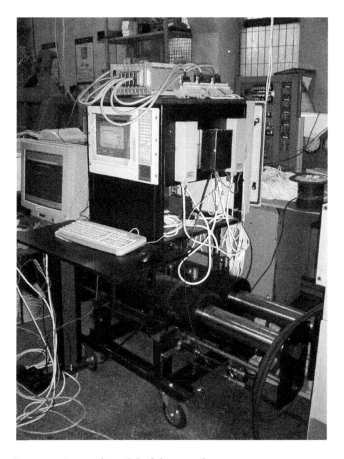

Figure 9.2.19 The first experimental model of the speed governor.

Section 1: Bibliography

1. Vasiliu, N. Contributions to the hydrodynamics of the radial-axial turbomachinery runners. PhD Thesis, University POLITEHNICA of Bucharest, Romania, 1976.
2. Anton, I. *Hydraulic Turbines*, Facla Publishing House, Cluj, 1979 (in Romanian).
3. Kundur, P. *Power System Stability and Control*, McGraw Hill, New York, 1994.
4. Mihoc, D. *Automatized Power Systems*. E.D.P., Bucharest, 1978 (in Romanian).
5. Voros, N.G., Kiranoudis, C.T., Maroulis, Z.B. Short-cut design of small hydroelectric plants. *Renewable Energy*, 19(4): 545–563, 2000.
6. Nicolet, C., Avellan, F., Simond, J.J. Hydro acoustic modelling and numerical simulation of unsteady operation of hydroelectric systems, LMH, LME, EPFL, 2007.
7. Lucero, T.L.A. Hydro turbine and governor modelling, Master Thesis, Norwegian University of Science and Technology, Trondheim, Norway, 2010.
8. Drăgoi, C. Modelling, Simulation and identification of the fast-electrohydraulic driving systems, with applications to the governing hydraulic turbines. PhD Thesis, University POLITEHNICA of Bucharest, Romania, 2001.
9. Călinoiu, C., Vasiliu, N., Irimia, C., Mihalescu, B. Numerical simulation of the operation of the servo systems with nonlinear servovalves. *6th International Conference on Energy and Environment CIEM*, 2013.

10. Ion Guă, D. D. Researches on the RTS of the hydraulic and pneumatic control systems. PhD Thesis, University POLITEHNICA of Bucharest, Romania, 2008.
11. Puhalschi, R., Vasiliu, D., Ion Guă, D. D., Mihalescu, B. 2012. Concurrent engineering by hardware-in-the-loop simulation with R-T workshop, *18th European Concurrent Engineering Conference–ECEC'2012, JW Marriott Hotel*, Bucharest, Romania, April 18–20, pp. 27–31, EUROSYS-ETI Publication, 2012.
12. Mitroi, M.A. Researches on the real-time modelling and simulation of the electrohydraulic control systems. PhD Thesis, University POLITEHNICA of Bucharest, Romania. 2013.
13. Vasiliu, D. Researches on the transients from the servopump and servomotors of the hydrostatic transmissions. PhD Thesis, University POLITEHNICA of Bucharest, 1997.
14. www.voith.com
15. Borel, L. Stabilité de réglage des installations hydroélectriques. Dunod et Editions Payot, Lausanne, Paris, 1960.
16. www.fcavalves.com/valvula-de-cono-fijo

Section 2: Bibliography

1. Vasiliu, N., Călinoiu, C. Electrohydraulic digital speed governor for hydropower units. Romanian Patent no.120101, 2003.
2. Vasiliu, N., Calinoiu, C. Technical specifications of the speed governors set up in Ramnicu Valcea hydropower station. University POLITEHNICA of Bucharest, Romania, 2003.
3. Călinoiu, C., Negoiţă, G. C., Vasiliu, D., Vasiliu, N. Simulation as a tool for tuning hydropower speed governors. *ISC'2011*, Venice, Italy, 2011.
4. BOSCH REXROTH AG. 4WRL 10…35, RE29086, 2009.
5. Imagine. Numerical challenges posed by modeling hydraulic systems. Technical Bulletin 114, 2001.
6. Lebrun, M., Vasiliu, D., Vasiliu, N. Numerical simulation of the fluid control systems by AMESim. *Studies in Informatics and Control with Emphasis on Useful Applications of Advanced Technology*, 18(2): 111–118, 2009.
7. LMS INTERNATIONAL. Advanced Modeling and Simulation Environment, Release 13 User Manual, Leuven, Belgium, 2012.
8. Popescu, T.C. et al. Numerical simulation–a design tool for electro hydraulic servo systems. In *Numerical Simulations, Applications, Examples and Theory*, INTECH PRESS, Zieglergasse 14, 1070 Vienna, Austria, 2011.
9. Union for the Co-ordination of Transmission of Electricity, *UCTE Operation Handbook*, Amsterdam, the Netherlands, 2002.
10. www.voith.com
11. www.boschrexroth.com
12. www.siemens.com
13. www.mathworks.com/products/simulink
14. www.famictech.com
15. www.festo.com
16. www.ni.com
17. www.adwin.de
18. www.analog.com

chapter ten

Numerical simulation of the fuel injection systems

10.1 Numerical simulation of common rail injection systems with solenoid injectors

10.1.1 Structure of the common rail fuel injection systems

The main features of the modern mechatronic concept of *common rail fuel injection systems* (CRFIS) are illustrated in this section using an injection system for a diesel engine [1–10] working with pressure in the range of 1000–2000 bar (Figures 10.1.1 and 10.1.2). The same concept has been used successfully promoted for gasoline engine injection systems, with pressures between 100 and 200 bar. The structure of these systems is almost identical to that of the diesel injection systems, except the number of stages of the injectors (1).

From a hydraulic point of view, a solenoid injector (Figure 10.1.3) is a two-stage electro valve with a smooth and stabile steady-state characteristic connecting the energizing time and the fuel injected quantity (Figure 10.1.4).

The overall dynamic stability of the engine is improved by the aid of an electrohydraulic proportional metering valve with the quasi-linear steady-state characteristic from Figure 10.1.5.

The time correlation between the solenoid current (A), the injection rate (g/s), and the needle lift (1/100 mm) for two series of injectors is shown in Figure 10.1.6. The good following accuracy of the digital input signal by the solenoid allows 4–9 partial injections on every piston stroke. The optimized injection map (Figure 10.1.7) creates the possibility of a significant reduction of the pollutants from the exhausted gases, and improves the engine power performances, helped by the huge progress in the field of automotive Electronic Control Units, running embedded software.

10.1.2 Simulation of a single injector in ideal conditions

The optimization of a common rail injection system needs a deep knowledge of the individual behavior of an injector running in ideal conditions in the first stage: constant environmental temperature, constant supply pressure from the common rail, and so on. The wide range of the operation temperature needs a wide study of all the system components. The injectors have to be tested in a refrigerating room at low temperatures, and in the hot room at extreme operating temperature. This huge amount of preliminary works lead to the optimal choice of all dimensions, clearances, and finally—to the controller tuning. Due to the high-operation pressure, the fuel compressibility becomes a dominant property in all the injection systems together with the fluid viscosity. Both properties are leading to cavitation problems, regarding especially the metering small orifices. The fatigue of

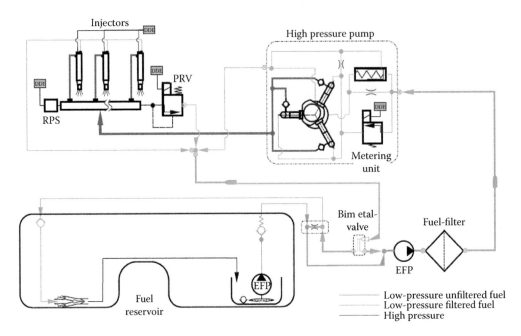

Figure 10.1.1 Structure of a typical common rail fuel injection for diesel engines.

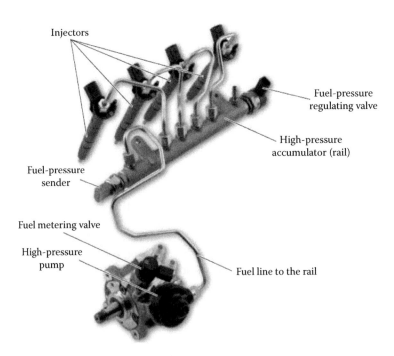

Figure 10.1.2 Overall view of a typical common rail system injection for diesel engines.

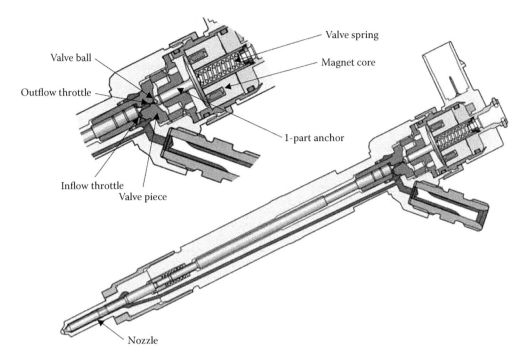

Figure 10.1.3 Main section of a solenoid injector (EURO 4).

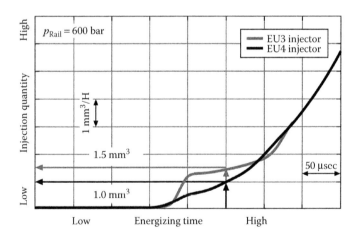

Figure 10.1.4 The steady-state characteristics of a solenoid injector.

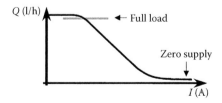

Figure 10.1.5 Steady-state characteristic (flow current) of the fuel metering valve.

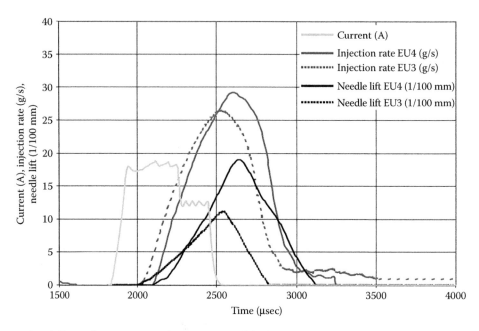

Figure 10.1.6 Typical parameters of a common rail fuel injection sequence.

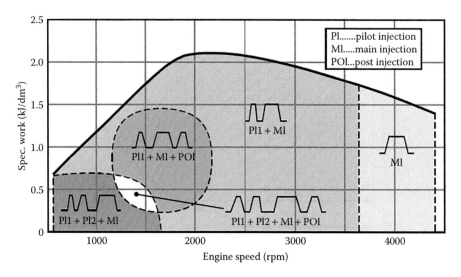

Figure 10.1.7 Typical optimized injection map of a common rail fuel injection system.

materials is also an important research field. The numerical simulations *in batch* can cut a lot of time from this long process. The schematic layout of a common rail system presented in Figure 10.1.8, suggests the above steps in the development of a new system for a given engine. The simulation model is shown in Figure 10.1.9. The next figures (10.1.10 through 10.1.22) can create a realistic image of the operation parameters of different injector electrohydraulic components.

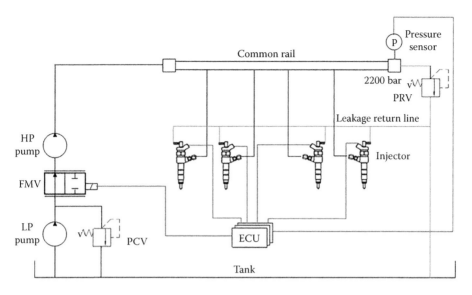

Figure 10.1.8 Schematic layout of the injection system.

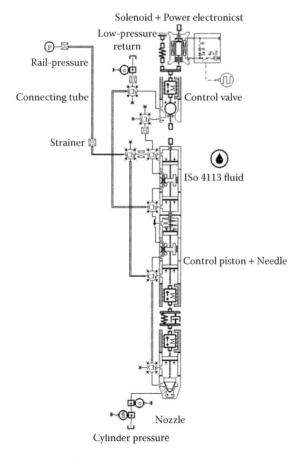

Figure 10.1.9 Simulation model of a solenoid injector constant pressure supplied.

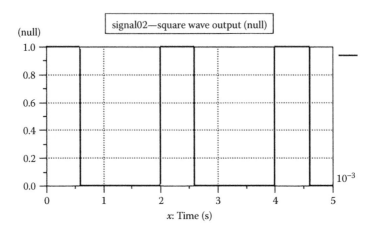

Figure 10.1.10 The input relative signal applied to the solenoid driver of the injector.

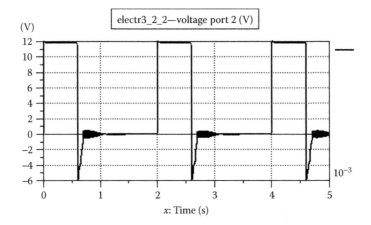

Figure 10.1.11 Voltage applied to the solenoid coil.

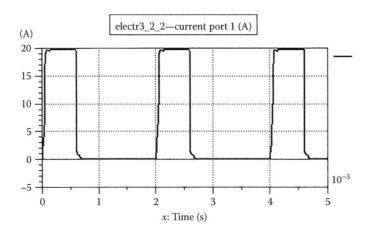

Figure 10.1.12 Current passing through the coil of the solenoid.

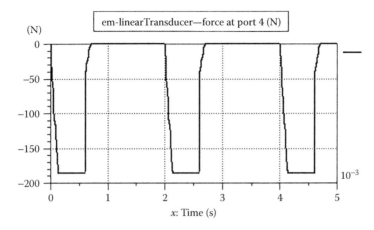

Figure 10.1.13 Force developed by the solenoid core.

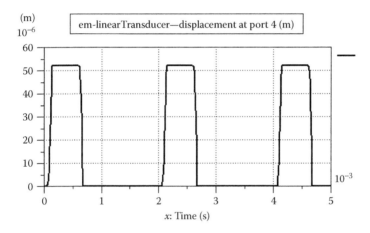

Figure 10.1.14 Solenoid core displacement.

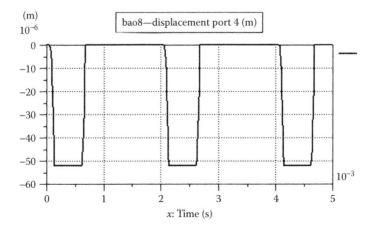

Figure 10.1.15 Pilot valve displacement.

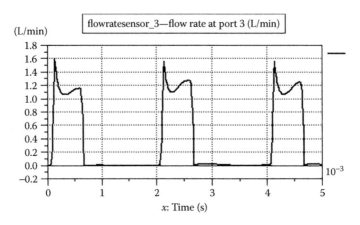

Figure 10.1.16 Flow rate passing through the pilot valve.

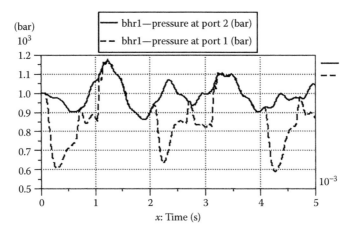

Figure 10.1.17 Pressure drop on the restrictor placed at the input of the injector and the control chamber of the needle motion.

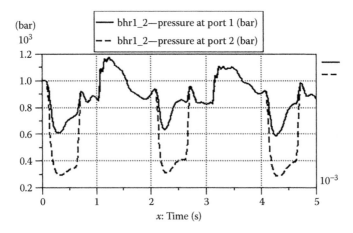

Figure 10.1.18 Pressure drop on the restrictor placed between the input of the injector and the input of the pilot valve.

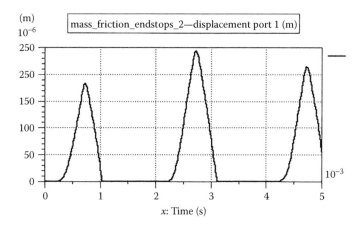

Figure 10.1.19 Displacement of the needle of the injector.

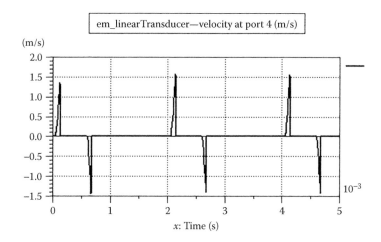

Figure 10.1.20 Speed of the needle of the injector.

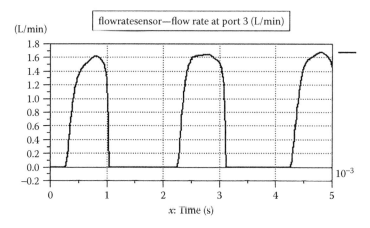

Figure 10.1.21 Flow rate delivered by the injector to the engine cylinder.

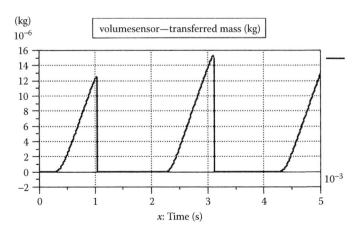

Figure 10.1.22 Mass introduced by the injector in the engine cylinder.

10.1.3   Using the Discrete Partitioning Technique in high-speed simulation of the common rail fuel injection systems

A successful tuning of a CRFIS needs a huge amount of simulation sessions, which alternate with software validation by HiL simulations. A simple way of reducing the overall computation time is the Discrete Partitioning Technique (DPT). The principle of this procedure is to divide the complete injection system into smaller subsystems and perform a co-simulation between the master and slave systems. This allows to each subsystem to be solved independently from one another. Each subsystem has its own integrator. Consequently, the time steps are typical of the local subsystem only. The division of the complete set of equations in subsystems having smaller Jacobians that saves time under the condition of using a high-speed bus. This takes advantage of multicore PCs since each subsystem runs independently on different cores.

The following simulations present the Amesim model of a typical Diesel CRFIS with a fuel-metering valve (FMV) placed on the suction side of the high-pressure pump. Such complex hydraulic system having high dynamics requires extremely long CPU times if a detailed complete modeling is needed. The aim of this example is to show the effectiveness of the DPT technique in terms of CPU time reduction and accuracy of the results. The high-pressure CRFIS modeled in this example from the language tutorial is constituted of the following components:

- A Common Rail System (CRS) with a pressure sensor
- Four solenoid injectors controlled by an electronic control unit (ECU)
- A low-pressure pump (LPP), including a pressure control valve (PCV)
- A high-pressure pump (HPP) having three radial pistons driven by an eccentric shaft
- A FMV at the suction side of the HP pump controlled by a PWM signal as a function of the error between the current pressure in the CR and the target pressure (a PI-type control)
- A pressure-relief valve (PRV) placed on the common rail, which opens in case of any overpressure in the system due to any malfunction. Figure 10.1.8 shows a schematic layout of the injection system. In order to facilitate the concept understanding, this numerical integration network (Figure 10.1.23) calls several other standard

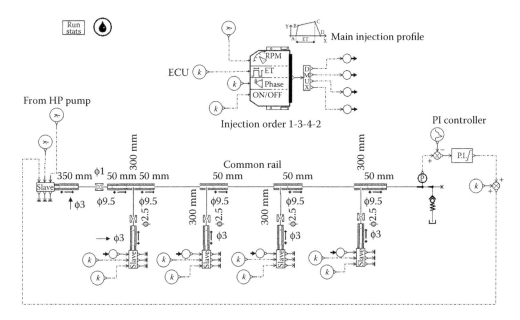

Figure 10.1.23 Overall simulation network used for Discrete Partitioning integration.

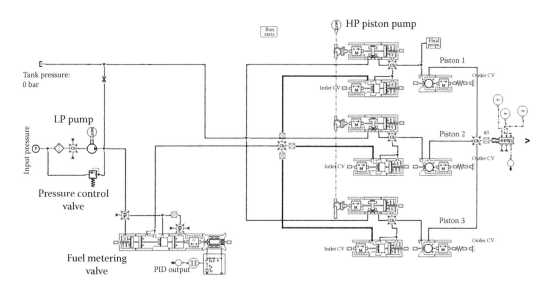

Figure 10.1.24 Simulation model of the high-pressure pump.

simulations: the HPP with regulation (Figure 10.1.24), the solenoid injector with connecting pipes (Figure 10.1.9), and the ECU. The co-simulation needs the following components in the Amesim sketch: special interface blocks available in the Discrete Partitioning library; CFD-1D lines, junctions, and components available in the Hydraulic library. The line sub model C:\libhydr\doc\html\submodels\ HLGCENTER00.html uses an internal solver to evaluate the 1D Navier–Stokes equation: the two-step Lax–Wendroff numerical scheme.

The DP technique is an effective solution to model in details the full dynamics of the complete assembly of complex hydraulic systems, such as an injection system *with reasonable CPU time.*

The following figures show some simulation results obtained with the following operating conditions: crankshaft, HP pump, and LP pump angular speed: 2000 rpm; set pressure point in the common rail: 1500 bar; energizing time of the injector's solenoid: 1 ms. The CPU time for 1 engine cycle is less than 10 min on a PC with 2 cores. This value depends very much on the processor and the number of processes running in parallel. A simple PI controller has been set up in order to regulate the desired set point of pressure in the common rail. The important low-frequency oscillations during the initial transient can be reduced with an improvement of the precontroller and the controller. The main results of the simulation experiments are presented in the following figures (10.1.25 through 10.1.32).

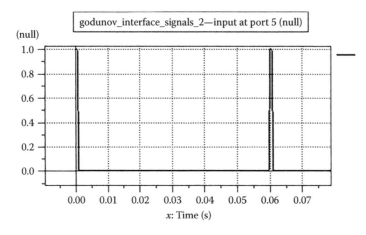

Figure 10.1.25 Sequence of input signals for controlling the injections needed to an engine for running at 2000 rev/min.

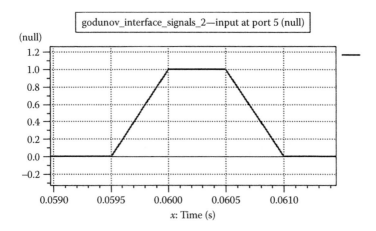

Figure 10.1.26 Relative input signal generating a fuel injection.

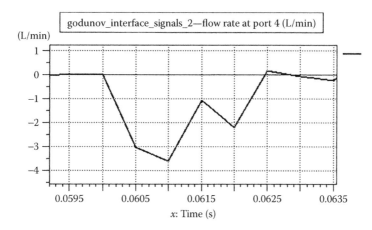

Figure 10.1.27 Typical flow rate variation during an injection.

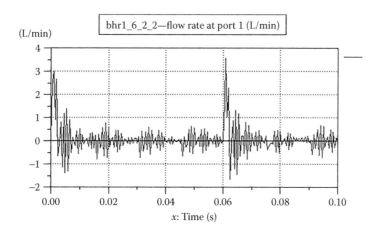

Figure 10.1.28 Flow rate variation at the input of the injector pipe in the first 100 ms.

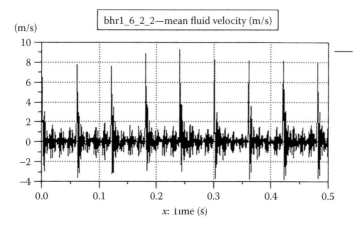

Figure 10.1.29 Mean flow velocity variation in the input pipe of an injector.

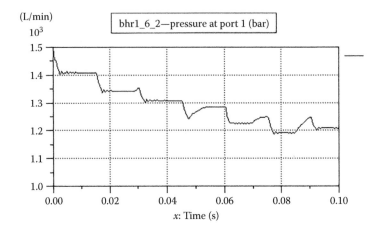

Figure 10.1.30 Pressure variation at the input of the common rail in the first 100 ms.

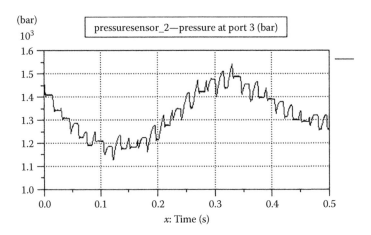

Figure 10.1.31 Pressure variation at the input of the common rail in the first 500 ms.

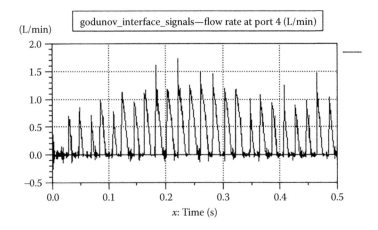

Figure 10.1.32 Flow rate variation at the input of the common rail in the first 500 ms.

10.1.4 Conclusion

The study of the whole system reveals the important role of the high number of nonlinear components, leading to a *rain* of pressure waves that can strongly disturb the overall steady state of the engine operation. The length and stiffness of all the pipes become very important from this point of view. *Nothing is happening in a smooth manner inside the system!*

The simulation results show high precision of the injection timing, rate, and duration offered by a two-stage solenoid actuated common rail diesel fuel injection systems. The Discrete Partitioning integration method promoted by Amesim can successfully replace a strong and expensive multicore computer. The huge number of differential and algebraic equations describing the fuel automotive fluid control systems (more than 272 equations) can be easy integrated by a PC's network with a minimum effort of dividing the whole simulation network in real independent subsystems well connected from a physical point of view. This computational tool can sustain the development of the new direct-drive piezoinjectors with a real potential for improving the fuel efficiency of the high-performance direct injection gasoline engines [10].

10.2 Dynamics of the piezoceramic actuated fuel injectors

10.2.1 Progress elements in fuel injection

Actuators based on piezoelectric ceramic materials prime movers (or piezoactuators) are finding broad acceptance in applications where precision motion and/or high-frequency operation is required. Piezoactuators can produce smooth continuous motion with resolution levels at the nanometer and the subnanometer level. This property makes them useful in precision positioning and scanning systems. The very fast response time, wide operating bandwidth, and high specific force may be beneficial for applications in fluid valve control, optical scanning, vibration isolation, and precision machining [1]. With proper design, piezoactuators have performance attributes and properties that can be valuable in precision positioning, vibration control, and scanning applications. Smooth, precise motion from the subnanometer to multiple-millimeter level is possible with a variety of solid-state actuation/amplification mechanisms. The application area of these actuators extends continuously in the field of high-pressure flow control valves. A typical example is presented in Figure 10.2.1a [2].

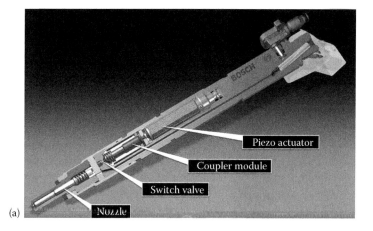

Figure 10.2.1 (a) Cut view of a modern piezoinjector. (Source: BOSCH.)   (*Continued*)

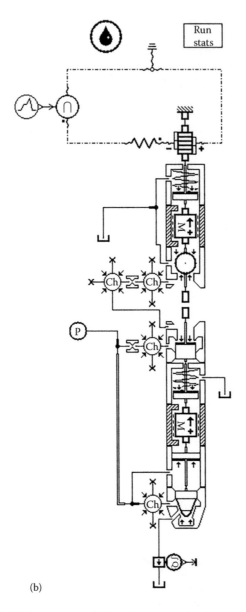

(b)

Figure 10.2.1 (Continued) (b) Amesim model for piezoceramic-actuated injector. (Source: BOSCH.)

10.2.2 Numerical simulations results

This chapter contains mainly the results of the numerical simulation performed by Amesim for a typical piezo inline injector developed by a few companies for high-quality diesel engines [3–10]. A complete Amesim mathematical model of such an automotive component is shown in Figure 10.2.1b [3]. The piezoelectric actuator was designed considering that a ball pilot valve of 1.5 mm diameter with a sharp seat of 1.0 mm diameter is used. The target is to obtain a force of about 150 N needed for compressing the spring of the control valve, and a stroke of about 36 μm, using a voltage power supply of 100 V.

This target can be reached using a stack of 700 wafers with a thickness of 0.05 mm and an area of 25 mm². This choice allows reasonable dimensions of the square piezo stack (5 mm × 5 mm × 35 mm). A restrictor of 0.2 mm diameter was introduced at the input of the control chamber of the needle. Another 0.3 mm diameter restrictor was considered between the control chamber and the pilot valve.

Some typical results concerning the injector dynamics are shown in Figures 10.2.2 through 10.2.19. Taking into account the high-speed response of the piezoactuator, a ramp-type input signal of 0.5 ms was considered. The same time was used for the negative slope ramp at the end of the control signal (Figure 10.2.2).

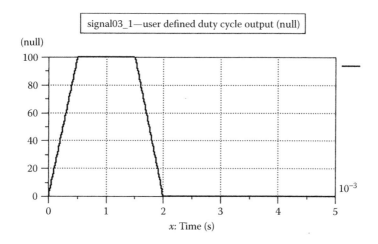

Figure 10.2.2 Duty cycle output considered in simulations.

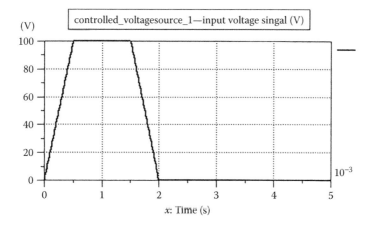

Figure 10.2.3 Input voltage signal used for the opening and the closing of the pilot valve.

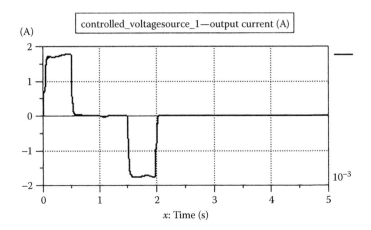

Figure 10.2.4 Current output of the controlled voltage source.

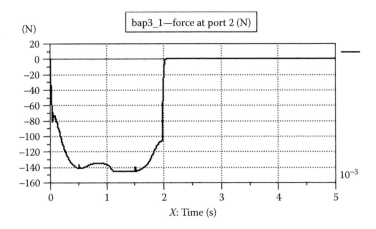

Figure 10.2.5 Force applied by the piezoactuator on the pilot valve spring.

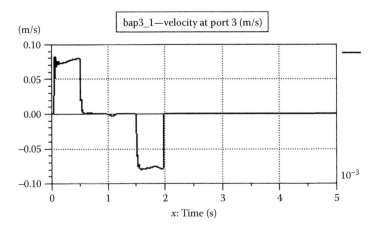

Figure 10.2.6 Velocity of the ball poppet during a control cycle.

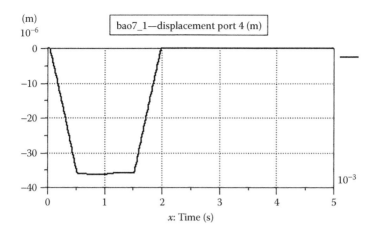

Figure 10.2.7 Ball displacement during the control cycle.

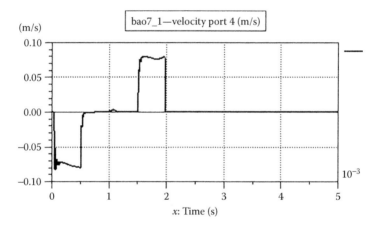

Figure 10.2.8 Velocity of ball poppet.

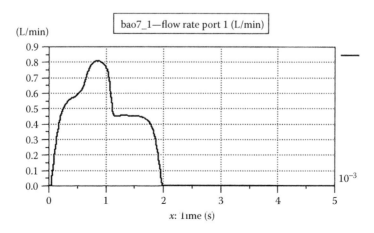

Figure 10.2.9 Flow rate of the pilot valve.

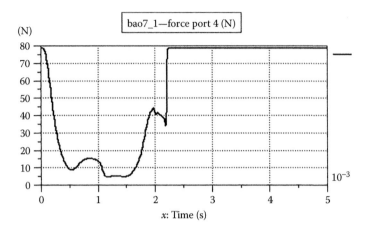

Figure 10.2.10 Force applied on the ball valve.

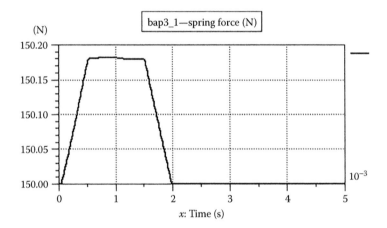

Figure 10.2.11 Force applied on the pilot valve spring.

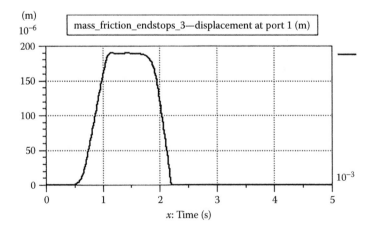

Figure 10.2.12 Needle displacement (maximum 0.19 mm).

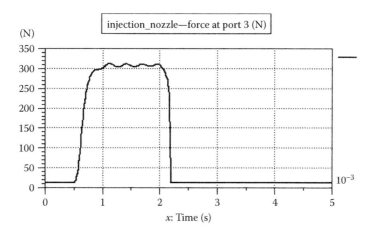

Figure 10.2.13 Overall force on the injector needle.

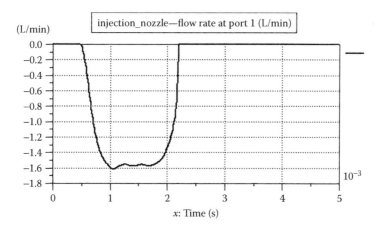

Figure 10.2.14 Flow rate injected in the engine cylinder (by 5 holes of 0.15 mm).

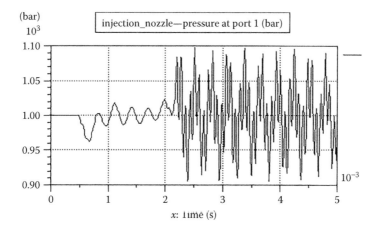

Figure 10.2.15 Pressure in the input port of the injector nozzle.

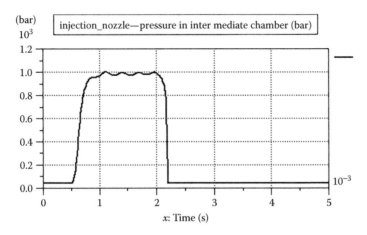

Figure 10.2.16 Pressure in the nozzle intermediate chamber.

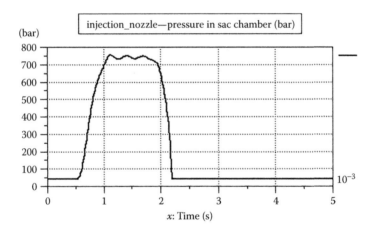

Figure 10.2.17 Pressure in the sac chamber of the nozzle (sac volume—0.2 mm^3).

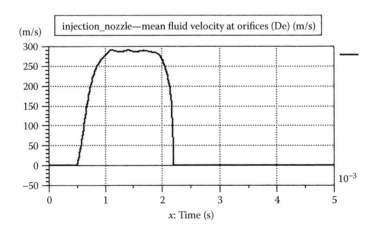

Figure 10.2.18 Mean fluid velocity at nozzle orifices.

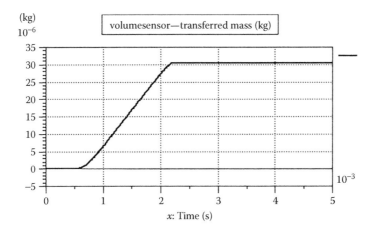

Figure 10.2.19 Transferred mass by the injector to the engine cylinder.

10.2.3   Conclusion

The simulation results show the capability of a piezoinjector to deliver 31 mg of fuel in 2.2 ms to the engine cylinder. This amount can be tuned with the engine demands [11,12], starting from a precise simulation of all the operation regimes.

Some new concepts have to be underlined in order to explain the high-level performances and reliability of such a two-stage flow-metering injection valve. First of all, the standard ball pilot valve used in the solenoid injectors was replaced by a conical poppet uncompensated pilot valve. The conical poppet valve is more reliable, and offers a better tightness. A coupled module between the piezoactuator and the needle is needed for thermal reasons. The main problem of this type of injectors remains the piezo material thermal stability.

10.3   Applications of Amesim in the optimization of the common rail agrofuel injection systems

10.3.1   Agrofuel problems

This chapter contains the comparative results of preliminary computer simulations of a mechanically and an electronically controlled injector running on diesel fuel and high oleic sunflower oil. The models are taking into account the specific viscosity, density, and bulk modulus of elasticity at fuel temperatures at 40°C and 95°C.

The aim of this study was to use a model of the two-stage electrohydraulic and a mechanical injection system in relation to two main parameters: (1) injection timing and (2) injection duration. The computer models and simulation models were provided by Simcenter Amesim software. The vegetable oil data was provided by Flower Power USA. The simulation models were obtained by reducing the initial complex model using different model reduction techniques such as activity index and state count. Different levels of injection valve models were considered in order to keep the time continuity and to evaluate the simplification techniques.

The results show a considerable distortion of injection timing and rate of fuel injection and demonstrate that even with fuel heated at engine coolant temperature, further injection system modifications are necessary in order to address injection timing and fuel injection fuel rate issues.

Pure plant oil (PPO) represents a fuel alternative option that, in some technical and environmental aspects, is superior to other alternative fuels for which its use entails a need for a specially designed engine (or modifications for current engines) and a need for a separate distribution infrastructure, thus it is assumed that there is little reason to see PPO as the primary fuel of the future. However, if used in agricultural equipment as fuel, coolant, lubricant, and hydraulic fluid, PPO (let us refer to it as AgroFuel or AGF) does have its benefits, and therefore should be given equal treatment as compared to other biomass-based, carbon dioxide (CO_2) neutral, biodegradable, and multipurpose fluids.

The vast majority of the current farm mobile equipment cannot run on ethanol, wind, solar, or plug-in electricity but AGF could be farm produced and used by farmers to continue to economically grow wholesome food as long as the Sun shines and regardless of the future of fossil fuels. AGF can be economically manufactured and used where it is needed the most, at the farm and for the farm. Unlike other biofuels, its production is simpler, requires much less energy than other biofuels, and it has no fire, explosion, and toxic hazards and waste disposal problems.

10.3.2   Agrofuel's sustainability

Using the sunflower meal as soil fertilizer, farm producing biofuel is a 100% sustainable activity. All we take from the land will be oil, a result of plant magic that tracks the Sun; absorbs the energy; and uses it to split the water and CO_2; releases most of the oxygen; and combines the remaining oxygen, hydrogen, and carbon into an energy-rich oil for later use (it is solar technology at its best). Designed by God, sunflower is the best natural solar panel and for long energy storage, accomplished in a process that is renewable and self-replicable.

A few generations ago, our forefathers dedicated 25% of their land to crops for their draft horses. Today, we know for a fact that with only 12% of land for sunflower, a farm can be fuel sufficient while continuing to produce safe, abundant, and affordable food, feed, fiber, and a better bottom line. That is, a 13% increase in land utilization for food and 100% decrease of backbreaking labor.

Good food growing is farmers' business and responsibility. As stewards of the land and frontline environmentalists, they will produce and utilize AGF purposely, so as not to compete with food production but to reduce production costs while improving the land.

The concept of using biomass-based liquid fuels, specifically vegetable oils (VO) as diesel fuel alternatives, is not new. Rudolf Diesel himself envisioned that his engine could run on VO. It appears that Rudolph Diesel himself originally thought that a high-efficiency engine and utilization of locally available fuels would enable independent countries and people to "be supplied with power and industry from their own resources, without being compelled to buy and import coal or liquid fuel" [7,8,9]. In 1912, he stated that: The diesel engine can be fed with vegetable oils … and as it can also be used for lubricating oil, the whole work can be carried out with a single kind of oil produced directly on the spot. Thus, this engine becomes a really independent engine.

Already tapping into our strategic reserves while at war to forestall disruption of our fuel supply and forced to send our sons and daughters to shed blood for foreign oil, AGF has the potential for improving that situation.

10.3.3   Agrofuel versus "Biodiesel"

Biodiesel is distinctly defined (only in the eyes of U.S. regulations, for tax and registration purposes) mono alkyl methyl ester. Ultimately, both biodiesel and AGF are used for power generation from biomass oil. Biodiesel is obtained from AGF as a feedstock for a chemical process that may require at least two times more (additional) energy input than that required for producing AGF alone. The main chemical input in biodiesel production is methanol (a fossil fuel-derived alcohol) and caustic ash, both toxic and hazardous. In the process, 10% of the AGF feedstock is lost as a low-value glycerol waste stream along with ample waste water, both requiring proper disposal.

In general, production of biodiesel on the farm is difficult and current production facilities are in essence chemical processing plants. Economy of scale, quality assurance, and public and environment safety dictates that large-scale, complex chemical process technologies be integrated. Biodiesel production involves the following:

- Fire, explosion, and toxic hazards and waste disposal problems
- Round the clock controlled dedicated personnel
- Quality assurance for the production, distribution, and usage
- Lengthy and costly zoning and permitting process

The finished product has its own additional problems that need to be addressed such as temperature and oxidation stability, fuel solvency, water and sediments, residual alcohol, completeness of catalytic reaction, and fuel systems material compatibility, and dedicated fleet for distribution.

Biodiesel is currently uneconomical to manufacture on a small scale, so additional road miles are needed to ferry the raw vegetable oil and chemicals to a centralized base. The finished Biodiesel must then be transported back to the various points of sale, including the farms where the vegetable oil originated. It is safe to say that biodiesel manufactured alone has a much larger environmental footprint than AGF production and use.

Biodiesel is not sustainable for a rural-local economy. Biodiesel plants tend to be expensive to set up, in order to comply with the vast array of legislation that governs chemical processing. It is not financially viable, for instance, for each town to have its own *micro* biodiesel processor. This cannot be said for AGF. A transportable seed crusher may be a shared facility that can be used by multiple communities for a genuine example of a local, *on the farm*, value-added economy.

Biodiesel is not safe for on-farm production. Unfortunately, even at industrial scale, it appears that due to negligence, incompetence, and just plain safety violations, biodiesel facilities tend to explode, burn to the ground, and in the process injure and kill people. Tragically, in the United States alone, in a three-year period (2006–2009) there were eight fires and six explosions that killed two and injured five people [6]. Taking into account all of the above-mentioned detail, it is obvious that it is difficult to produce a commercial-grade biodiesel on a family farm. With biodiesel at higher cost than off-road fossil fuel diesel, and no availability in rural areas, farmers continue to use polluting high sulfur fuel that are bad for farmers and the environment. In addition, there are no incentives or benefits for farmers to sell their crops at wholesale prices and then buy back (in the form of fuel and animal feed) at the retail prices. By contrast, advantages of the local production of vegetable oil include the following:

- A simple process as the only steps are cold pressing and filtering, therefore the production could be decentralized and create jobs in rural areas.
- Decentralization of production allows a minimization of raw material transport, an optimization step crucial for overall energy balance.
- Low-energy consumption in optimized production, for instance, compared to energy utilization for the production of fossil fuel (13%), biodiesel (26%), and AGF (12%).
- AGF is biodegradable and in Germany, for instance, it is classified as no hazard to ground or surface water, whereas biodiesel, a water hazard, is classified the same as heavy oil.
- Producing and using their fuel and meal could reduce farm input, thus lowering the cost of farming.

10.3.4 Agrofuel versus fossil diesel fuel

Fossil diesel fuel is a blend of nasty chemicals (aromatics, sulfur etc.). From the standardization point of view, the fuel is not a pure chemical but an amalgam with characteristics that vary with its geographical origin. The current fossil fuel standard describes only its characteristics in fuel terms but not its contents. The fossil diesel fuel has been poisoning the air and waters for the past hundred years, and we still do not know what we pour into our engines and what we get out. By contrast, AGF is a valuable alternative, which could open the biofuel business opportunity to farmers and help the farmers survive rather than. AGF could provide the fuel for farming to continue if, God forbid, a bomb goes off somewhere and the price of oil jumps sky high.

Plant oil has been embraced by humans for a millennium for food, fuel, medicinal, and spiritual use, and there are no known adverse effects (humans adapted and evolved with it), thus it is safe. Among oilseed plants, sunflower is the best solar panel and energy storage designed by God. It is self-replicating year after year and it produces fuel, food, and fodder where it is needed the most, at the farm. Sunflower oil is intrinsically safe and may be stored at home, kids and pets can drink it, and it is not going to set the house or car on fire. If AGF is made from only one seed variety and refined with identical technology by trained farmers, fuel variance is drastically minimized and the quality is built in naturally. The sunflower plant can only produce what is genetically programmed for.

In order to explore the suitability of using AGF in diesel engines, the authors have conducted a simulation of two injection parameters, timing and rate of injection as applicable for mechanical and common rail solenoid-controlled injectors with fossil diesel fuel and AGF at 40°C and 95°C.

10.3.5 Materials and methods

Short season high-oleic sunflower (HOSF) cultivars were grown and harvested in Washington and Oregon States under nonirrigated and irrigated conditions and as a single (spring planting) or double (summer planting) crop by utilizing existing farm infrastructure and agricultural equipment (specific corn or wheat farming). The yields were specific to those under similar conditions in traditional sunflower regions (Dakotas, Kansas, and Texas). The oil was extracted by cold pressing in mechanical press as those typically in farm operations. The seeds moisture level was below 12% and the oil was clarified by mechanical settling and fine filtration. As is, the oil meets and exceeds most of the specifications of the of DIN V 51 605 rape oil fuel standard as follows:

Property	Results	Specification limits	Units
Density (15°C)	915–915	900–930	kg/m³
Flash point	201–270	min. 220	°C
Kinetic viscosity (40°C)	39.42–39.63	max. 36	mm²/s
Calorific value, lower	37,149–37,30	min. 36,000	kJ/kg
Cetane number (DCN)	49.6–62.8	min. 39	—
Carbon residue	0.16	max. 0.40	% (m/m)
Iodine value	84	95–125	g I/100 g
Sulfur content	1< to 1.2	max. 10	mg/kg
Total contamination	11.33–45.0	max. 24	mg/kg
Acid value	7.45–11.33	max. 2.0	mgKOH/g
Oxidation stability 110°C	5.6–12.2	min. 6.0	h
Phosphorus content	13.8–21.1	max. 12	mg/kg
Earth alkali (Ca + Mg)	2.9–22.4	max. 20	mg/kg
Ash content	<0.001–0.001	max. 0.01	% (m/m)
Water content	495–595	max. 750	mg/kg

For reference, here Diesel Fuel Oils as per ASTM 675 standard specifications are as follows:

Property	Specification/grade 2D	Unit
Density 40°C	849	kg/m³
Density 65°C	869	
Density 95°C	893	
Flash point	52	°C
Kinetic viscosity (40°C)		
minimum	1.90	mm²/s
maximum	4.90	
Calorific value lower	42.0	kJ/kg
Cetane number, min.	40.0	—
Carbon residue	0.35	on 10% mass max
Sulfur	2.00	% mass max (m/m)
Ash	0.10	% mass max (m/m)
Water sediment	0.05	% vol. Max

The HOSF oil under investigation has a ultra high content of monounsaturated fatty acids (>90%) and a low content of polyunsaturated fatty acids (5%<), thus the viscosity–temperature relationship can be predicted based on the amount of polyunsaturated PUFA and monounsaturated (MUFA) fatty acids (FA) present. A mathematical equation developed in Reference [5] relates absolute viscosity (μ) to temperature and mass fraction (y) of monounsaturated fatty acids as follows:

$$\mu = A \exp (B/RT) + y \exp (C/RT) \tag{10.3.1}$$

Where R is the universal gas constant (8.314 kJ/kg mol K) and T is the absolute temperature (K) with the values of constants A, B, and C as 3.31×10^{-5}, 3.55×10^4, and 5.17×10^3, respectively. By using the method described in Reference [5], it was determined that for plant oil having 76% MUFA and 7% PUFA, the kinematic viscosity–temperature relationship (correlation coefficient 0.82 to .90) is as follows:

Temperature (°C)	40	50	65	80	95
Viscosity (mm²/s)	39.92	27.18	18.07	12.57	9.45

By extrapolation, the results from Reference 5, we can predict HOSF's kinematic viscosity and temperature relationship as follows:

Temperature (°C)	40	50	65	80	95
Viscosity (mm²/s)	39.53	26.91	17.89	12.45	9.36

This is in a good correlation with HOSF's kinematic viscosity measured at 40°C of 39.42–39.63 (mm²/s). For comparison, nr.2 diesel fuel's kinematic viscosity varies from approximately 2.61 mm²/s at 40°C to approximately 0.86 mm²/s at 95°C as determined according to ASTM D 445 procedure.

For oil-cooled engines specific to some off-road applications, the coolant temperature could be as high as 140°C, thus the minute fuel volume confined in the injector's body and ready to be injected will reach a temperature of at least 150°C. Under those conditions, for instance, diesel fuel kinematic viscosity could not be sufficient to assure adequate lubrication and minimal leakages in the injector's needle body assembly.

Taking into account that at 95°C HOSF's viscosity is of the same order or magnitude as of the diesel fuel at 40°C, it is conceivable that under those circumstances and from hydraulic viewpoint only, we assume that HOSF is a viable diesel fuel alternative. Beside viscosity, fuel density and bulk modulus of elasticity are assumed to have an important role in fuel injection timing and fuel metering.

Density of vegetable oils (VO) is decreasing linearly with increasing temperature. In general, VO have molecular chains mostly of 18 carbons, thus for all practical purposes their density and its variation with temperature are pretty much the same. According to data available in the literature [10] for VOs, HOSF' density can be predicted as follows:

Temperature (°C)	15	40	50	65	80	95
Density (kg/m³)	914	896	891	883	870	862

This is in a good correlation with HOSF's density of 915.0–915.8 kg/m³ measured at 15°C. For comparison, diesel fuel's density varies from approximately 893 kg/m³ at 40°C to approximately 849 kg/m³ at 95°C.

Bulk modulus of HOSF is higher than that of diesel fuel. It decreases with rise of temperature but it increases with increase of pressure and it is by far more dependent of VOs' air content than fatty acids profile. From Reference 11 for VO at 25°C, bulk modulus of elasticity dependence of pressure is as follows:

Pressure (MPa)	25	50	100	150
Bulk modulus (MPa)	1900	2100	2500	2750

From Reference 11 for VO at 100 MPa, bulk modulus of elasticity dependence on temperature is as follows:

Temperature (°C)	40	50	65	80	95
Bulk modulus (MPa)	2370	2342	2076	1884	1796

For comparison, diesel bulk modulus has approximately 1550 MPa at 40°C and approximately 1250 MPa at 95°C.

The fuel-specific values for viscosity, density, and bulk modulus of elasticity were introduced as parameters into the fluid module of the Amesim injection simulation software to determine injection timing and injection fuel rate of HOSF and diesel fuel at low (40°C) and high fuel temperatures (95°C).

The Amesim fuels injection simulations were made for the following set of values:

	HOSF fuel		Diesel fuel	
Temperature (°C)	40	95	40	95
Viscosity (cSt)	39.92	9.45	2.61	0.86
Density (g/cm3)	0.896	0.862	0.893	0.849
Absolute viscosity (cP)	3576	814	222	77
Bulk modulus (MPa)	2370	1796	1553	1250

10.3.6   Main results of the numerical simulations

The mechanical injector simulation models (general and detailed) are shown in Figures 10.3.1a and 10.3.2. The main results of the simulations are shown in Figures 10.3.3 through 10.3.10.

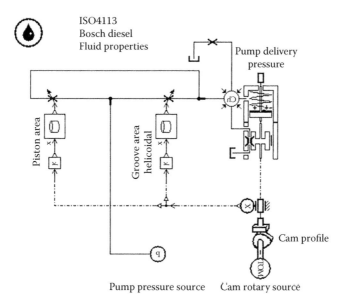

Figure 10.3.1 Amesim general model for mechanical actuated injector.

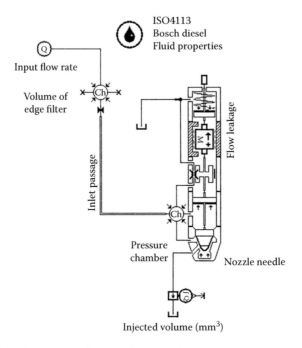

Figure 10.3.2 Complete Amesim model for solenoid injector.

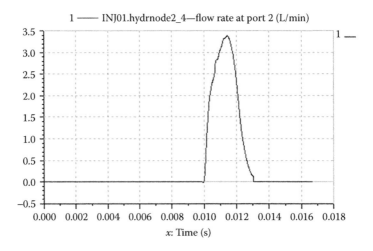

Figure 10.3.3 40°C Diesel fuel mechanical injection flow rate.

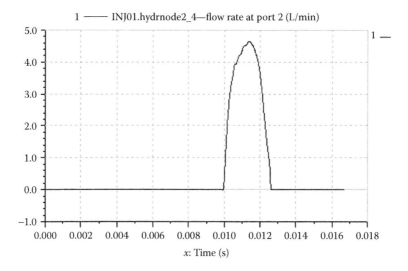

Figure 10.3.4 95°C Diesel fuel mechanical injection flow rate.

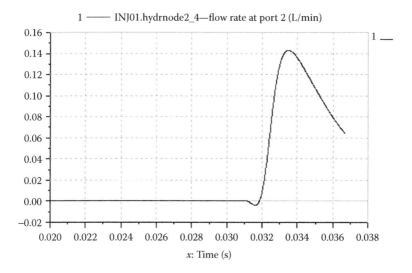

Figure 10.3.5 40°C HOSF fuel mechanical injection flow rate.

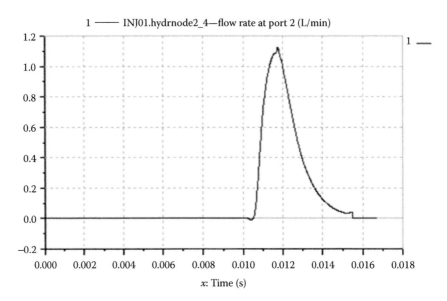

Figure 10.3.6 95°C HOSF fuel mechanical injection flow rate.

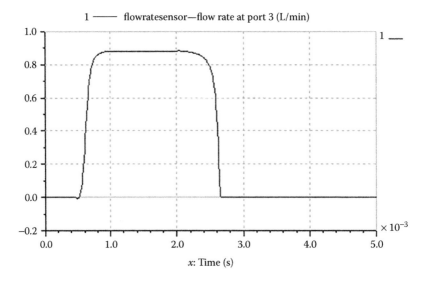

Figure 10.3.7 40°C Diesel fuel solenoid CR flow rate.

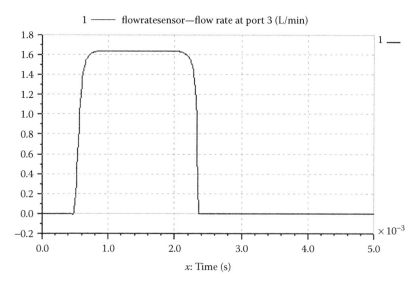

Figure 10.3.8 95°C Diesel fuel solenoid CR flow rate.

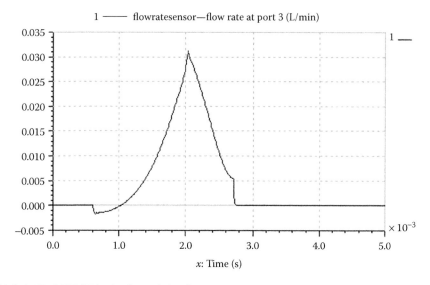

Figure 10.3.9 40°C HOSF fuel solenoid CR flow rate.

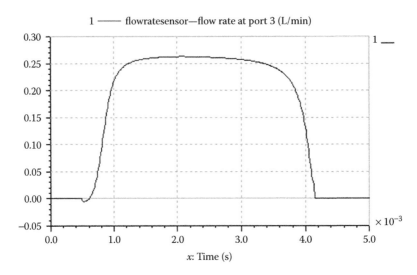

x: Time (s)

Figure 10.3.10 95°C HOSF fuel solenoid CR flow rate.

10.3.7   Conclusion

The simulation results show substantial differences on the injection timing and rate and duration of injection that could be the root cause of catastrophic failure of diesel engines not adapted to run on straight vegetable due to late and incomplete combustion, lubricating oil dilution, plugging of injectors, and piston rings sticking due to oil polymerization, and low-power and higher fuel consumption.

However, a dramatic improvement has been observed by heating the fuel from 40°C to 95°C and it is expected that further simulations at higher fuel temperatures (150°C) will reveal the performance that will get closer to that of diesel fuel injection [14].

Section 1: Bibliography

1. Robert Bosch GmbH. *Diesel Engine Management*, 4th ed. John Wiley & Sons, Chichester, UK, 2005.
2. Aird F. *Bosch Fuel Injection Systems*, HP Books, New York, 2001.
3. Breitbach H. *Fuel Injection Systems Overview*. Delphi Corporation, Gillingham, UK, 2002.
4. Jost K. 1998. New common-rail diesels power Alfa's 156, Automotive Engineering, January 1998, pp. 36–38.
5. Knecht W. Some historical steps in the development of the common rail injection system. *Transactions of the Newcomen Society*, 74, 89–107.
6. Petruzzelli A.M. A story of breakthrough. The case of common rail development, *35th DRUID Celebration Conference*, Barcelona, Spain, June 17–19, 2013.
7. Smil V. *Prime Movers of Globalization: The History and Impact of Diesel Engines and Gas Turbines*. MIT Press, Cambridge, MA, 2010.
8. www.delphi.com.
9. www.siemens.com.
10. www.continental-automotive.com.

Section 2: Bibliography

1. www.dynamic-structures.com.
2. https://au.bosch-automotive.com/en/parts_and_accessories/motor_and_sytems/diesel/
3. lmsimagine.lab/ame_dir/demo/Solutions/Automotive/Powertrain/Engine/FuelSystem/
4. http://www.amtgarageforum.nl/public/topics/603481-funtion_hdi_siemens.pdf
5. https://www.elsevier.com/books/fuel-systems-for-ic-engines/978-0-85709-210-6
6. https://www.dieselnet.com/tech/diesel_fi_common-rail.php
7. http://alternativefuels.about.com/od/dieselbiodieselvehicles/a/diesel crd.htm
8. http://pmmonline.co.uk/technical/common-rail-diesel-fuel-systemissue-peugeot-307-hdi
9. http://www.repairmanual.net.au/car-repair/common-rail-diesel
10. http://www.delphi.com/docs/
11. Hardenberg H., *Die geometrischen Strömungsquerschnitte von Lochdüsen für Direktein spritzmotoren*, MTZ 45 (1984) 10, S. 427–429, 1984.
12. De Groen O., Kok D., *Rechenprogramm zur Simulation von Hochdruckeinspritzsystemen für Nutzfahrzeuge*, Motortechnische Zeitschrift MTZ – 57, 1996.

Section 3: Bibliography

1. Martin Rauber M., Russ W., Winthuis N., Werkmeister R., Meyer-Pittroff R., The influence of temperature and mixing ratio on the kinematics viscosity of mixtures from rape seed oil and diesel fuel. *The International Conference on Hydraulic Machinery and Equipments*, Timisoara, October 16–17, 2008, Romania, Romanian.
2. Cigizoglu K. B. Alternative Diesel Fuel Study on Four Different Types of Vegetable Oils of Turkish Origin, in "Turgut Ozaktas. Energy Sources, Part A: Recovery, Utilization, and Environmental Effects", Istambul, 1997.
3. Hersey M.D. Viscosity of diesel fuel oil under pressure, National Advisory Board for Aeronautics Technical Note no. 315, September 3, 1929.
4. Rodenbush C.M., Hsieh F.H., Viswanath D.S. Density and viscosity of vegetable oils. *Journal of the American Oil Chemists Society*, 76(12): 1415–1419, 1999.
5. Fasina O.O., Hallman H., Vraig-Schmidt M., Clements C. Predicting temperature-dependence viscosity of vegetable oils from fatty acid composition. *Journal of the American Oil Chemists Society*, 83(10), 899–903, 2006.
6. Azian M.N., Mustafa Kemal A.A., Panau F., Ten W.K., Viscosity estimation of triglycerols and some vegetable oils, based on their triglycerols composition. *Journal of the American Oil Chemists Society*, 78(10): 1001–1005, 2001.
7. Diesel R. The Diesel oil-engine, Engineering, 93, pp. 395–406 (1912), Chemical Abstracts No.6, 1984 (1912).
8. Diesel R. The diesel oil-engine and its industrial importance particularly for Great Britain. *Proceedings of the Institution of Mechanical Engineers*, pp. 179–280 (1912). Chem. Abstr., 7, 1605 (1913).
9. Diesel R. *Die Entstehung des Dieselmotors*. Verlag von Julius Springer, Berlin, Germany p.115, 1913.
10. Davis J.P., Dean L.O., Faircloth W.H., Sanders T.H. Physical and chemical characterization of normal and high-oleic oils from nine commercial cultivars of peanut. *Journal of the American Oil Chemists Society*, 85: 235–243, 2008.
11. Varde K.S. Bulk modulus of vegetable oil-diesel fuel blends. *Fuel*, 63: 713–715, 1984.
12. Yang H., Briker Y., Szynkarczuk R., Ring Z. Prediction of density and cetane number of diesel fuel from GC-FIMS and piona hydrocarbon composition by neutral network, The National Centre for Upgrading Technology, Devon, AB, Canada, T9G 1A8.
13. Vasiliu N., Vasiliu D., Calinoiu C., Manea I. Modeling and simulation of the biofuel electro hydraulic injection systems by AMESIM. *2010 European Simulation and Modelling Conference ESM'2010*, October 25–27, 2010, Hasselt University, Belgium.
14. Manea I. Researches on the agrofuel hydraulic injection systems. PhD Thesis, University POLITEHNICA of Bucharest, Romanian, 2017.

chapter eleven

Numerical simulation and experimental validation of ABS systems for automotive systems

11.1   Development and validation of ABS/ESP models for braking system components

11.1.1   Models and libraries used in the modeling of the road vehicles

From more than 100 years, the cars exist and run on the public roads, and during this time the cars developed and continuously improved. The fundamental transformations on the automobiles began to appear along with the draconian restrictions imposed at the end of the second millennium and at the beginning of the third millennium. Moreover, the prospects shown by the adoption of new legislation related to road safety and emission compliance made the car manufacturers to think more in advance about the cars that will run on public roads, in the near future. It is obvious that these rules, mentioned earlier, and which are related to environmental protection and conservation of mineral resources on one hand, and reducing costs, improving performance and driving pleasure, on the other hand, lead to essential changes of the automobile both in its subassemblies' conception as well as in the strategy and the management of their control.

Considering the developments of the last decade, we can observe that the automobile is becoming increasingly complex. In fact, it becomes a complicated system that incorporates modern technologies from many technical fields. A very affordable way to characterize these complex systems is the construction of mathematical models of relevant processes and the use of these models in a systematic way.

As a first step, you can use simple mathematical models that describe the behavior of the vehicle. These models are then used in an optimized sequence to synthesize certain optimal configurations for a vehicle or, for example, for implementing various energy management strategies.

It should be noted that the number of elements that are used in these systems is growing. Starting from this remark we should add that optimal results can be obtained only by optimizing the system as a whole and not by optimizing it element by element. However, optimizing the whole system can be problematic and the only approach is the development of mathematical models for each component and use of numerical methods for optimizing the system structure and the control algorithm. These models should be able to extrapolate the behavior of the system as a whole. We should also keep always in mind that we need to get simple models, functional, that meet the specified technical definition needs.

But let us not forget that a model must be simple yet complex enough to capture all physical phenomena that are representative for the modeled system.

Given all these considerations presented previously, it is logical that with the evolution of cars, systems come to the need of evolving also for the modeling and the simulation tools used.

The main objective of this chapter is to present the development and test of road vehicles' hydraulic and pneumatic systems. A strong base to reach this objective is to split the research activity into three main categories as follows:

- Physical modeling
- Numerical methods
- Computer science and programming

Interaction between the three activities is an important factor to ensure consistency of the final results.

To have a more accurate understanding of these three categories, a small list of details is presented for each of the category.

1. *Physical modeling*: For this category, the development will be focused on:
   a. Simplified models for hydraulic circuit design
   b. Simplified global models (mechanical & hydraulic part) for strategy development with Real-Time (RT) capabilities
2. *Numerical methods*: In this category, the development is focused on:
   a. Model reduction techniques, to eliminate high dynamics that are unnecessary for the model fidelity: index of activity, state count, and modal shapes' analysis
   b. *Computer science and programming*: To have a first overview of the more complex model that will include almost all the vehicle systems, a sketch is presented in Figure 11.1.1. This model represents a virtual system simulation platform for cars.

In order to build this platform, a series of new LMS Imagine components and models have been developed, tested, and integrated in applications on virtual car system simulation:

- Mechanical 1D models' development
- Hydraulic component design Real Time (HCD RT)
- Power train models (3D engine model, Gearbox model, four-wheel drive models)
- Steering system
- Braking system
- Driver

It is important to understand that a limited number of new components will be developed in the first phase for almost all the subsystems shown in Figure 11.1.1 but the main research and development of this chapter will remain focused on the development and integration of the antilock braking system/electronic stability program (ABS/ESP) system. The other components are developed for the author to better understand the *rules* and the specificity of developing of virtual models for road vehicles subsystems and for better understanding of those subsystems' interactions in a virtual environment.

In all research and development stages, the general idea held in mind was that the final models have to be simple yet sufficiently accurate to obtain representative results.

The aim of developing new 1D mechanical component is to build and develop a structure of Amesim model simplification for Real-Time applications.

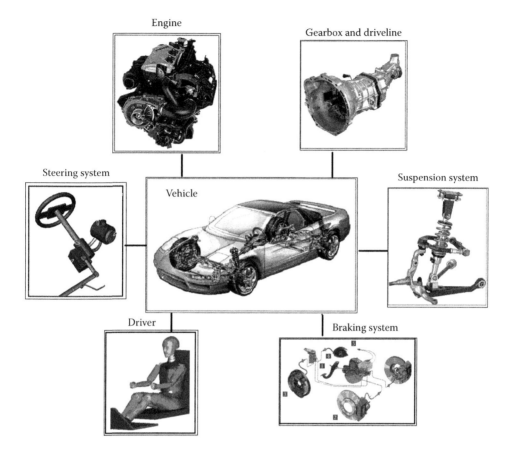

Figure 11.1.1 Vehicle with subsystems overview.

Before each development of a complex model or component for virtual simulation platform, a simple analysis was made to establish the basic components that are necessary for model simplification. Starting from this analysis, a series of new components and libraries were developed and included in Amesim.

Several models have been established and presented as follows.

For the Mechanical library, some evolutions of the *dampers models* were made. Starting from the existing models, new dampers submodels are now available in the Mechanical library (Figure 11.1.2).

The output force is the same as the input force but the output velocity is calculated using the damper rating. The displacement at port 1 is a state variable. The derivative of this variable is equal to the velocity at this port.

Figure 11.1.2 New damper model for Mechanical library.

Figure 11.1.3 New transformers models for Mechanical library.

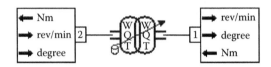

Figure 11.1.4 WTW01A transformer submodel for Mechanical library.

An upgrade of the transformers from Mechanical library was made along with the creation of new *transformer models*. The upgrade of the transformers consists of the creation of new models, starting from the existing ones but with reversed causality (Figure 11.1.3).

An example of a new model of transformer is presented in Figure 11.1.4. WTW01A is a submodel of a modulated transformer between two rotary shafts. In other words, it is a variable ratio gear train.

The user must set a data file table to provide angular displacement at port 1 as a function of input angular displacement at port 2.

To have all the elements for Real-Time applications, a new library of *endstop* models was created (Figure 11.1.5). This new library will be included in the next version of Amesim in the Mechanical library.

RSTP01 (Figure 11.1.6) is a submodel of a simple endstop between two bodies capable of rotary motion. The gap or clearance in mm is calculated as an internal variable. A zero or

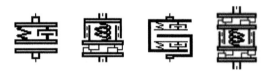

Figure 11.1.5 Components of endstop library models.

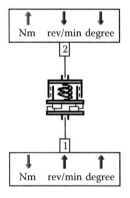

Figure 11.1.6 RSTP01 endstop submodel for Mechanical library.

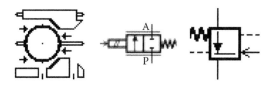

Figure 11.1.7 New specific library for braking system.

negative gap implies contact. Power variables are computed corresponding to the spring effect. This is integrated to the given activity. The aim of this library development was model simplification. In fact, the endstop models without damper in contact were developed to eliminate the algebraic loops that would appear in the case of using free inertia mass components.

Along with the endstop presented earlier, some other models were developed for real-time applications. Those models are used for specific application such as the braking system (Figure 11.1.7) and are not included in the Amesim Libraries.

The starting models for this development are from multiple Amesim Libraries: Mechanical, Hydraulic, or HCD. From these libraries, only some of the components were considered for development, and a little specific library was created to be used for building a model of an ABS/ESP braking system/electronic stability program system.

The specificity of the components of this library is the possibility to run in Real-Time applications and those components can be the base for future development, if requested, for all the elements of the libraries listed previously. At this point, it can be seen that many components from mechanical library were modified and developed for compatibility with Real-Time applications.

11.1.2   Basic layouts of the ABS/ESP systems

After all the functionalities of the simulations programs have been emphasized, some models have been created and some library models have been updated, the aim of this chapter is to build and then to simplify a real model of an ABS/ESP system. The final goal of this study is to be able to include the resulted system in a HiL phase. The system will be modeled using Simcenter Amesim software, and validated by using sets of experimental data.

This study was made in collaboration with Honda and Nissan. All the experimental data were produced by these companies.

In the first phase, each component of the global system will be validated. After the validation, the components are to be integrated in a global model of the braking system. This global model will be first validated and will be reduced in the second phase to be able to run as a Real-Time application. In order to obtain a good system simplification, the next hypotheses of a proper model have to be taken into account. Thus, a proper model is the one that

- Has physically meaningful parameters.
- Has physically meaningful state variables.
- Has the minimum complexity required for addressing the modeling target.

In this first phase, a complete hydraulic braking system will be modeled, simplified, and validated with experimental data. This first model will be a high frequency one.

Before building the entire system model, each component of the braking system will be modeled, simplified, and validated to be included into the global system.

Into the system presented in Figure 11.1.8a and b, only the left part will be modeled in Amesim (this part includes Front Left and Rear Right calipers and the hydraulic block that controls the two calipers). The right part is identical and can be modeled in the same way.

11.1.3 Model inputs and outputs

In order to make the system validation, Honda provided the control laws for all the valves and the results obtained from the test bench.

The inputs of the system are the Master Cylinder pressure (Figure 11.1.9) and the piloted force of each valve (the characteristic $F = f$ (Current, Displacement)) (Figure 11.1.10). Figure 11.1.11 presents the displacement of the valve and the force.

Having the force characteristic and the input current and using the displacement of the valve, we can obtain the force acting on the valve. This type of characteristic was used inside Amesim models by adding directly the table data with the exact data provided.

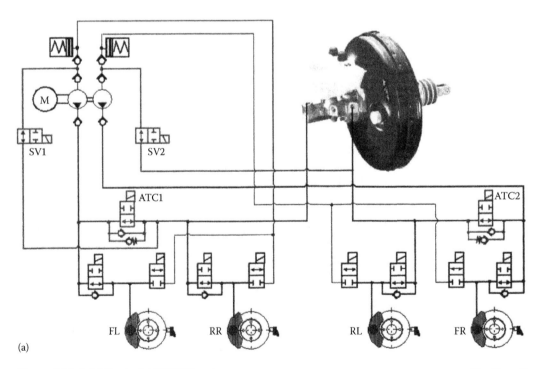

(a)

Figure 11.1.8 (a) Typical ABS/ESP system. (*Continued*)

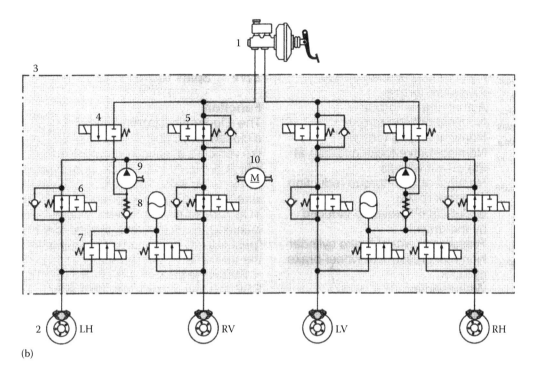

(b)

Figure 11.1.8 **(Continued)** (b) Schematic diagram of the antilock braking system/traction control system (ABS/TCS) hydraulic circuit for X brake-circuit configuration.

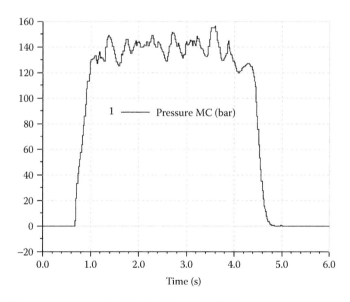

Figure 11.1.9 Different pressure cycles for the Master Cylinder.

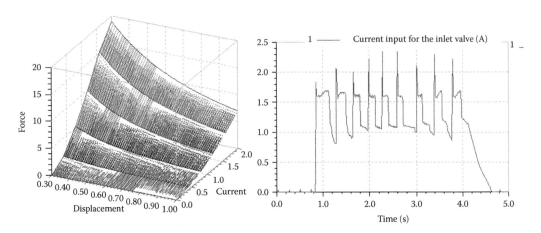

Figure 11.1.10 $F = f$ (current, displacement) characteristic and the input current for the valve.

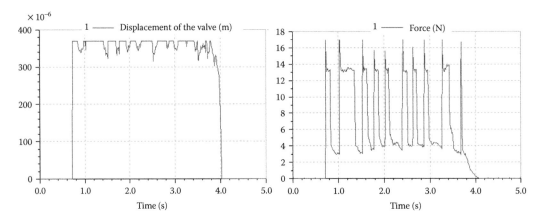

Figure 11.1.11 Displacement of the valve and force.

All these characteristics were built and with this input data, all the ABS/ESP system valves were validated statically and dynamically.

All these validations and testing are integrated in the global ABS/ESP system. The outputs of the system (Figure 11.1.12) are Caliper pressure (it is the main variable to be considered for system validation). This pressure is a very important output of the system because based on this output, all the validations of the system will be made. From all the sets of test bench data, we have chosen one set to do a detailed validation of the components and of the entire ABS/ESP system.

11.1.4   *Modeling system components*

In order to have a real physical system representation, the next fluid properties were taken into consideration in the modeling process: density, compressibility, cinematic viscosity, aeration level, and bulk modulus. Having all those fluid characteristics, we are able to use in Amesim a model of fluid properties that gives a very good representation for the real physical system. A simple model will be built to correlate all fluid properties in a single

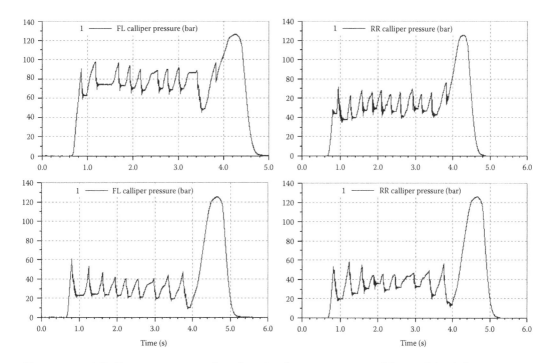

Figure 11.1.12 Different variations of test bench caliper pressure used for model validation.

Amesim fluid properties icon. Since it is difficult to measure the air content in a hydraulic braking circuit (American Society for Testing and Materials [ASTM] norm), we establish an air content of 0.8%. This value was set in to be able to make a closer approximation of the bulk modulus.

The justification of the 0.8% aeration level has been done experimentally to better understand the used fluid properties model and his parameters: Using the Amesim simulation we have run some test and we saw that a value greater than 0.8 does not fit the experimental bulk modulus. We can observe the difference between the two aeration levels in Figure 11.1.13.

We can see that a variation of bulk modulus that is closer to the experiment is for 0.8% aeration and this is the level that will be used for simulation. For the first phase of the project, all the information about the hydraulic block is needed. The internal volume and pressure drops can influence the global dynamics of the braking system, so in the first phase we are going to consider all the elements to see how big is that influence. All the information about the hydraulic network that connects the different components of the hydraulic bloc is available and will be introduced as parameters in the Amesim blocks.

The hydraulic block has two types of valves: (1) regulation valves (Inlet and Regulation valves) and (2) ON/OFF valves (Outlet and Suction valves). A detailed model of each valve will be built using HCD components. The models of valve require physical and geometrical parameters such as mass of the spool, the diameter of the piston, the diameter of internal fixed orifices, the angle of cones, spring stiffness, and preload. Those parameters will concur to the modeling of dynamics for each valve from the braking circuit.

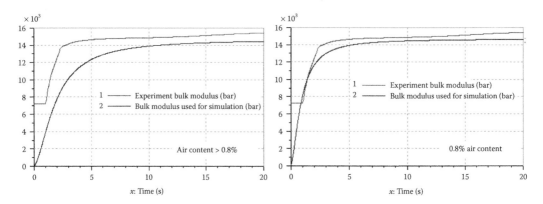

Figure 11.1.13 Bulk modulus comparison between experiment and simulation.

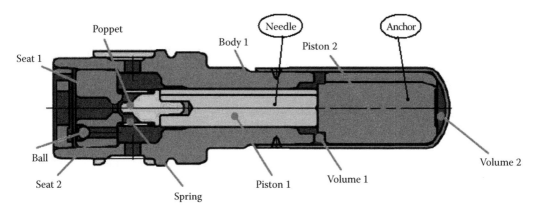

Figure 11.1.14 Common type regulation valve design.

To control the valve, we will use the characteristic $F = f$ (Current, Displacement). In this phase of the project, we will not consider the hysteresis of the solenoid because we do not use an electric circuit inside the model. Having the Regulation Valve sketch (Figure 11.1.14) and all the parameters, we can build in Amesim, using HCD components, the model of this valve (Figure 11.1.15). In a first stage, we build a complex model of the valve taking into account all the elements and hydraulic lines that appear on the sketch. We are going to use elements from Mechanical and Hydraulic libraries also.

Having this first level of the model, we can run tests to see some results for this valve. The aim of the model simplification is to obtain, for the simplified model, results that are very similar with that of the complex model or even the same.

First, we establish the inputs for this valve: the pressure from Master Cylinder, the pressure given by the Inlet Valve, and the Input current to pilot the solenoid force

This type of valve is normally a pressure regulation valve. It imposes a pressure control not a flow rate. To be able to begin the simplification of the model we have to do some simulations using some Amesim facilities that will allow us to understand where to work on the model to be able to simplify it. The goal of this simplification is to reduce the simulation time but to have always the same simulation results.

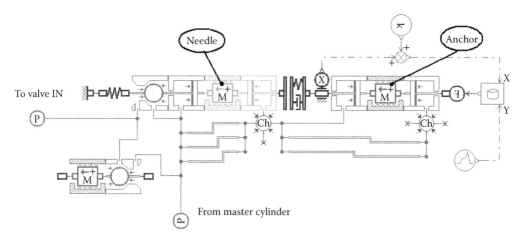

Figure 11.1.15 Regulation valve complex model.

11.1.5   Using Amesim facilities for simplifying the models

For this model, we will use two Amesim facilities: activity index calculations and State Count. The ACTIVITY INDEX facility is a powerful analysis tool based on energy transfer in the components of a system. Using activity index, we can identify the most energy-active components of a system and the most energy-passive components. It can be used to simplify complex systems and this can be done by eliminating, where possible, the components associated with low-power flow that makes small contribution to a dynamic behavior of the system. After a simulation, we can check the activity index of each valve element. The two elements that represent the one-dimensional motion of a ball poppet valve with conical seat (BAP24) are the most energy-active elements (Figure 11.1.16).

Using this facility may induce some wrong decisions if the power in the main circuit is very large compared to the power used to pilot the valve. It is essential to identify the main element that represents the main power flow and the pilot stage from these elements. One dynamic element should stay in the model to represent each power level.

Before formulating any conclusions on model simplification, let us use now the other Amesim facility: State Count.

The main purpose for the State Count facility is to identify the reason for a slow run. For the same simulation, we will open the State Count to analyze which of the state variables are slowing down the simulation.

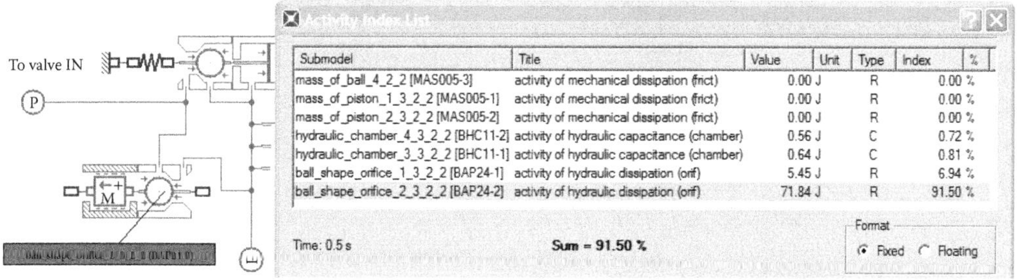

Figure 11.1.16 Activity index for all elements of valve.

Checking the elements that induce the biggest effort for the integrator, we can see that those are all the hydraulic chambers, all the hydraulic lines and the two masses of the piston (Figure 11.1.17). We will do the simplification step by step to check every time that we do not have wrong results. For the first level of simplification, we try to eliminate the hydraulic chambers, lines, and the pistons around the anchor (Figure 11.1.18.). We keep the two masses and the pistons from the left, as we established earlier.

If we check the plot, we can see that the pressure from anchors back has disappeared (we have eliminated the piston from anchors back and also the volume chamber). All the other variations are identical with the variations from the complex model.

To be able to perform the second level of simplification, we have to use again the activity index and the state count facilities. We can see that all the masses have light power exchange and the mass of the needle is the most CPU time consuming. Yet, we are not going to eliminate those masses. We will eliminate all the pistons and also replace the check valve with a simple model from the hydraulic library. We saw that this check valve is the most energy-active element

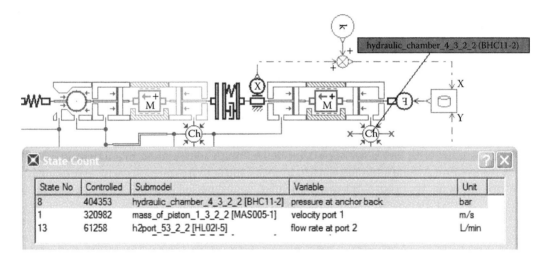

State No	Controlled	Submodel	Variable	Unit
8	404353	hydraulic_chamber_4_3_2_2 [BHC11-2]	pressure at anchor back	bar
1	320982	mass_of_piston_1_3_2_2 [MAS005-1]	velocity port 1	m/s
13	61258	h2port_53_2_2 [HL02I-5]	flow rate at port 2	L/min

Figure 11.1.17 State count window for the valve model.

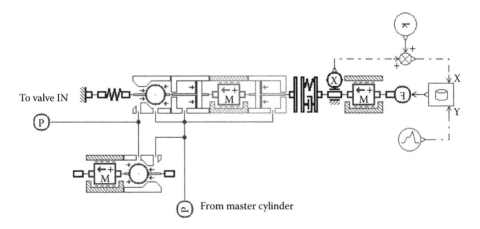

Figure 11.1.18 Regulation valve—first level of simplification.

for the main flow part but is also energy-passive from the pilot point of view (ball mass). The simple check valve from the hydraulic library will keep the main flow of the replaced check valve. We try this new model, look at the results, and then we can formulate a conclusion about a new simplification. The plotted results are the same. The CPU time lowers from 114.5 s for the complex system to 0.328 s for the second level of simplification. The model of the valve has reached a level that can be used in system simulation (Figure 11.1.19). If it is required, the model will be more simplified if the CPU time model validation is too high. Clearly, for a Real-Time application this model will probably be more simplified.

To simplify the check valve model and to keep the same characteristic for the new component we build a simple system, and using the characteristic $Q = f(P)$ we can set the new valve parameters. Using the flow rate characteristic (Figure 11.1.20), we can prescribe

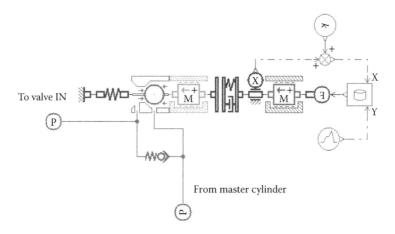

Figure 11.1.19 Regulation valve—second level of simplification.

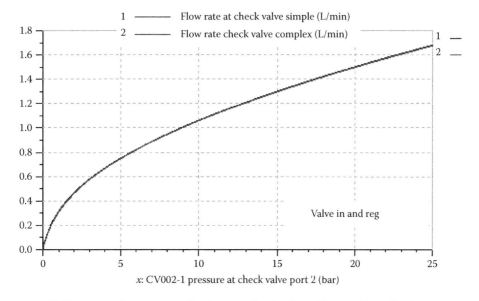

Figure 11.1.20 Flow rate characteristic for the two check valves (Reg and In valves).

the two parameters for the new valve: nominal flow rate fully open valve and corresponding pressure drop.

To model and to simplify this valve, we are going to use the same algorithm as for the Regulation valve. We start in the same way by building a complex model of the valve (Figure 11.1.21) that contains all the elements that appear in the valve sketch (hydraulic lines, pistons, hydraulic chambers, masses).

Using the activity index and state count, we are going to simplify the model (Figure 11.1.22) and verify that the simulation results do not change.

As the model has to be able to run in Real-Time application, this is not the last level of simplification. In the second phase of the project, this valve model will be more simplified and will be able to work within a system that runs under a Real-Time platform.

The Outlet valve is an ON/OFF valve (Figure 11.1.23). As for the other valves, the solenoid activation/deactivation dynamics will be neglected. Only the dynamics linked to the mechanical response of the valve will be included. We are going to use a simple input with a pulse-width modulation (PWM) signal. The relative displacement of the valve is not important here. We need to know only the values of the end stops to be able to say when the valve is fully open or fully closed.

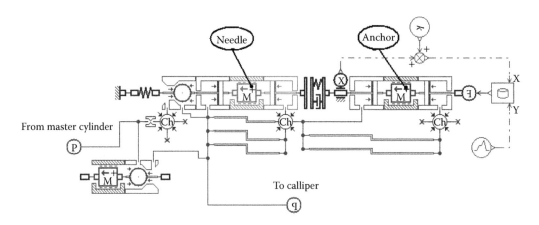

Figure 11.1.21 Inlet valve complex model.

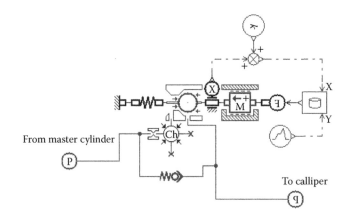

Figure 11.1.22 Inlet valve simplified model.

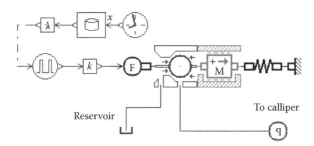

Figure 11.1.23 Outlet valve simplified model.

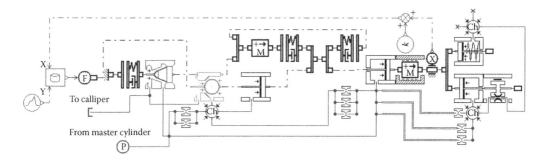

Figure 11.1.24 Suction valve complex model.

Similar to the Outlet valve, the Suction valve is ON/OFF (Figure 11.1.24). It is a specific one because it has a relative displacement inside. Thus, this valve is a little more complex in comparison with the others. In fact, this valve has two phases for opening: (1) Phase 1—The Ball is lifting from the Second Seat and (2) Phase 2—The Second Seat is lifting from the First Seat. In the first stage, the valve will be modeled with all the pistons, internal chambers, leakages, and masses. This complex model of the Suction valve will consume a lot of CPU time for simulation. Thus, we have to simplify it by eliminating the elements that do not influence the low-frequency results of the simulation, however, have a bad influence on the CPU time. To have a simple model, all the internal chambers, orifices, and leakages should be eliminated.

Using the Activity index and the State count facilities leads to a simpler model (Figure 11.1.25). The simplified model is free of some elements that consume a lot of CPU

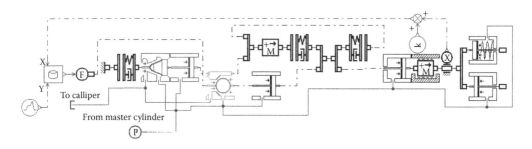

Figure 11.1.25 Suction valve first level of simplification.

time and is not of big importance from the power exchange point of view. We have to remember that this valve is the most complex component of the ABS/ESP system and even with this simplification, the model is not yet complete. Additional work will be done to improve the Amesim model.

Inside the braking system, there are two piston pumps that are driven by an electric motor (one for the primary circuit, the other for the secondary circuit). The control of the electric motors is done through PWM current input. The electric motors that drive the pumps will not be modeled. Instead, an equivalent signal corresponding to the rotary velocity of the pump will be used as input.

From the sketch presented in Figure 11.1.8a and b, we can see that an accumulator (reservoir) is included at the pump inlet. It is a spring accumulator and, similar to the pump, will be first modeled with HCD components and further it will be simplified to be used for Real-Time applications.

In order to simplify the system more, we have chosen not to model the Master Cylinder. The pressure coming from the Master Cylinder chamber will be considered as the source (a simple model for pressure source from Hydraulic library will be used). The booster also is not to be modeled because it is not of interest for the ABS system.

Similar to the valve, the calipers will be modeled using components from the HCD library (Figure 11.1.26). The caliper model from Amesim is of a big importance to have the same characteristics as the real caliper. Having the characteristic of required volume, we are going to build a simple Amesim model with a spring and an end stop inside. We are going to set the parameters for the spring and the end stop to reproduce the required volume characteristic.

The experimental characteristic and Amesim-simulated characteristic are almost identical (Figure 11.1.27). Now we have a validation of the caliper models and we can use those models further in our project.

It is well known that a hydraulic braking circuit has long pipes and hoses and those elements have a big influence for the dynamic response of the system. All the pipes and hoses from the hydraulic unit to the wheels will be modeled with components from Hydraulic library and will include the compressibility effect of the fluid and the wall expansion. Also depending on the boundary conditions, the inertia can be important and will be included in the modeling aspects (in the first phase of the model). Of big importance is the wall expansion for the hoses. To model this wall expansion, a very important parameter is the equivalent bulk modulus of the hose as a function of pressure. In order to obtain the

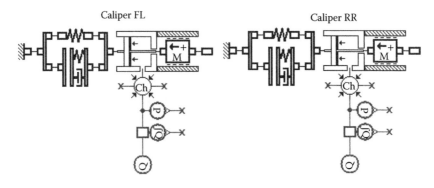

Figure 11.1.26 Amesim system for validation of calipers.

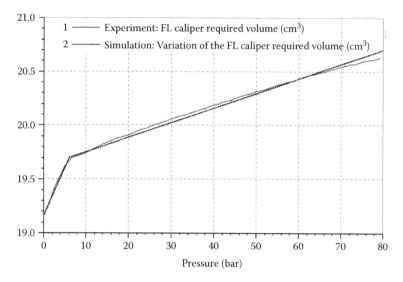

Figure 11.1.27 Calipers compressibility.

linear characteristic of the Inlet valve, we are going to build a simple model that contains the elements from the test bench. We set the input pressure to a constant value of 140 bars and we plot the pressure difference between the input pressure and the caliper pressure for different values of the electrical current. If we look at the plot that shows the two variations (experimental and from Amesim) (Figure 11.1.28), we can see that the two curves are very similar. We can say that statically the Amesim model of the Inlet valve has the same results as the valve from test bench.

For the Inlet valve dynamic validation, we have to check if the pressure increase rate matches with the experimental data. Using a model developed for this type of validation, a simulation was made with a pressure of 140 bars and a sudden increase in the current to 2.5 A so that the solenoid force opens the valve. We can see that after 60 bars the model pressure rise is slower (Figure 11.1.29).

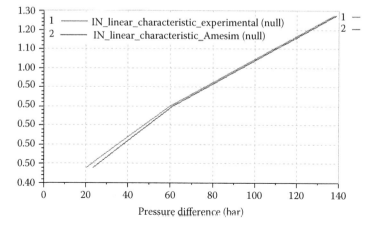

Figure 11.1.28 Comparison of linear characteristics of the Inlet valve.

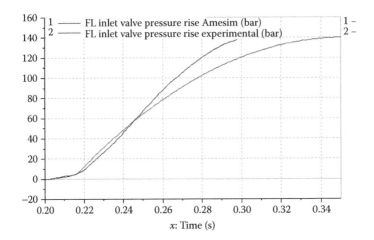

Figure 11.1.29 Comparison experimental versus Amesim pressure rise for Inlet valve.

For the elements involved in this *time response* we can say that the response looks like a first-order lag response and thus a remote control (RC) system or a L/R system. In our case, we can say that the contribution is a RC system with the R term coming from the valve variable orifice and the C term coming from the caliper compressibility.

In this stage of the project, a global model of the braking system will be developed. This model includes all the elements and all the parameters described earlier. The input for Input and Regulation valves will be a 2D table that represents the dependency $F = f$ (Current, Displacement). For the Outlet and Suction valves (ON/OFF valves), a PWM type signal will be used. The output of the model will be the caliper pressure as a function of time.

Now that we have put all the elements together in a global model (Figure 11.1.30) we can begin the validation of this model by using the test bench data. The test bench data represent a number of tests for different road surfaces: brake on dry asphalt from initial speed of 100 km/h, brake on the compact snow surface from initial speed of 50 km/h, brake on ice surface from initial speed of 100 km/h, brake on mix surface from initial speed of 50 km/h (right side on dry asphalt, left side on frozen snow), and acceleration on mix surface from initial speed of 50 km/h (right side on dry asphalt, left side on frozen snow).

First we have to set the inputs of the test and also to set what outputs to monitor to make the model validation. We are going to use model inputs such as: the master cylinder (MC) pressure has a predefined cycle, and for the Model outputs: Calipers pressure.

There are two types of used input data: the (1) 1 ms file and (2) 10 ms file. Note that we do not have a file sampled at 1 ms for the command of the valves and we have to use the commands from the 10 ms file. Analyzing the caliper pressures from the simulations (curves 3 and 4 from Figure 11.1.31), we can see that there are some differences between them. We used the same Amesim model with the same valve inputs. The only difference is the MC pressure input.

As a conclusion, we can say that every little modification of the MC pressure input results in a different simulation result.

In this first ABS/ESP system construction stage for integration in a loop phase, we have made the work hypothesis, we have validated each valve model, and also we have made a part of the entire system validations in the given conditions.

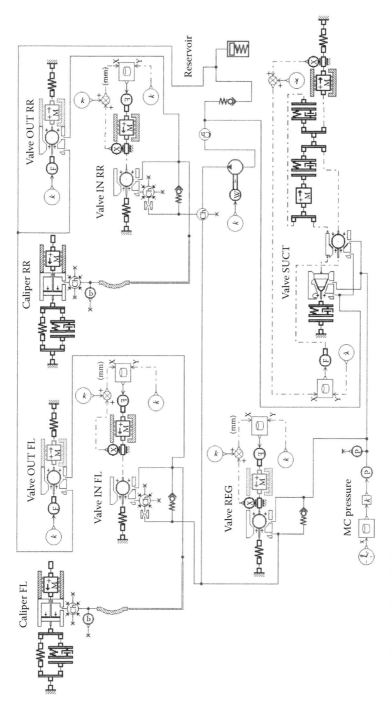

Figure 11.1.30 Model of the ABS global system.

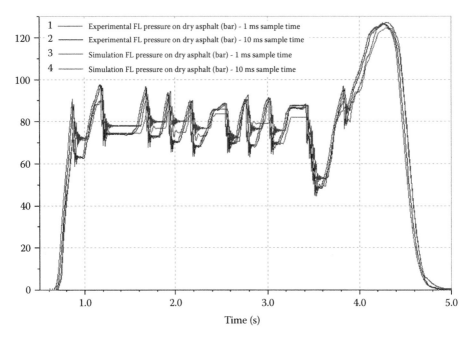

Figure 11.1.31 Comparison between the FL caliper pressure.

We saw that some differences appear between the simulation results for slightly different pressure inputs. Some parameters (such as caliper gap and air content) can influence the system functioning a lot and this can be seen in the simulation results.

After the first set of tests, the general conclusion is that the results are encouraging and we can continue the development of the global ABS/ESP model for Hardware-in-the-Loop (HiL).

11.1.6   Conclusion

The synthesis of such a complex digital control system such as ABS/ESP needs a huge effort of writing proper mathematical models and then simplifying them to run in real time. The successful researches developed in by Lebrun M., Dragne F.D., Alirand M., Baţăuş M., and Ion Guţă D.D. [13–16] have shown the way to follow in each similar complex case such as common rail fuel injection systems, variable valve timing (VVT), and so on using the Amesim facilities such as State Count. However, the *Real Time simulation* (RTS) models' validation has to cover all the possible operation situations connected to the temperature drift of all the components, the transient occurring in extreme conditions, and so on.

11.2   Brake system model reduction and integration in a HiL environment

11.2.1   Problems of reduction of the ABS/ESP system for HiL

In this second phase of this study, the model will be reduced and validated to be able to run in real time. In order to be able to do the simplification of the global model, each component must be reduced step by step to use it in a Real-Time system simulation.

The results are sufficiently accurate but the model is too complex. To have a first idea on the complexity of the system, we have to look at the number of state variables—43 state variables. This number is too high and as a consequence, the model is not suitable for real time. Another clue of the model complexity is the CPU time—this model is consuming too much CPU time. As a conclusion, the model has to be reduced to be able to use it for real time. In a first step, a simplification of each component has to be done. The last level of simplification, for each component from the first phase of the project, will be reused to start the second phase of simplification.

The Outlet valve was simplified and validated to be used in the global model. We retake this model and try reducing it to a stage that can be used in real time. This model takes into account the mass of the spool. This mass connected to the springs leads to a mode that is too high to be considered for Real-Time simulation.

In order to reduce the Outlet valve model, a simple force input will be used and a pressure that is in fact the pressure from the caliper in a normal working process will be used. The pressure and the spring force act to close the valve. The command force acts to open the valve. This model of valve has two state variables in the MASS model (Figure 11.2.1).

Using submodels for BALL POPPET, MASS, and SPRING that are specially created for real time, the complexity of the model will be reduced. Checking the results of the two valve models, it can be seen that the two flow rates are the same (Figure 11.2.2).

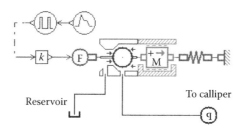

Reservoir   To calliper

Figure 11.2.1 Outlet valve simplified model (for reduction tests).

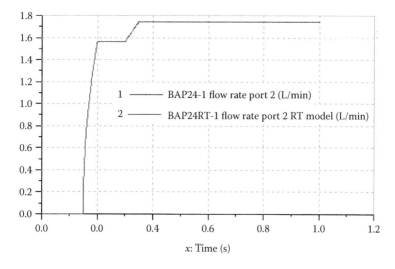

Figure 11.2.2 Flow rate comparison.

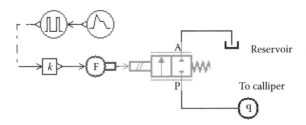

Figure 11.2.3 Outlet valve modeled with Hydraulic library components.

As a conclusion, this model of valve can be used inside a model for real time. It is not to forget that there is still a state variable inside the model. To eliminate this state variable, a new valve specific model is build. This new model does not contain any state variables (Figure 11.2.3).

This new model has different parameters from the HCD model. The correct value of the parameters for the new valve model should be introduced so that the simulation results have to be the same. The values for the maximum flow coefficient and the critical flow number are the same for the two models. To set the saturation input value, the maximum force value from the solenoid tables has to be considered. The maximum force is 36 N, so the value of the saturation input valve parameter will be set to 36. For this value, the valve will be fully open.

The area used for flow rate computation is calculated with the following formula Equation 11.2.1:

$$\text{area} = F_{\text{input}} \cdot \frac{\text{area}_{\text{max}}}{36} \tag{11.2.1}$$

To set the value for the equivalent cross-sectional area at maximum opening (that is in fact the maximum area from the aforementioned formula) in correlation with the area of the HCD model, some calculation has to be done using the parameters of the HCD valve model.

To compute the maximum flow area of the valve (Figure 11.2.4), we use the next following formula:

$$\text{area}_{\text{max}} = \pi \cdot x_{\text{max}} \cdot (db + x_{\text{max}} \cdot \cos(\alpha)) \cdot \sin(\alpha) \cdot \cos(\alpha) \tag{11.2.2}$$

where:
   area_{max} is the maximum flow area [mm²]
   db is the ball diameter [mm]
   α is the ($\pi/2$)—seat semi-angle [rad]
   x_{max} is the maximum poppet lift [mm]
   $db = 1.582$ mm
   $\alpha = \pi/2 - \pi \cdot 59.5/2$
   $x_{\text{max}} = 0.17$ mm

Using the nonlinear Equation 11.2.2, the calculated value of the area will be: $\text{area}_{\text{max}} = 0.40369$ mm². To verify that this area is correct, we run a simulation of the valve model with HCD components and we look at the variables list of the HCD valve.

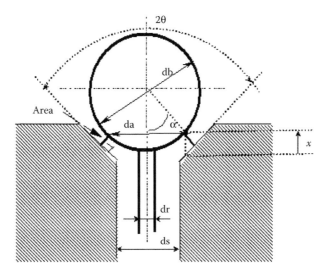

Figure 11.2.4 Valve model sketch.

The value of the flow area (0.264208 mm²) is different from the calculated flow area (0.40369 mm²). Now we make another computation and we calculate the area of the seat hole:

$$\text{area}_{\text{hole}} = \frac{\pi \cdot d_s^2}{4}$$

(11.2.3)

where:
   $\text{area}_{\text{hole}}$ is the seat hole area [mm²]
   d_s is the seat diameter [mm]

Using this formula, we will find: $\text{area}_{\text{hole}} = 0.2642$ mm². This value corresponds to the value that we find in the variable list of the HCD valve model. In conclusion, the maximum flow rate through this valve is set by the valve orifice and not by the poppet lift. When we set the parameters for the new valve model, we have to be very careful and we have to verify that the correct values of the parameters are introduced.

The Outlet Valve has reached a level that gives exactly the same results as the valve with geometry model (HCD model) and is faster from the CPU point of view. In conclusion, this model can be used for real time. In the first phase of the project, the Inlet valve was simplified (Figure 11.2.5) and validated to be used in the system simulation. This model will be reused and reduced to a stage that can be used in Real-Time system simulation.

This valve model has three state variables: one state variable in the hydraulic chamber model and two state variables in the mass model. After a simulation run, using the State Count facility, it can be seen that the hydraulic chamber is the one that consume the most CPU time. In order to reduce the number of state variables, the hydraulic chamber will be eliminated. Let us not forget that if the hydraulic chamber will be eliminated, the orifice model and the valve model cannot be connected directly. The conclusion is that the orifice should be eliminated too.

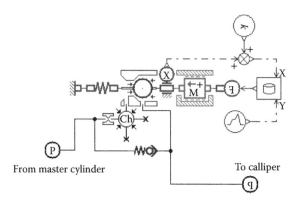

Figure 11.2.5 Inlet Valve simplified model (from global model).

To build a new Inlet valve without hydraulic chamber or orifice, a comparison (Figure 11.2.6) between the model with chamber and orifice and the new model without chamber and without orifice will be done. The results are the same (Figure 11.2.7) and in conclusion, this new model of the valve will be used from now on.

By eliminating the orifice and the chamber, one state variable was eliminated (the one that consumes the most CPU time). There are still two state variables in the mass model. To reduce the number of states more, some especially Real-Time designed submodels are used. By choosing those new submodels, the number of state variables is reduced to 1. The Inlet valve has reached a level with no orifice and only one state variable.

The regulation valve is almost the same as the Inlet valve. It is a simpler valve because it has no orifice or hydraulic chamber. The only difficulty comes from the mass model that has two state variables. By choosing the right submodel for the mass (Real-Time submodel), the number of state variables of the regulation valve will be reduced to 1.

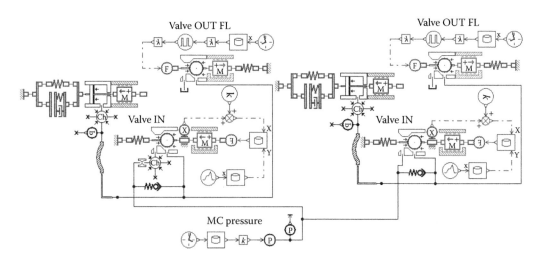

Figure 11.2.6 Comparison model for simplification of the Inlet valve.

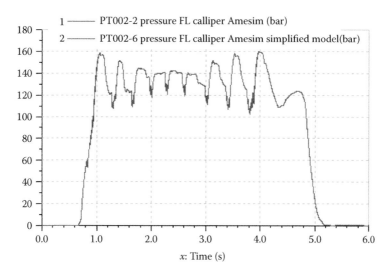

Figure 11.2.7 Caliper pressure comparison.

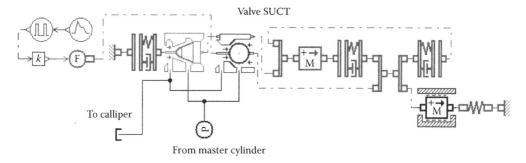

Figure 11.2.8 Suction Valve second level of simplification.

For the reduction of the Suction valve (Figure 11.2.8), in a first step all the pistons of the model will be eliminated. The minimum number of components that are necessary to model the valve geometry will be kept. The pistons can be eliminated because a valve model with relative displacement—BRP024 can model the difference between the areas inside the valve. This difference has to be modeled because the areas in front of the anchor and areas behind the anchor are not the same.

After the compilation, it can be seen that the model has four state variables: two state variables for the simple mass model and two state variables for the model of mass with end stops. In order to simplify the valve model and to reduce the number of state variables, Real-Time submodels will be used for the masses, valves, and end stops (submodels designed for Real-Time applications with no mass inside). The end stops used here contain only the spring. As there is no mass inside the models, there is no need to damp the oscillations in the end stop. The new model of Suction Valve (Figure 11.2.9) will contain some new specific models and has only two state variables. To see if the new model of valve gives the same results, a comparison of the flow rates of the two valves was made (Figure 11.2.10).

The valve model has reached a level that can be used in real time. There are still two state variables and there is nothing that can be done to eliminate them and to keep the

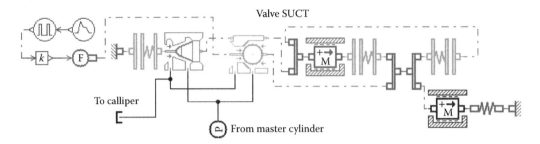

Figure 11.2.9 Suction Valve model for RT applications.

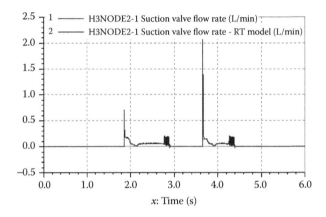

Figure 11.2.10 Comparison of the flow rate for the Suction Valve models.

geometry of the valve. Knowing that this valve is an ON/OFF valve, a simple model from the Hydraulic library can be used. The valve model is the same that was developed for the Outlet valve. This new model does not contain any state variables or the valve geometry (Figure 11.2.11).

The simplest Suction Valve model for RT applications has the structure shown in Figure 11.2.12.

To set the parameters for the new model, a calculation of the maximum opening area for the suction valve HCD model has to be done. Remember that this valve has two valves inside: (1) a ball poppet with conical seat and (2) a poppet with conical seat. To calculate the maximum opening area for the Suction valve, a calculation of areas for the two valves has to be done. To calculate the area, we are using the relation Figure 11.2.13:

$$\text{area}_1 = \left(D_E - h \cdot \sin\frac{\alpha}{2} \cdot \left(\cos\frac{\alpha}{2} + \sin\frac{\alpha}{2} \cdot \tan(\beta_1) \right) \right) \cdot \frac{\pi \cdot h \cdot \sin\frac{\alpha}{2}}{\cos(\beta_1)} \qquad (11.2.4)$$

where:

area_1 is the maximum flow area [mm²]

h is the poppet maximum lift [mm]

D_E is the seat hole diameter [mm]

α is the poppet angle [rad]

β_1 is the angle with the perpendicular (Figure 11.2.13)

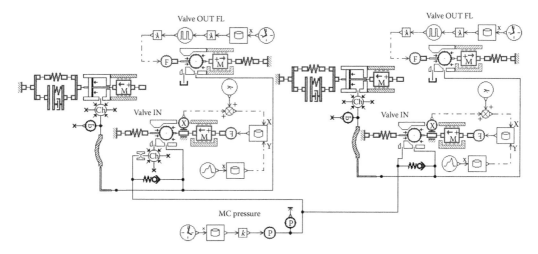

Figure 11.2.11 The new model without any state variables or valve geometry.

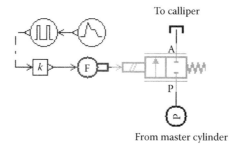

Figure 11.2.12 Simplest Suction Valve model for RT applications.

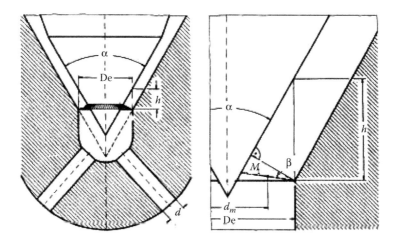

Figure 11.2.13 Conical poppet with conical seat.

$$\tan(\beta_1) = M_1 - \sqrt{M_1^2 - \frac{1}{2}} \tag{11.2.5}$$

Here,

$$M_1 = \frac{1}{4}\left(\frac{D_E \cdot \sin\frac{\alpha}{2}}{h \cdot \left(\sin\frac{\alpha}{2} + m''\right)} \cdot \left(1 + \frac{1}{\tan^2\frac{\alpha}{2}}\right) - \frac{1}{\tan\frac{\alpha}{2}} \right) \tag{11.2.6}$$

where m'' is the *correction of Hardenberg*:

$$m'' = \frac{(D_A - D_E) \cdot \cos\frac{\alpha}{2}}{2 \cdot \tan\frac{\sigma}{2}} \cdot \left(\tan\frac{\alpha}{2} - \tan\frac{\sigma}{2}\right) \tag{11.2.7}$$

To calculate the areas, a complex calculus for the poppet with conical seat has to be done. This calculus is made inside the Amesim HCD model of poppet with conical seat valve. A simpler solution is to plot the Amesim simulation results for the flow area and read the maximum value of the flow on the plot. To compute the maximum flow area, the two area values will be added:

$$\text{area} = a_1 + a_2 \tag{11.2.8}$$

where:
  $a_1 = 4.95481 \text{ mm}^2$
  $a_2 = 0.125664 \text{ mm}^2$
  $\text{area} = 5.08047 \text{ mm}^2$

The saturation input value is set to the maximum value of the solenoid force that acts on this valve. The area calculated before will be the value of the equivalent cross-sectional area at maximum opening Figures 11.2.14 and 11.2.15.

   This way, a simple Suction valve model was built. The model does not keep the valve geometry but has no state variable inside. It can be seen that the results are the same as for the model with valve geometry. As a conclusion, this model can be used for to build systems for real time.

11.2.2   Reduction of the ABS/ESP global model

In the previous section, each valve was reduced to reach a level suitable for real time. The complexity of the global system will be reduced by using some of the Amesim facilities for model simplification. The model has 43 state variables and the necessary CPU time for a simulation is very long and thus is not suitable for Real-Time system simulation.

   To be able to simplify the system, State count and Activity index facilities will be used (a presentation of the two facilities was made in the first phase of the project).

   Using the two facilities, it can be determined which components of the model can be eliminated to reduce the complexity but keeping the same results. Remember that the

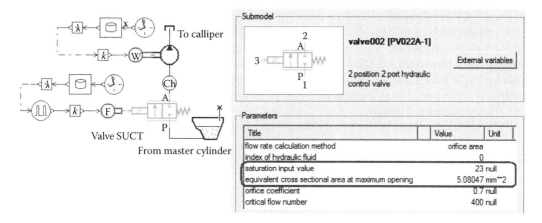

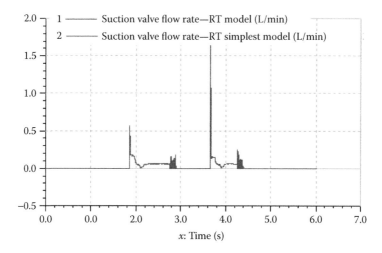

Figure 11.2.14 Suction Valve simplest model and parameters list.

Figure 11.2.15 Comparison of the flow rate for the Suction Valve RT models.

Activity Index and the State Count should be used in parallel: one confirms the results of the other.

From the Activity Index point of view, the most energy active components are the Inlet Valves and the calipers (end stops and springs). This facility confirms that the compressibility effects from the hydraulic chambers should be removed (those elements are energy passive).

The first phase of the global model simplification consists of:

- Eliminating the compressibility effect of the calipers hydraulic chambers and the compressibility effect of the Inlet valves hydraulic chambers.
- Replacing the Outlet and Suction valves with the simplest models developed in the previous section.

To eliminate the compressibility of the hydraulic chambers of the Inlet Valves, the chambers and the orifices will be eliminated. The dead volume of those chambers is very small and it has no impact on the compressibility of the system. Care should be taken when setting the parameters for the Inlet valves. For the caliper chambers, the hydraulic stiffness has to be taken into account because the volume of those chambers is important. The solution to eliminate the compressibility of the hydraulic chamber is to bring this chamber into the mechanical domain.

The hydraulic chamber is in series with the spring and the end stop of the caliper. To compute an equivalent mechanical stiffness for the Front Left caliper, the formula presented in the following will be used:

$$k_{mec\_equiv} = \frac{k_{mec} \cdot S^2 \cdot k_{hydr}}{k_{mec} + S^2 \cdot k_{hydr}} \qquad (11.2.9)$$

where:

k_{mec_equiv} is the equivalent mechanical stiffness
k_{mec} is the mechanical stiffness
k_{hydr} is the hydraulic stiffness
S is the area of the piston

The mechanical stiffness for the spring and for the end stop is known. To compute the hydraulic stiffness, an average value of the pressure has to be considered and the bulk modulus for that pressure has to be determined. To set the value for the equivalent bulk modulus (Bequiv), an average value of the pressure for the calipers has to be considered. The maximum value of the caliper pressure is around 140 bars. Using the bulk modulus variation function of pressure, the value of the bulk modulus at an average pressure of 70 bars can be found (Figure 11.2.16). To compute the equivalent mechanical stiffness for both spring and end stop, the formulas below will be used (equivalent stiffness of the spring):

$$k_{mec\_equiv}^{spring} = \frac{k_{spring} \cdot S^2 \cdot k_{hydr}}{k_{spring} + S^2 \cdot k_{hydr}} \qquad (11.2.10)$$

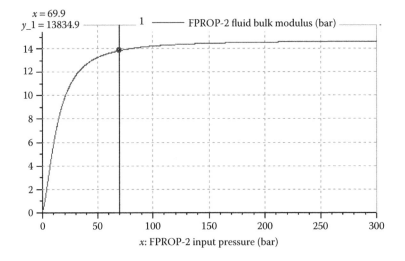

Figure 11.2.16 Bulk modulus at the average pressure of 70 bars.

After the calculation, the next values were obtained:

$$k_{mec\_equiv}^{spring} = 7.38 \cdot 10^6 \, N/m \tag{11.2.11}$$

Knowing the equivalent stiffness of the caliper components, a simple model can be built to verify that the results for the caliper compressibility are the same (Figure 11.2.17).

For the caliper models without hydraulic chamber, some specific end stop models are used (those models of end stops do not include a damper—same approach as that of the Suction valve). A comparison will be done between the old system with hydraulic chamber and the new system. For the new system, two models will be built. One model with the analytic computation of the equivalent stiffness (presented earlier). The other model with the equivalent stiffness will be determined by simulation.

The hydraulic chamber was eliminated. In order to have the equivalent volume consumption to increase the pressure, the caliper gap has to be modified. The value for the caliper gap will be set to 0.26 mm (from 0.21 in the initial end stop model) to compensate the loss of volume consumption (Figure 11.2.18).

Bringing the chamber in the mechanical domain generates some little differences between the compressibility of the two models (Figure 11.2.19). Those differences appear

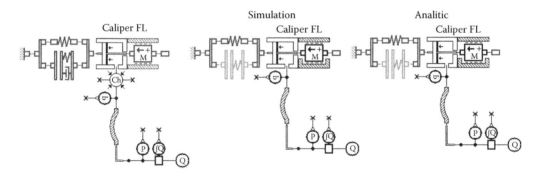

Figure 11.2.17 Model for validation of the equivalent stiffness (chamber in mechanical domain).

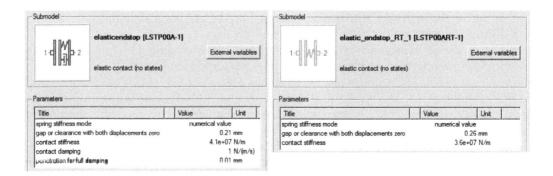

Figure 11.2.18 Parameter list for the old and new end stop models.

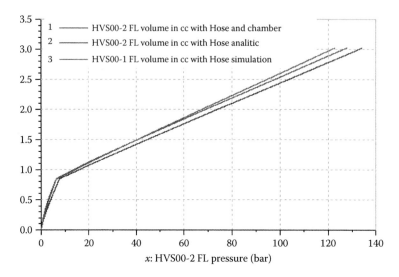

Figure 11.2.19 Comparison of FL caliper compressibility.

if an analytic calculation of the contact stiffness is done. This could be due to the fact that the bulk modulus was approximated: A bulk modulus was set for an average pressure of 70 bar.

In order to replace the Outlet and Suction valves with simplified models, in the previous sections of this chapter, the valves were simplified and reduced to a level that allows their utilization in Real-Time system simulation. Although each model has less elements and less state variable, it is able to keep the same simulation results. The last levels of simplification for the Outlet valve and Suction valve (remember that those are ON/OFF valves) will be used to simplify the global model.

After eliminating the caliper hydraulic chambers and using the Outlet and Suction valves as well as Inlet and Regulation valves simpler models, the complexity of the system was reduced by 14 state variables. If a comparison is made between the old global model and the first level of simplification, it can be seen that the CPU time is reduced from 84 s to 20 s (the simulation runs four times faster).

Looking at the plotted results (Figure 11.2.20), it can be seen that the results are still similar and we can say that the simplified models are validated for use in the global model.

The second phase of the global model simplification consists of

- Bringing the hydraulic stiffness of the hose into the mechanical domain.
- Eliminating the masses from the calipers and valves.
- Eliminating the lines and setting the parameters of the remaining hydraulic chambers to have a model compatible with Real-Time applications.

To bring the hose into the mechanical domain, the same technique will be used as for bringing the hydraulic chamber into the mechanical domain. In this case, only the simulation model will be used to determine the stiffness of the caliper.

As introduced in a previous section, for Inlet and Regulation valves, the mass icon model can include a mass or not. Here the model that does not include any mass parameter will be used. For the calipers the mass icon will be eliminated. To be able to do this, a specific submodel from the HCD library—BAP0RT will be used. This is a model with reversed causality.

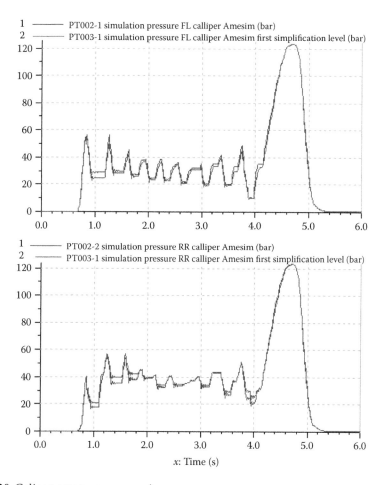

Figure 11.2.20 Caliper pressures comparison.

Instead of calculating a flow rate from a velocity input, the model calculates a velocity from an input flow rate.

Now all the components used for this new model will have associated submodels developed for real time (valves, springs, masses, end stops, etc.). This new system has only 12 state variables but it still has the hydraulic lines models. The elimination of the hydraulic lines models is the object of next section. The next step is to eliminate the hydraulic lines. There is no need to pass the hydraulic stiffness of the lines into the mechanical domain because the hydraulic stiffness of the two lines is much bigger than the hydraulic stiffness of the hoses (that were already passed into the mechanical domain), and the influence on the global stiffness of the calipers comes from the hoses and not from the lines. The lines will be eliminated without any other action. This last model (Figure 11.2.21) has reached level of simplification that should be able to run in real time regarding the number of state variables (this final model has only 8 state variables in comparison with the initial model that has 43 state variables).

Figures 11.2.22 through 11.2.31 contain the main result of simulations obtained by this model.

To be able to prove that the model is suitable for real time for sure, an Eigen values analysis will be done. In order to do this analysis, a set of linearization times will be set.

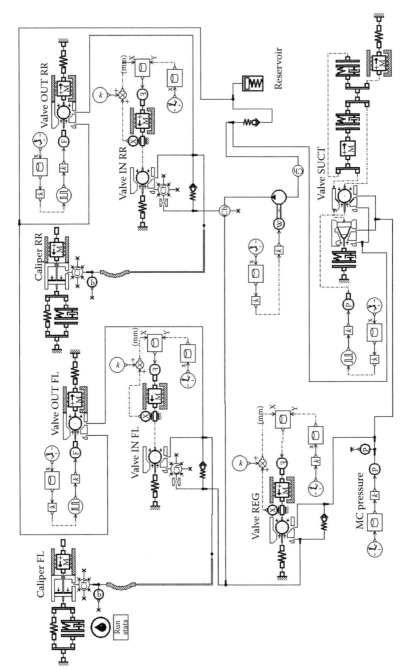

Figure 11.2.21 Global model of an ABS system for a braking circuit.

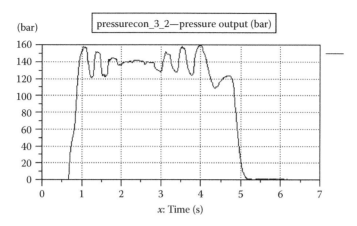

Figure 11.2.22 Pressure at the output of the master cylinder.

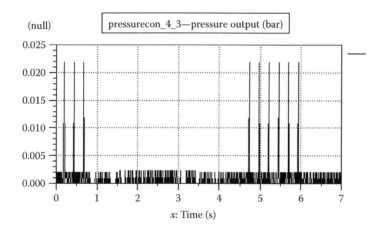

Figure 11.2.23 Output signal from the ABS controller.

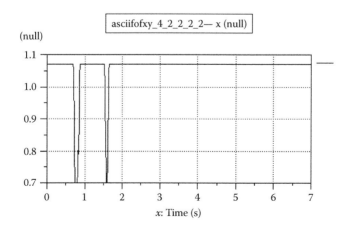

Figure 11.2.24 Error signal for the position loop of the valve regulation solenoid armature.

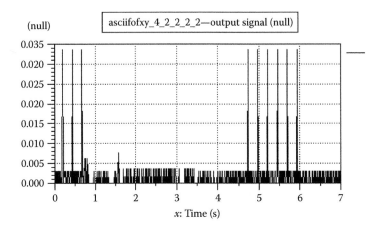

Figure 11.2.25 Output signal from the valve regulation controller.

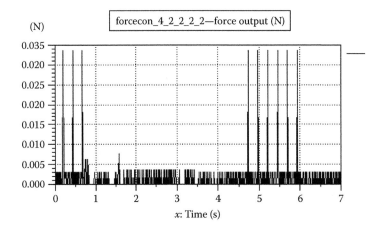

Figure 11.2.26 Electromagnetic force applied to the pressure control valve.

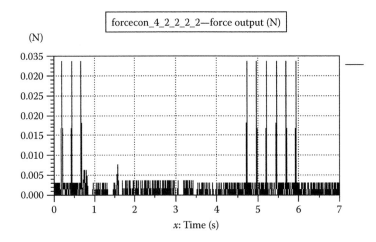

Figure 11.2.27 Force applied by the solenoid to the armature of the pressure control valve.

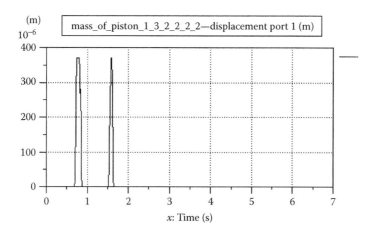

Figure 11.2.28 Mobile armature of the control pressure valve displacement.

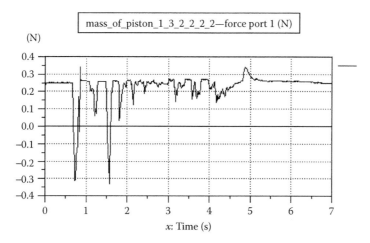

Figure 11.2.29 Force applied by the solenoid mobile armature on the pressure control valve ball.

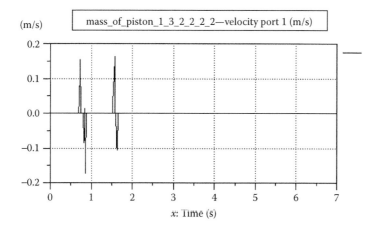

Figure 11.2.30 Pressure control valve mobile solenoid armature velocity.

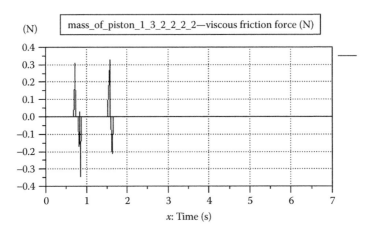

Figure 11.2.31 Pressure control valve mobile solenoid armature viscous friction force.

The mode of the hydraulic chamber (of the pump outlet) is the highest (9605 Hz). The real part of the Eigen value of this frequency shows that this chamber is too stiff. In this case, the model will not be able to run in with a fixed step integrator. In order to reduce the hydraulic stiffness of the chamber, a bigger value for the volume of chamber has to be set. The new value of the chamber volume should be a compromise: The value has to allow the model to run real time and the value should not influence the simulation results. The new value for the volume of chamber is set to 100 cm^3.

With this new value, the mode of the hydraulic chamber changes and now the value allows to use the model for real time. A quick comparison of the CPU times between this model and the initial one (with 43 state variables) shows a big improvement as follows: The CPU time drops from 80 s to 5.4 s. This simulation has been done with a variable step. A new simulation is done with a fixed step of 0.5 ms and by using Euler as an integration method. The results are very encouraging. Of more importance is now to see a comparison between the simulation results of this Real-Time model and the experiments' results (Figure 11.2.32).

It can be seen that the results for the final model are good enough. There are still some little differences for the Rear Right caliper but those differences are of the same order as for the initial complex global model.

11.2.3   Conclusion

As a first conclusion, after these simplifications, the model has reached a level that can be used for real time with simulation results that are very close to the experimental results. In this chapter, the simplification process for each valve was completed. Using some Amesim facilities for model simplification (State Count and Activity Index), some reduced models have been obtained. Those models have a minimum number of state variables with components still including physical parameters that are able to run with a reduced CPU time.

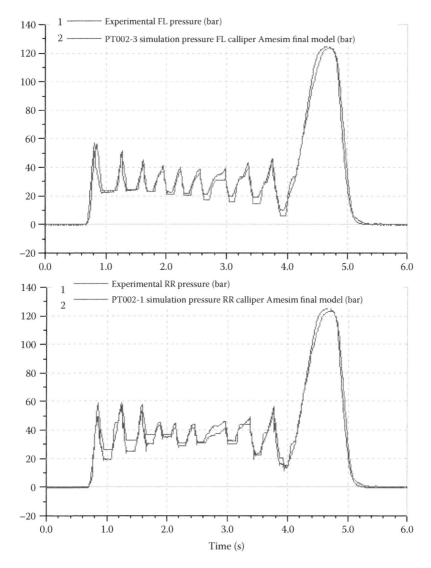

Figure 11.2.32 Comparison between initial model and final model (fixed step simulation).

In a second step, the valve models were integrated in a global model and all the steps for global model simplification were presented. The same Amesim facilities for model simplification (as for the valves) were used as well as Linear Analysis.

For each step of the simplification, there was an analysis of the simulation results to verify that the results are close to the expected ones. A very important aspect of model simplification is the difference between the main circuit and the pilot. Both should be simplified independently. One power element of these two parts should remain to give access to the right behavior of the system. The finalization of each system components allows us to continue the model final integration and to do the tests and validations required for the final conclusion of the work. Different methods based on physical or energy considerations [1–6] can lead to similar conclusions, useful for the HiL simulation.

11.3   Validation of the Real-Time global model by comparison with the experimental data

11.3.1   Validation with dry surface (asphalt) experimental data

In this case, the ABS is active. The brakes are operated in condition of dry asphalt, and the valves are piloted with the data from the dry asphalt experiments (Figure 11.3.1).

In the previous chapters, we have seen that the input data for the Master Cylinder pressure has a big influence on the caliper pressure. The experimental data had to be adapted to be able to use them as inputs for the model. There are two types of experimental data, depending on the acquisition sample time: (1) a file with 1 ms and (2) a file with 10 ms acquisition sampling time.

The model had to be also modified by adding some delay models for the commands of the valve (the data from the 1 ms file have a delay of 0.19 s in comparison with the data from 10 ms file). The delay model is added for all the valves of the model. The results found using the 1 ms sampling file and the delay model are more accurate. The simulation was made with a variable step because the delay models have a bad influence on the simulation results with a fixed step.

11.3.2   Validation with compact snow experimental data

In this case, the ABS is active. The brakes are operated in condition of compact snow and the valves are piloted with the data from the compact snow experiment (in this case, only the 10 ms file with engine control unit is used).

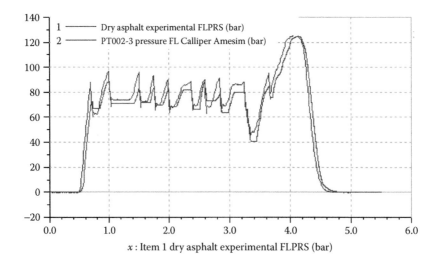

Figure 11.3.1 Validation with dry asphalt data.

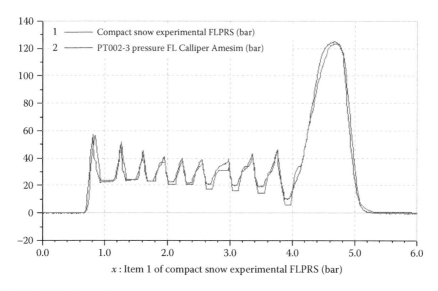

x : Item 1 of compact snow experimental FLPRS (bar)

Figure 11.3.2 Validation with compact snow data.

The results of the simulation are very close to the experimental results. This simulation with fixed step confirms a bigger difference between the simulation and the experiments, for the Rear Right caliper—the same difference as for the validation complex models (Figure 11.3.2).

11.3.3   Validation with very frozen snow experimental data

In this case, the ABS is active. The brakes are operated in condition of very frozen snow. The valves are piloted with the data from the very frozen snow experiments (Figure 11.3.3).

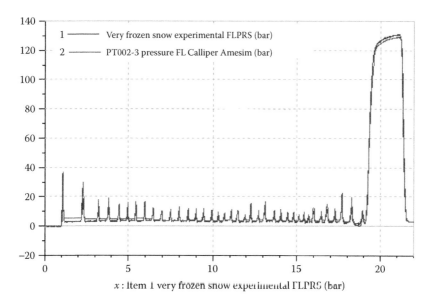

x : Item 1 very frozen snow experimental FLPRS (bar)

Figure 11.3.3 Validation with very frozen snow data.

11.3.4 Validation with mixed surface brake experimental data

In this case, the ABS is active. The brakes are operated in condition of dry asphalt for the rear right wheel and frozen snow for the front left wheel. The valves are piloted with the data from the mix braking surface experiments (Figure 11.3.4).

11.3.5 Validation on mixed surface acceleration experimental data

In this case, the ESP is active. The brake pedal is not operated. Only the pump is used to increase the pressure in the calipers. The rear right wheel is on dry asphalt and the front left wheel is on frozen snow (Figure 11.3.5).

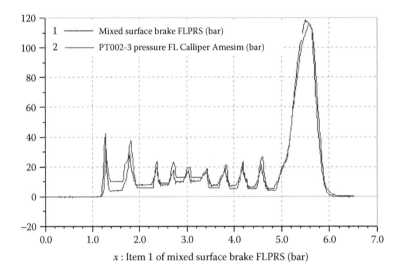

Figure 11.3.4 Validation with mixed surface brake experimental data—FLPRS.

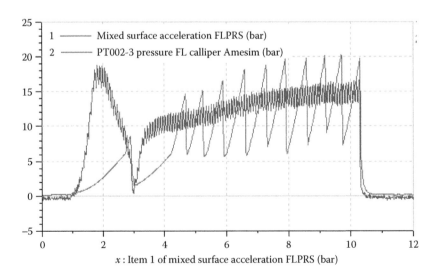

Figure 11.3.5 Validation with mixed surface acceleration data—FLPRS.

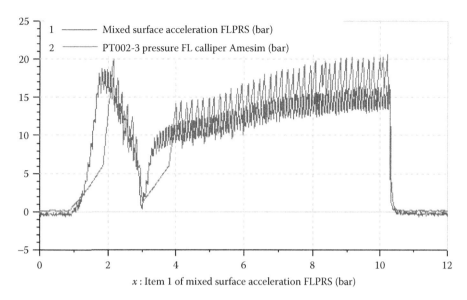

Figure 11.3.6 Validation with mixed surface acceleration data.

The comparison of the results does not look so good. It can be seen that in the first phase, the pressure does not increase as it should do. Remember that in the first phase of the project, there was a mention about the influence of the caliper gap for the caliper pressure rise in this case. There was also a mention about the influence of the air content on the pressure rise. In this case, there is another factor that has a direct influence on the pressure rise—the volume for the hydraulic chamber (from the pump outlet) that was imposed to 100 cm^3. If this volume is reduced to 1 cm^3 and a simulation is made with a variable step (the fixed step is no longer a solution here because the system becomes stiff), as it can be seen in Figure 11.3.6 this volume has a big influence on the pressure rise. Let us not forget that the value of this volume has to be set as a compromise for the model to be feasible for real time.

Another factor with influence on the pressure rise is the model of the pump. In the experimental data, the input for the pump model is represented by a percentage of the pump motor duty. Additional information is needed for the inputs of the pump to be able to develop a model that can be feasible for this case of simulation.

11.3.6   Conclusion

The synthesis of such a complex digital control system such as ABS/ESP needs a huge effort of writing proper mathematical models and then simplifying them to run in real time. The successful researches developed in by Lebrun M., Dragne F.D., Alirand M., Baţăuş M., and Ion Guţă D.D. [1–5] have shown the way to be followed in each similar complex cases such as common rail fuel injection systems, VVT, and so on using the Amesim facilities such as State Count. However, the RTS models' validation has to cover all the possible operation situations connected to the temperature drift of all the components, the transient occurring in extreme conditions, and so on.

Section 1: Bibliography

1. Alirand, M., Orand, N., Lebrun, M., Model simplification for nonlinear hydraulic circuits, IMECE2005-79329, *International Mechanical Engineering Congress & Exposition* November 5–11, 2005.
2. Alirand, M., Jansson, A., Debs, W., A first idea of continuity in model simplification for hydraulic circuit—an automatic gearbox real time application, *The Ninth Scandinavian International Conference on Fluid Power*, SICFP'05, June 1–3, 2005, Linköping, Sweden.
3. Alirand M., Favennec G., Lebrun M., Pressure component stability analysis: A revisited approach. *International Journal of Fluid Power*, 3(1): 33–47, 2002.
4. Băţăuş, M., Cancel, L., Dragne, F.D., Oprean, I.M., Vuluga, B., The study of the gearshift process in automated mechanical transmissions—EAEC, European Automotive Congress, Belgrad, Serbia & Muntenegru, 2005.
5. Băţăuş, M., Dragne, F.D., Oprean, I.M., The analysis of the gearshift process in an automated mechanical transmission–CAR 2005, Pitesti, Romania, 2005.
6. Băţăuş, M., Gallo, F., Ripert, P.J., Real-time simulation of detailed powertrain models, *Advanced Transmissions for Low CO2 Vehicles*, June 4, Paris, France, 2008.
7. Băţăuş, M., Maciac, A.N., Oprean, I.M., Vasiliu, N., Automotive clutch models for real time simulation, *Proceedings of the Romanian Academy, Series A: Mathematics, Physics, Technical Sciences, Information Science*, No. 2/2011.
8. Băţăuş, M., Maciac, A.N., Oprean, I.M., Vasiliu, N., Real time simulation of complex automatic transmission models, Proceedings of *Virtual Powertrain Creation, 2010*, ATZ Live, München, Germany, 2010.
9. Băţăuş, M., Maciac, A.N., Oprean, I.M., Andreescu, C., Vasiliu, N., Real time simulation of drivetrain launch devices. *The 11th International Congress CONAT 2010*, Vol. 2, paper CONAT20101078, pp. 229–236. ISSN 2069-0401.
10. Călinoiu, C., Vasiliu, N., Vasiliu, D., Catană, I., Modeling, simulation and experimental identification of the hydraulic servomechanisms. Technical Press House Bucharest, 1998 (in Romanian).
11. De Groen, O., Kok, D., *Rechenprogramm zur Simulation von Hochdruckeinspritz systemen für Nutzfahrzeuge*, Motortechnische Zeitschrift MTZ – 57, 1996.
12. Dragne, F.D., Băţăuş, M., Oprean, I.M., The selection of an appropriate gear ratio in an automatic city-bus transmission. *The 10th International Congress CONAT 2004*, The Automobile and Future Technologies, Brasov, Romania, 2004.
13. Dragne, F.D., Alirand, M., Oprean, I.M., Vasiliu, N., ABS valve model reduction by AMESIM. *Proceedings of the Romanian Academy, Series A: Mathematics, Physics, Technical Sciences, Information Science*, 10(2): 189–196, ISSN 1454-9069.
14. Dragne, F.D., Băţăuş, M., Oprean, I.M., The possibilities of selecting the gear ratios in an automatic transmission according to the driving conditions. *Review Mechanisms and Robotics*, 4(2): 19–24, 2005.
15. Dragne, F.D., Cancel, L.A., Oprean, I.M., Vasiliu, N., Valve model simplification and validation using tools from Imagine.Lab—AMESim Simulation software, ESFA International Conference Bucharest, Romania, 2009.
16. Ion Guţă, D., Vasiliu, N., Vasiliu, D., Călinoiu, C., Basic concepts of real-time simulations (RTS). *U.P.B. Scientific Bulletin*, Series D, 70(4): 291–300, 2008.
17. Lebrun, M., Vasiliu, D., Vasiliu, N., Numerical simulation of the fluid control systems by AMESim. *The 15th International Conference on Control Systems and Computer Science CSCS15*, Bucharest, Romania, 2005.
18. Louca, L.S., Stein, J.L., Hulbert, G.M., Sprague, J., Proper model generation: An energy-based methodology. *3rd Int. Conference on Bond Graph Modeling and Simulation*, Phoenix, AZ, January 1997, pp. 1–6.
19. Merritt, H.E., *Hydraulic Control System*, 1st ed. John Wiley & Sons, New York, 1967.
20. Oprean, I.M., Băţăuş, M., Dragne, F.D., The impact of the automobiles on the green-house effect, *CIEM 2005*, Bucharest, Romania, 2005.
21. Oprean, I.M., *The Modern Car*, Romanian Academy Press House, Bucharest, 2003 (in Romanian).

22. Panait, T., Bățăuș, M., Dragne, F.D., Oprean, M., Car Recycling for a cleaner Future Car, Pitesti, Romania, 2005.
23. Popescu, T.C., Vasiliu, D., Vasiliu, N., Numerical simulation—a design tool for electro hydraulic servo systems, *Numerical Simulations, Applications, Examples and Theory*, Intech press, Zieglergasse 14 1070 Vienna, Austria.
24. Vasiliu, N., Vasiliu, D., Mare, J.Ch., Using SIMULINK and ACSL for the simulation of a hydraulic power systems. *Proceedings of the European Simulation Multiconference*, Lyon, France, pp. 185–189, 1993.
25. Vasiliu, D., Vasiliu, N., Catană, I., Validating the transfer function of a hydraulic servomechanism by numerical simulation, *Bulletin U.P.B.*, Series C, 56(Nr.1–4): 69–75, 1994.
26. Vasiliu, N., Călinoiu, C., Vasiliu, D., Ion Guță, D., High accuracy electro hydraulic servo mechanisms with additional derivative feedback. *Recent Advances in Aerospace Actuation Systems and Components*, June 2007, Toulouse, pp. 183–188.
27. Vasiliu, N., Vasiliu, D., *Fluid Power Systems*, Vol. I, Technial Press Bucharest, 2005 (in Romanian).
28. Ye, Y., Youcef-Toumi, K., Modeling in the physical domain: An optimization-based approach, *2002 ASME International Mechanical Engineering Congress & Exposition*, New Orleans, LA, November 2002, pp. 1–8.

Section 2: Bibliography

1. Alirand, M., Orand, N., Lebrun, M., Model simplification for nonlinear hydraulic circuits, IMECE2005-79329, *International Mechanical Engineering Congress & Exposition*, November 5–11, 2005.
2. Alirand, M., Favennec, G., Lebrun, M., Pressure component stability analysis: A revisited approach. *International Journal of Fluid Power*, 3(1): 33–47, 2002.
3. Dragne, F.D., Alirand, M., Oprean, I.M., Vasiliu, N., ABS valve model reduction by AMESIM. *Proceedings of the Romanian Academy, Series A: Mathematics, Physics, Technical Sciences, Information Science*, 10(2): 189–196, ISSN 1454-9069.
4. Dragne, F.D., Cancel, L.A., Oprean, I.M., Vasiliu, N., Valve model simplification and validation using tools from Imagine.lab—AMESim Simulation software. *ESFA International Conference*, Bucharest, Romania, 2009.
5. Ion Guță, D., Vasiliu, N., Vasiliu, D., Călinoiu, C., Basic concepts of real-time simulations (RTS). *U.P.B. Scientific Bulletin*, Series D, 70(4): 291–300, ISSN 1454-2358, 2008.
6. Louca, L.S., Stein, J.L., Hulbert, G.M., Sprague, J., Proper model generation: An energy-based methodology. *3rd International Conference on Bond Graph Modeling and Simulation*, Phoenix, AZ, January 1997, pp. 1–6.

Section 3: Bibliography

1. Dragne, F.D., Alirand, M., Oprean, I.M., Vasiliu, N., ABS valve model reduction by AMESIM, *Proceedings of the Romanian Academy, Series A: Mathematics, Physics, Technical Sciences, Information Science*, 10(2): 189–196, ISSN 1454-9069, 2009.
2. Dragne, F.D., Bățăuș, M., Oprean, I.M., The possibilities of selecting the gear ratios in an automatic transmission according to the driving conditions. *Revista Mecanisme și Manipulatoare*, 4(2): 19–24, 2005.
3. Dragne, F.D., Cancel, L.A., Oprean, I.M., Vasiliu, N., Valve model simplification and validation using tools from Imagine.lab – AMESim Simulation software, *ESFA International Conference*, Bucharest, Romania, 2009.
4. Ion Guță, D., Vasiliu, N., Vasiliu, D., Călinoiu, C., Basic concepts of real-time simulations (RTS), *U.P.B. Scientific Bulletin*, Series D, 70(4): 291–300, ISSN 1454-2358, 2008.
5. Lebrun, M., Vasiliu, D., Vasiliu, N., Numerical simulation of the fluid control systems by AMESIM. *The 15th International Conference on Control Systems and Computer Science CSCS15*, Bucharest, Romania, 2005.

chapter twelve

Numerical simulation and experimental tuning of the electrohydraulic servosystems for mobile equipments

12.1 Structure of the electrohydraulic servosystems with laser feedback used for ground leveling equipments

This part presents the use of numerical simulation and experimental identification of the electrohydraulic position servomechanisms with laser feedback set up on mobile equipments that are used for improving the quality of the lands in hydropower dams, agricultural lands, civil engineering works, and so on [1–5]. This type of servomechanism is manufactured by the Japanese Topcon Corporation [6,7] having a modular structure, which is specific for all the fluid-powered land leveling equipment (with scoop or blade). The specific tuning of this kind of servosystems needs the following activities:

- Numerical simulation of a servomechanism with laser position feedback
- Experimental identification of the operating parameters of a real complete system
- Laboratory identification activities on a device that reproduces the operation conditions of the electrohydraulic servomechanism mounted on a typical leveling equipment

Longitudinal dikes, which prolong transversal dams on riverbeds (Figure 12.1.1), are similar to land dams, the only difference between them is the parallel disposal with the river bed. These dikes are permanent, being subjected to the water action in most of their lifespan, having the function of limiting the flooded area by generating an accumulation lake.

The dam body is made of rocks taken from loan pits sited near the dam. The ground must have a certain granulometry with well-defined fractions for particles of different diameters. For a proper density repartition the ground must be laid in layers of a certain thickness, so that compaction performed before the next layer is laid down to be evenly done on the entire breadth. If the optimum breadth of the material laid down in layers in the dam body is not properly performed, according to the type of compaction equipment, it will lead to an over compaction of the dam crowning, which may generate water flood and infiltrations through the dam body.

For preventing this to happen, the dikes or dams' builders adjust the breadth of the ground layer before its compaction by leveling it with a special equipment that was created purposefully for this operation, provided with Topcon laser-controlled modular systems, manufactured in Japan or other similar systems such as Spectra Lasers Corporation, the inventor of the rotating laser [8].

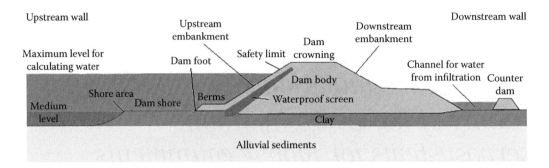

Figure 12.1.1 Main section through an earth dam or dike.

The objectives of this part of the book are listed as follows: a brief presentation of the Topcon laser feedback electrohydraulic modular servosystems, created especially for the leveling equipment [1]; a presentation of the device that simulates the operational conditions of Topcon [2]; the numerical simulation and the experimental identification of the Topcon laser feedback electrohydraulic modular servosystem set up on the test device [3], which includes two electrohydraulic servomechanisms [4], and the assessment of the experimental performance of the industrial system uses by an antegrade.

The leveling technology using laser (Figure 12.1.2) implies a leveling performed by a complex installation, which is able to perform work from two passes, a rough leveling, and a fine one at finish, with deviations from the reference plane of maximum 2.5 cm on the entire leveled surface and with a significant reduction of the tracking, transposition, and materialization process during the leveling project.

The modular mechatronic system with laser, electronic, and electrohydraulic components that allows reaching this leveling technology may be mounted on any land-leveling equipment whose work bodies, scoops, or blades are hydraulically powered. It is conceived as an additional option of the land-leveling equipment, which offers the

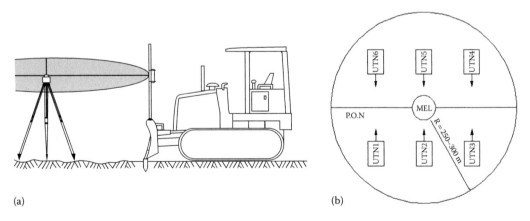

(a)   (b)

Figure 12.1.2 The laser-leveling technology: (a) setup of the laser modules transmitter and receiver and (b) automatic leveling after an optical leveling plan (PON) performed simultaneously by six land-leveling equipments UTN.

Figure 12.1.3 Leveling machine.

possibility of leveling land automatically to it, without any human-error occurrence in what regards precision.

In the classic acception, land leveling controlled by laser systems implies a modular system with the following structure:

- *The laser transmitter* placed in the center of the surface to be leveled above a point with a known quote mark on a tripod that may be adjusted vertically, emitting a laser beam in its rotation movement. This generates the laser reference plane or the optical reference plane (with programming options for the longitudinal and transversal slopes in the forward direction), which will be followed by the work body of the equipment during leveling. After setting the slope needed at leveling, the laser transmitter positions itself automatically.
- *The laser receiver,* whose support is connected to the work body of the land-leveling equipment, intercepts the laser fascicle generated by the laser transmitter and transmits altimetric information, namely the position of the work body accountable to the laser reference plane, to an electronic control and monitoring module, placed in the cabin of the land-leveling equipment.
- *The electronic monitoring and control module* that collects and amplifies the information supplied by the laser, compares it with a prescribed position specified for the leveling quote value, computes the error, and sends a reference input to an electrohydraulic drive system for cancelling the error.
- *The electrohydraulic system,* controlled by the electronic module has the role of driving the hydraulic cylinders of the blade for maintaining the work body in the leveling plane set by the leveling project, plane that is parallel to the laser reference plane. Figure 12.1.3 presents two types of land-leveling machines equipped with laser modular systems.

12.2   Test bench for simulation of the real operational conditions of the laser module on the equipment

Figure 12.2.1a shows the hydraulic part of the laboratory test device that simulates the real behavior of the Topcon laser-controlled modular system that is purposefully created for equipping the automatic land-leveling machines in horizontal plane. Figure 12.2.1b shows the test bench assembly [1]. The scheme of the simulator is presented in Figure 12.2.2.

On the rod of the upper hydraulic cylinder, the optical receiver of the laser beam is fixed that is vertically actuated by the upper cylinder, the bottom cylinder, or by both the hydraulic cylinders. The simulator for testing the laser-controlled equipment includes two electrohydraulic servomechanisms: (1) The first one simulates the real behavior of the hydraulic cylinders of the blade of the land-leveling machine and (2) the second one simulates the profile of the land to be leveled.

The first servomechanism contains a hydraulic cylinder, which is similar to the one mounted on the machine, supplied from the hydraulic delivery block Topcon depending on the level of detection of the laser reference plane and generated by a rotary laser transmitter Topcon.

The second servomechanism consists of a hydraulic cylinder controlled by a proportional valve with integrated electronics (OBE), by means of a data acquisition board, an industrial process computer (IPC), and the data acquisition software Test Point [9].

The Topcon electronic block receives the electric signal from the laser receiver, which is placed on the rod of the upper cylinder of the simulator. The signal size depends on the level of detection of the optical reference plane generated by the rotary laser transmitter. The control signal sent to the proportional valve of the Topcon hydraulic kit is proportional to the level error. According to this input, the rod of the bottom cylinder pulls or pushes the body of the upper cylinder in reverse direction to that of the displacement of the cylinder rod.

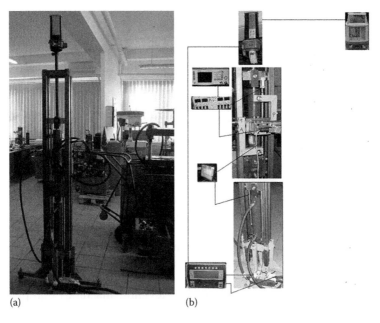

(a)   (b)

Figure 12.2.1 Simulator for testing the Topcon laser-controlled modular system: (a) hydraulic module and (b) test bench assembly.

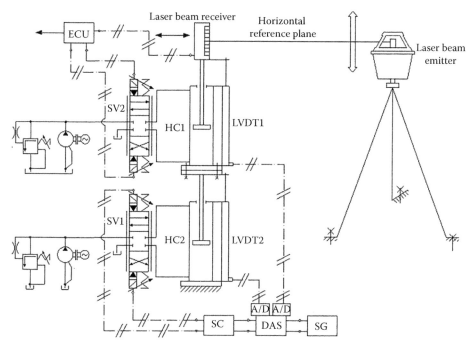

Figure 12.2.2 Scheme of the simulator for testing the laser controlled modular system. TOPCON: SV1, SV2—industrial OBE servovalves; HC1, HC2—hydraulic cylinders; LVDT1, LVDT2—displacement transducers; ECU—TOPCON electronic control unit; SG—signals generator; DAS—data aquisition system; A/D—analog/digital converters.

The upper cylinder is controlled in close loop by means of a servocontroller and by the action of a signal generator for simulating various profiles for the uneven land. The two inductive transducers of linear displacement of the cylinders are connected by means of a data acquisition board to an IPC with Test Point Das software.

12.3 Numerical simulation and experimental identification of the laser-controlled modular systems for leveling machine in horizontal plane

The dynamic behavior of the laser-controlled modular servosystem was simulated by Amesim using the realistic model, which is shown in Figure 12.3.1. All the components of the simulation network are based on mathematical models validated by practice. The algorithm of numerical integration of the set of algebraic and differential equations is chosen automatically. If the model is not correct or the inner and outer parameters are not properly chosen, the program does not work. Usually this happens because the differential equations are incompatible or undetermined.

The simulation model includes two electrohydraulic servomechanisms for controlling the position rod of the upper cylinder according to the feedback. It includes an inner control loop and an outer control loop.

The first inner loop is set at the level of the hydraulic servomechanism of simulation for uneven land, which is excited with step, sine, composed, and variable frequency sine signals.

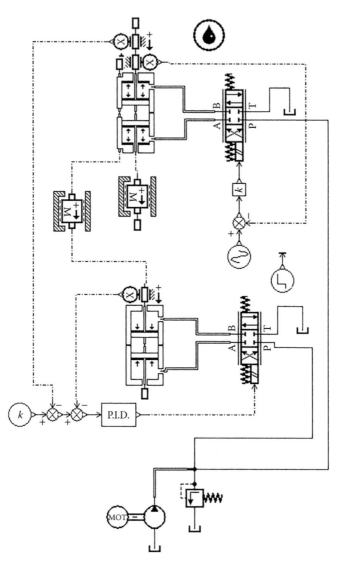

Figure 12.3.1 Amesim simulation model of a Topcon laser-controlled modular system set up on a testing device.

The second inner loop is set at the level of the servomechanism of monitoring with laser control, which is similar to the Topcon laser-controlled modular system.

The main control loop is realized between the output of the first servomechanism and the input of the second one.

A realistic input signal from the land was used for the servomechanism that generates the profile of uneven land: a composed signal with an amplitude of 0.14 m and a frequency of 0.033 Hz in an interval of 30 s. One curve from Figures 12.3.2 and 12.3.3 represents the displacement of the rod of the servocylinder that generates disturbances by moving the laser beam receiver (position transducer), and the other curve represents the displacement of the monitoring servocylinder rod coupled with the body of the disturbance generator servocylinder.

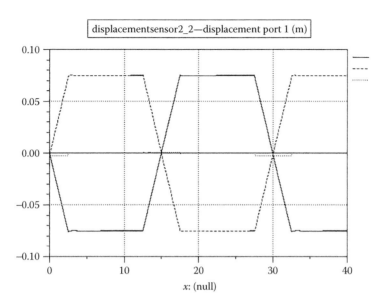

Figure 12.3.2 The response of the two servomechanisms for a composed input signal.

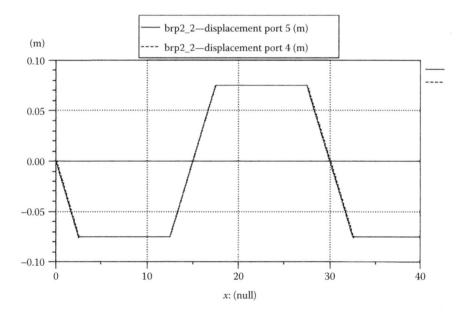

Figure 12.3.3 The response of the two servomechanisms for a composed input signal superposed by changing the output sign for pointing out the small following error.

Figures 12.3.2 through 12.3.8 show some of the significant numeric simulations as follows. Starting from the previous experience in the domain, a realistic test signal is composed by two ramps (up and down) and a constant sequence. Consequently, the simulations presented in Figure 12.3.2 show the system response for such signals with an amplitude of 0.14 m and a period of 30 s running 40 s. Changing the output sign and superposing the responses of the two servosystems are easier to point out the small following error of the whole system (Figure 12.3.4). The deviations of the profile of the leveled land from the optical horizontal reference plane have a maximum value of 0.8 mm.

Figure 12.3.5 presents another realistic case when the servomechanism generating the profile of uneven land is excited with a constant sinusoidal signal with an amplitude of 0.15 m and a frequency of 0.05 Hz, running 50 s. The meaning of the curves 1 and 2 is the same as that of the curves shown in Figure 12.3.2. The maximum following error is very small: about 1.8 mm, as shown in Figure 12.3.6.

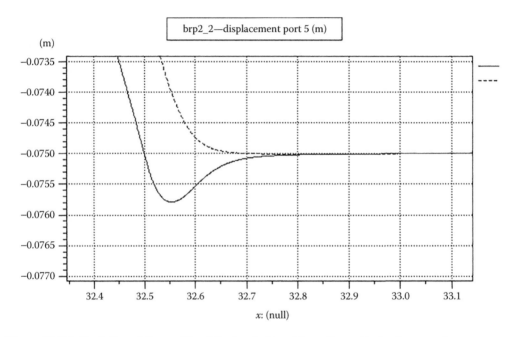

Figure 12.3.4 Deviation of the profile of the leveled land from the optical reference plane.

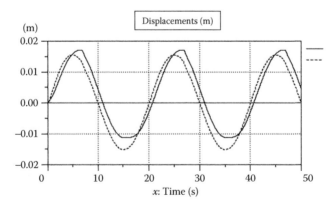

Figure 12.3.5 The answer of the laser-monitoring mechanism (continuous line) excited with constant sinusoidal signal (dot line).

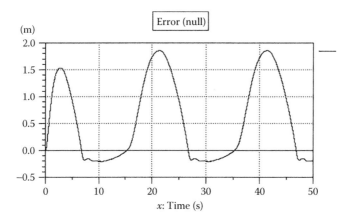

Figure 12.3.6 Deviation of the profile of the leveled land from the optical reference plane for a constant frequency sinusoidal input signal (maximum 1.8 mm).

A more realistic test needs the use of a variable frequency sinusoidal input signal with variable frequency from Figure 12.3.7. The simulation result is presented in Figure 12.3.8.

The graph shown in Figure 12.3.9 results from the algebraic sum of the curves shown in Figure 12.3.8, which leads to the deviation of the profile of the leveled land from the optical reference plane shown in Figure 12.3.10. For practical purposes, the control deviations can be neglected because the maximum values are below 8 mm.

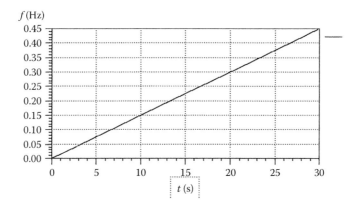

Figure 12.3.7 The variation of the frequency of the sinusoidal signal.

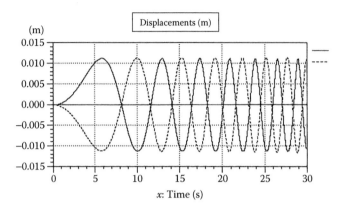

Figure 12.3.8 The answer of the laser-monitoring servomechanism at exciting the servomechanism generator of profile with variable frequency sinusoidal signal.

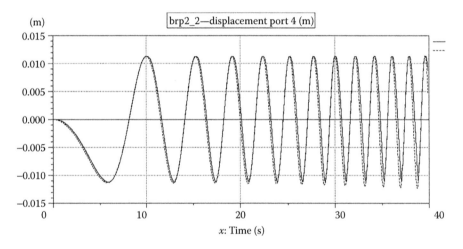

Figure 12.3.9 Dynamic frequency response of the two servomechanisms.

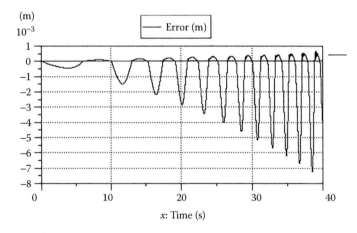

Figure 12.3.10 Deviation of the profile of the leveled land from the optical reference plane for a variable frequency sinusoidal input signal (maximum 8 mm).

12.4   Experimental identification

The results of the experimental identification of the Topcon laser-controlled modular system mounted on test devices are shown in Figures 12.4.1 through 12.4.6.

Figure 12.4.1 shows the dynamics of the laser-controlled hydraulic monitoring system when a constant sinusoidal signal with a frequency of 0.025 Hz and an amplitude of 0.072 m is applied at the input of the hydraulic mechanism generator of uneven land profiles. The test took 50 s and it proved a proper dynamic of displacement of the monitoring servosystem toward the generator of uneven land profile.

The graphics shown in Figure 12.4.2 were obtained by repeating the test with the same frequency of the sinusoidal signal of excitation 0.025 Hz but with a higher amplitude 0.080 m. The test time was 46 s and the results have shown a proper behavior of the monitoring servomechanism with laser control again.

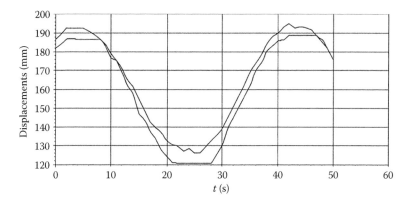

Figure 12.4.1 The answer of the laser-monitoring mechanism at the excitation of the servomechanism generator by a constant sine signal of 0.025 Hz and 72 mm amplitude—upper curve: input signal and lower curve: output signal.

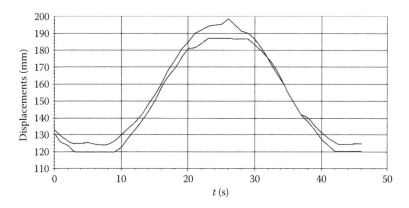

Figure 12.4.2 The answer of the laser-monitoring mechanism at the excitation of the servomechanism generator by a constant sine signal of 0.025 Hz and 80 mm amplitude—upper curve: input signal and lower curve: output signal.

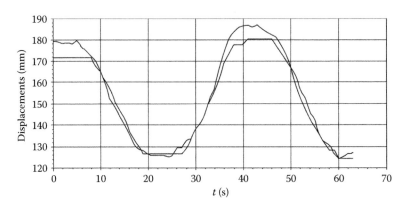

Figure 12.4.3 The answer of the system to the excitation of the servomechanism generator of profile with triangular signals of 0.025 Hz and amplitude of 60 mm—upper curve: input signal and lower curve: output signal.

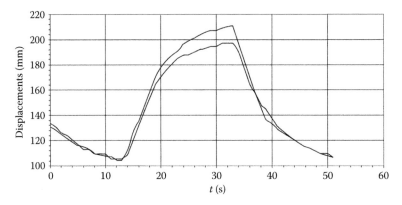

Figure 12.4.4 The answer of the laser control-monitoring mechanism at the excitation of the servomechanism generator of profile with rectangular frequency signals of 0.025 Hz and amplitude of 105 mm—upper curve: input signal and lower curve: output signal.

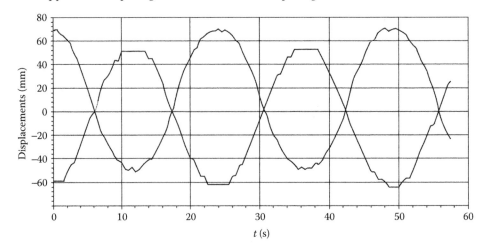

Figure 12.4.5 The answer of the two servosystems for the excitation of the generator of profiles with a constant sinusoidal signal of 0.020 Hz and 65 mm amplitude, and response amplitude of 55 mm.

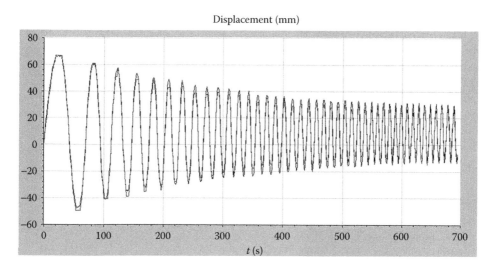

Figure 12.4.6 The frequency response of an industrial laser feedback servomechanism set up on an autograder—maximum error: 25 mm.

Figure 12.4.3 shows the dynamic of the monitoring hydraulic servosystem with laser control, when a constant triangular signal with a frequency of 0.025 Hz and an amplitude of 0.060 m, which takes 63 s is applied at the input of the hydraulic servosystem generating uneven land profiles. The test proves the proper work of the laser-controlled servomechanism. Figure 12.4.4 shows the dynamics of the hydraulic servosystem with laser control when a constant rectangular signal with a frequency of 0.025Hz and an amplitude of 0.105 m is applied at the input of the hydraulic servomechanism generator of uneven land profiles. The test time was 51 s.

During all the tests presented previously, the inductive transducers of linear displacements of the hydraulic cylinders were connected in such a way that the two graphs are overlapped for easy assessment of the dynamic behavior of the hydraulic servomechanisms. Figure 12.4.5 shows the dynamics of the hydraulic system with laser control when a constant sinusoidal signal with a frequency of 0.020 Hz and an amplitude of 65 mm is applied at the input of the hydraulic mechanism generator of uneven land profiles. The test needed 115 s and a constant sinusoidal one was used as an excitation signal; the inductive transducers of linear displacement of the hydraulic cylinders were set so that they can offer information regarding the real direction of displacement of the cylinders.

Figure 12.4.6 shows the frequency response of a real electrohydraulic servosystem with laser feedback for a sinusoidal input signal sited in the frequency range of 0.01–0.10 Hz. The amplitude range limits are 115–34 mm for a time test of 694 s [4].

12.5   Conclusion

The laser leveling of the land layers laid down when making the dam and the land dikes from the hydropower systems represents a safe and efficient solution for providing optimum breadth with maximum errors of 2.5 cm on the entire surface of the laid layer. This kind of leveling performed before compaction of each land layer provides a proper and homogenous compaction of the dam and represents an optimum solution for reducing infiltrations and falling of the crowning, which may lead to water flood.

The laser-controlled modular systems such as Topcon or similar ones are now present in enterprises not only for the most modern land-leveling machines. They may be mounted on any kind of hydraulic power systems of a land-leveling machine, no matter of the degree of wear or the manufacturer.

The dynamics and steady-state performances obtained by a comparative simulation and laboratory test of a Topcon laser-controlled modular system set up on an autograder that performed an automatic leveling and then on a test device connected to a simulator that is properly equipped with electrohydraulic drives are comparable. The match of the numerical simulation and the laboratory validation are very useful tools for the preliminary tuning of any leveling system before setting up on any leveling machine.

Bibliography

1. Popescu T.C. 2008. *Researches on the Synthesis of the Hydraulic Drive Systems*, PhD Thesis, University Polytechnica of Bucharest, Bucharest, Romania (in Romanian).
2. Popescu T.C., Vasiliu D., Vasiliu N., Calinoiu C. 2011. Applications of the electrohydraulic servomechanisms in management of water resources. In *Numerical Simulations, Applications, Examples and Theory*, Intech Press, Vienna, Austria, pp. 447–472.
3. Popescu T.C., Vasiliu D., Vasiliu N. 2011. Numerical simulation—A design tool for electro hydraulic servo systems. In *Numerical Simulations, Applications, Examples and Theory*, Intech Press, Vienna, Austria, pp. 426–446.
4. Popescu T.C. et al. 2008. Numerical simulation and experimental identification of the laser controlled modular system created for equipping the terrace leveling installations. In *Proceedings—Reliability and Life-time Prediction, 2008 31st International Spring Seminar on Electronics Technology*, IEEE, Budapest, Hungary, May 7–11, 2008, pp. 336–341.
5. Popescu T.C., Sovaiala Gh., Nita I. 2008. Device for testing laser controlled modular systems, Patent Demand CBI no. A/00586-2008.
6. http://www.topcon.com.sg/survey/rl01.html.
7. http://www.topcon.co.jp/en/positioning_top/.
8. http://www.spectralasers.com/en/products/laser-levels.html.
9. https://www.dataq.com/products/test-point/tstpnt.htm.

chapter thirteen

Using Amesim for solving multiphysics problems

13.1   Real-Time systems and Hardware-in-the-Loop testing

Hardware-in-the-Loop (HiL) has been used extensively used mainly in the aerospace industry and military applications due to the high cost needed for obtaining the required computing power. Recent advances in electronics and computer science have shown that a cost-effective solution for running complex systems simulation is available, thus widening the field of applications for HiL testing. This chapter presents the basic tools and principles that are used in the development of such a platform, which enables the study of electrical power trains, especially those designed for road vehicles. A general-purpose test rig was developed, whose preliminary structure is briefly described.

Computer simulation of mathematical equations has a long history and a well-established place in the engineering design. After computing machines became powerful enough to be able to control real equipment, a new field of engineering emerged. At the beginning, processing units were based on analog circuitry and ran faster than the systems that are being controlled. The command signal rates were limited to adapt them to the processes' evolution rate. With the advent of digital computing systems, the controllers were commonly slower than the supervised plants. The notion of real time emerged, referring to the computer's generic reaction time relation with the process that it controls.

It is hard to define what a Real-Time system is, given the variety of computing platforms and applications that exist today. Two main characteristics are common to all real-time systems and could define them: (1) They are bounded to a process that evolves at its natural rate and (2) they respond to inputs in an absolutely determined manner. The core of every Real-Time system is a computing platform, which is composed of a hardware part and a software part. The software can be further splitted into two entities: (1) the operating system and (2) the user-defined programs. A Real-Time operating system is the brain that manages every aspect of the system. It is usually generic, meaning that the same basic functionality can be ported to multiple platforms. Users can develop custom programs that implement specific functions using low-level routines that are hardcoded into a kernel; the advantage is that the probability of malfunction is decreased dramatically as the hard-tested kernel takes care of the resource management.

Common users for Real-Time systems are satellites, military equipment, avionics, medical devices (noninvasive scanners), command and control for power plants (nuclear mostly), airbags, and even mobile phones. Some of the used cases imply a strict no-error policy, whereas others can tolerate occasional less-than-critical malfunctions. Based on the complexity and scale of the damage produced, if the system fails, there are hard and soft Real-Time systems. However, this is not a proper classification as it lacks a measure for estimating the *softness* of a system. A soft Real-Time system is the one that runs as fast as the corresponding process but either lacks a Real-Time operating system or cannot guarantee the response time. For example, video decoding is said to be a Real-Time operation because it must be present to the viewer in a frame for every 1/25th of a second, but if there is a problem and a few frames are skipped, then the error is not a critical one and the system's functionality is not compromised. In case of a nuclear power plant, even a generic error could generate a planetary disaster, so it is difficult to give a complete definition of a Real-Time system.

HiL testing fits very well in the context of Real-Time systems. The concept is not new, but the tools used for it have evolved to the point where they became widely available as off-the-shelf components. There is of course a lot of work to be done for integrating them into the final product. Looking at the V-shaped diagram of product development from Figure 13.1.1, one can see that HiL is a process that connects the component development level with the system integration and testing level.

A HiL system intrinsically requires a Real-Time operating system as it must drive a specified equipment in the same way in which it would be driven in the product for which it was designed. By imposing timing constraints and enough priority levels, it can be guaranteed that even with a generic platform, the obtained results are accurate and the functionality can be reproduced on the target device. HiL started with engine control unit (ECU) testing as it did not have moving parts or different subsystems, but later it started to incorporate devices such as shock absorbers, actuators, hydraulic systems, and so on. This was the transition from *Model-in-the-Loop* or *Software-in-the-Loop* to system-wide testing and integration. This evolution shortened the development cycle with up to 75%. HiL systems can be regarded as hybrid or multiphysics systems, because they involve components from different industries and integrate them into a fully functional mock-up of the studied equipment.

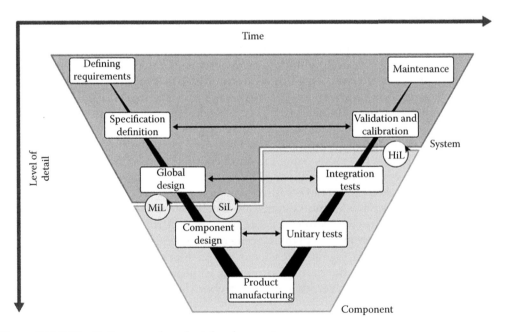

Figure 13.1.1 The V-diagram of product development.

13.2 Objectives of the Hardware-in-the-Loop simulation of the road vehicles electrical power train

This part of the book presents an original methodology to deal with specific problems in Real-Time simulation and testing (HiL) illustrated by a practical solution patented by Cătălin Vasiliu [1] to a technical problem of great interest: the study of electric transmissions used for traction. To meet market requirements, industry needs innovative design techniques that allow shortening development cycles of new products, and yet they are flexible enough to be reused for several projects. This section describes a modular architecture (Figure 13.2.1) that includes both hardware and software. To prove the concept, a test bench has been designed, which contains two vector-controlled electric motors, three Real-Time units from National Instruments' PXI range, and an industrial PC. The electric motors are mechanically coupled to the same shaft. Real-Time units perform three individual tasks: (1) acquisition of electrical parameters, (2) electric vehicle command and control, and (3) the simulation of the mathematical model. The mathematical model is developed using Simcenter Amesim includes all the subcomponents of a car. It is

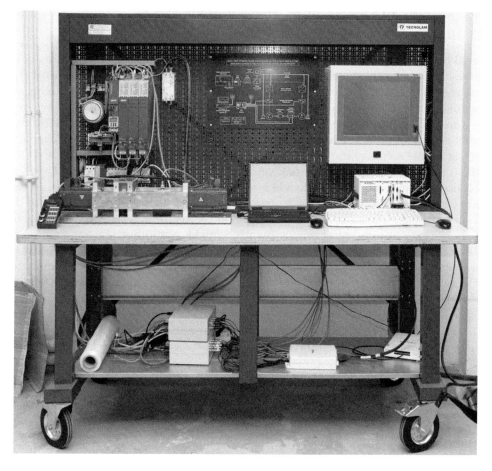

Figure 13.2.1 Test bench for electric power train from the Fluid Power Laboratory of U.P.B. (Courtesy of Catalin Vasiliu.)

integrated using partially a LabVIEW toolkit, a graphical user interface (GUI), and a controller being created using the same software. Due to the complexity of the mathematical model, a fuzzy controller was chosen because it offers superior performance to the classical proportional–integral–derivative (PID). The test results gathered during a number of test runs with different scenarios showed that the proposed solution is viable. As modularity is a key concept, adaptation of electric motors of different types or sizes is as simple as replacing them and changing their physical parameters in the software.

Also, by developing innovative methods of converting mathematical models from Amesim, a high degree of independence between creating the models and Real-Time testing was achieved.

By using commercial off-the-shelf components, the proposed architecture avoids extensive research to design and build new devices that are required for simulation. The same principle can be applied to other engineering devices, with minor changes, proving its versatility.

The keywords of this work are Real-Time, simulation, testing, power train, fuzzy, electrical brushless motors, electric vehicle, and data acquisition.

13.3 Specific tools used in the development of a test bench for electric power train

The goal of the design is to create a HiL system for testing electric power trains of road vehicles. It should be able to test three main subsystems but not all simultaneously: (1) the electric motor, (2) the controller for the car, and (3) the car model. The motor can be tested as it would be mounted on a full-scale vehicle model, and the car model would be driving a real load. This opens the way to a multitude of scenarios: tuning of the inverters, testing different geometries of the actual electric machine, car controller development, parameter tuning, or model accuracy improvement. Although the same tests could be done on a real car, the time lost in getting the car ready, transporting systems in other locations, or the protection of confidential data makes HiL testing a much more cost-effective alternative and not only from an economical point of view. The HiL system includes a Real-Time operating system, data acquisition hardware, simulation software, actuators and, of course, the component being tested. The chosen software provides a multiphysics engineering platform that can handle almost any simulation task. In this work, Amesim from Siemens PLM Software (former LMS International [2,3]) and LabVIEW [4] from National Instruments are used as basic software components.

Amesim is a simulation environment that, as many others, offers an advanced user interface for building representations of engineering systems. What sets it apart is that it is based on bond graphs. Bond graphs are a manner to represent physical dynamical systems by modeling the energy transfer between components, based on the power conservation laws.

Bond graphs require that each component should have a real physical representation, and the greatest advantage is that by simply monitoring the power flow, enough information for validation is available. If other programs transmit information between blocks that have no intrinsic meaning, in Amesim, every signal will have a physical unit and significance attached, so it will be very hard to make mistakes.

The philosophy behind *LabVIEW* is that by using standard blocks, one can build virtual instruments that resemble their actual counterpart, neglecting the build of all the subsystems that are common to specific categories. By integrating drivers and utilities for communicating with the hardware boards made by National Instruments, it offers a solid base for data acquisition and processing.

With its Real-Time extension, LabVIEW can run complex HiL systems with minimal effort and hardware. It can import .dll files in a specific format, so that external models can be incorporated if necessary. The big advantage over other platforms is that it integrates everything needed in one package and does not rely on any third-party software. All other commercial Real-Time platforms, such as dSPACE [5], use MATLAB®/SIMULINK® to run in Real-Time.

13.4   Amesim simulation environment features used for Hardware-in-the-Loop

Amesim has one more notable characteristic besides the one already described: It allows multidomain engineering system simulation. A device can be modeled starting from the very basic structure up to whole systems of similar components (e.g., the common-rail fuel injection system of a car).

The graphical representations are International Organization for Standardization (ISO) symbols or intuitive icons that facilitate the building of self-explanatory diagrams. There are many ways to analyze a system to get meaningful and accurate results. There is a possibility to run the system to get the equilibrium values for the state variables: Use these values further for dynamic runs, hold the inputs constant and see if there are oscillations, and change the time step or method anytime during the simulation.

There is also a linear analysis mode, which can pinpoint models that have an unusual behavior.

The energy-based solving method also can indicate where the most CPU time is consumed by counting the number of submodel calls and by calculating an activity index, which corresponds to the importance of a certain submodel for the entire system. All these tools help to improve the solving of big systems to integrate them on the Real-Time platforms. When CPU time is an issue, every simplification counts and for the model of the car used, this is very important, which leads to a better overall performance.

For creating Real-Time capable models, there are two restrictions that cannot be overlooked: The model must not contain implicit state variables, and the model must be able to be integrated with a fixed-step solver with a time step as close as possible to the one being used in the Real-Time system. The solver built in Amesim uses a complex scheme of choosing the proper method to obtain the desired solution, making some systems run much faster than if using a fixed-step solver; however, variable-stepping is not implemented yet for Real-Time applications.

There are intensive research programs that focus on faster integration designed for real-time platforms using variable-step solvers or other techniques as model-partitioning; it is possible that in the future versions of Amesim, some of them will be applied. From the car models that are offered as demos in Amesim, targeted at different aspects such as handling, comfort, noise and vibration, and thermal or fuel management, a suitable model was chosen, one that can be run with a fixed-step solver and that can offer an easy interpretation of the data. The model encompasses all the mechanical subsystems of a car, being used for the study of passenger comfort and maneuverability. By replacing the torque source from the schematic with the real electrical motor, a very powerful HiL system was obtained. Of course, there are a lot of changes needed to make the model to work in Real-Time, of which most are related to replacing models with simpler ones or with models without implicit variables.

The big advantage of this example is that it can simulate the wind drag resistance and the road rolling resistance accurately and dynamically. These can be used as perturbations in the final system, making the identification of the desired parameters possible. Using the Amesim-LabVIEW interface, it is possible to run the Amesim model in LabVIEW and then to use the inputs and outputs on the virtual car to drive a virtual user interface and the real electric motors. In this setup, the Amesim model is solved using the fixed-step methods from LabVIEW and an intricate system of VI's to step through the equations while maintaining stability altogether. From Amesim's point of view, the procedure is simple enough to be done in a very short time.

13.5   Vehicle modeling in Amesim

Amesim offers a multitude of models for vehicles, ranging from simple to very complex. Two main aspects are being studied: dynamics, by analyzing parameters such as acceleration, power, speed, and comfort, which requires a structural analysis.

The chosen model is the most complex model available. Starting from this model, by adding and removing components, a model that is suitable for Real-Time simulation will be created. The model represents the entire vehicle structure, including dynamic, structural, and environment elements (road profile, wind, etc.). The model includes the following subsystems: chassis aerodynamics, front and rear suspension, elastocinematic elements of suspension for each wheel, advanced tire modeling, road profile, and power steering brakes (simple model). Even though not all the above values are displayed or used, the system still calculates them, which demonstrates the capabilities for simulation that Amesim possesses. It can be said that the model is completely a representative for a vehicle in all the fields except the internal combustion engine. If needed, however, the engine (as well as any other system) can be added without much difficulty. Figures 13.5.1 through 13.5.8 show the complete model used and the variation of some typical parameters during a simulation.

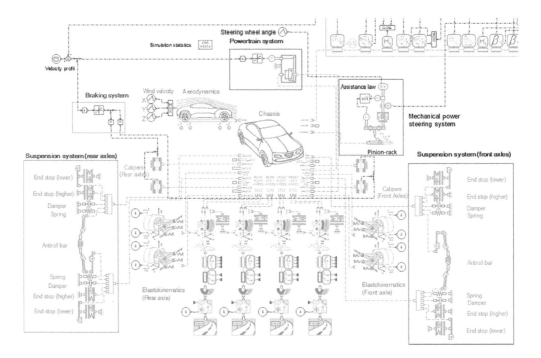

Figure 13.5.1 Complete vehicle model used in Amesim (partial view).

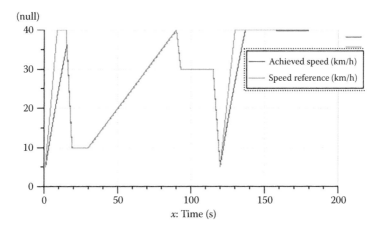

Figure 13.5.2 Achieved versus reference vehicle speed.

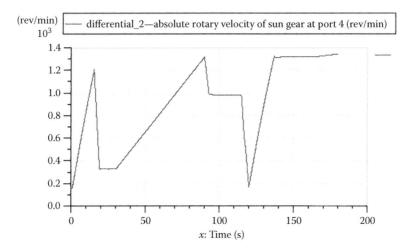

Figure 13.5.3 Engine RPM.

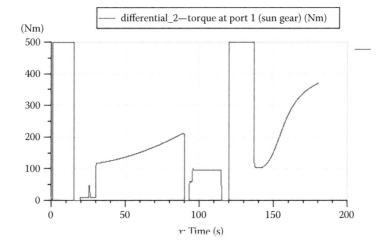

Figure 13.5.4 Engine torque.

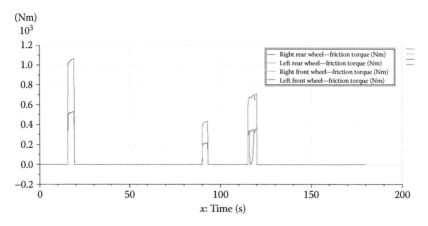

Figure 13.5.5 Brake torque.

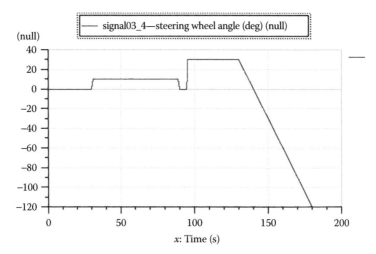

Figure 13.5.6 Steering angle reference.

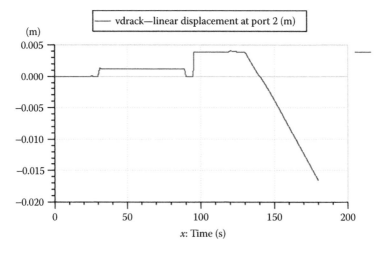

Figure 13.5.7 Power steering displacement.

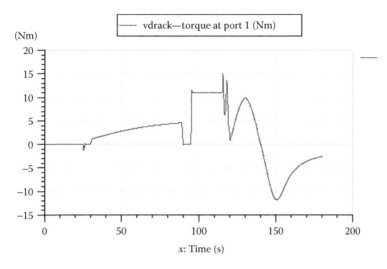

Figure 13.5.8 Power steering torque.

13.6   Connecting the real electrical motor to the virtual model

Connecting exterior elements is done through special blocks, which transmit the input value to the simulation model (Figure 13.6.1). Direct connection such as other simulation programs is not possible in Amesim. Amesim uses blocks that transform the input signals into actual physical measurements, the validity of the transformation being the responsibility of the user. The gearbox of a classical vehicle offers numerous issues, starting from the complicated structure, high cost, hard to model reliability, up to weight, response time, and effectiveness in a wide variety of environments. We considered it would be interesting to eliminate this component when analyzing the simulation results. Therefore, the input torque from the electrical motor will be connected to a simple gear reducer, then to a differential, and to the input of the chassis block. The motor axle speed is computed into the simulation model and then transmitted to the exterior.

A full diagram of how different components interact is presented in Figure 13.6.2.

The torque developed by an electrical motor is applied to the gear reducer-differential assembly, out of which a scaled torque is the output. This scaled torque is applied to the wheels from which, due to the contact with the road, a force is obtained. The force is further applied to the chassis and through vehicle dynamics calculations, a velocity is obtained. Through the wheels, this is transformed into angular velocity, which is applied to the gear reducer. This scales it and sends it further into the exterior. The chassis model is extremely complex with several 100 parameters and over 5000 lines of code. It has 15 degrees of freedom, as indicated in Figures 13.6.3 and 13.6.4.

Detailed theoretical considerations used in building the chassis model are beyond the scope of this paper; for further information about the equations used in the modeling, Reference [6] can be consulted. This also shows that Real-Time simulation is a collaborative process, which allows the user to access varied resources without the need of understanding each of them in detail. Next it will be presented the main two ways of introducing disturbances to the system will be presented in the following: (1) changes in aerodynamics and (2) changes in the road type.

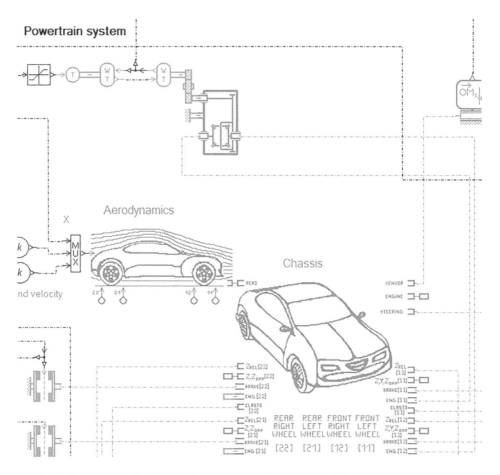

Figure 13.6.1 Connecting the electrical motor to the Amesim model.

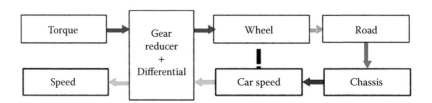

Figure 13.6.2 Interaction between model components.

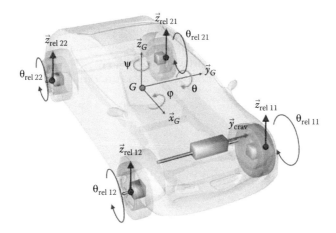

Figure 13.6.3 Chassis degrees of freedom.

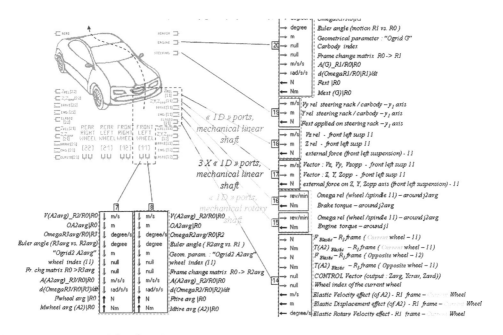

Figure 13.6.4 Some of the chassis parameters.

13.7   Modeling aerodynamic parameters

Usually, the aerodynamic force is assumed to act in a point on the same vertical axis as the centre of gravity. If we add to this a force generated by an air current on a direction different from that of the direction of travel, the chassis will be acted upon by a three-dimensional system of forces (Figure 13.7.1) The origin point of the aerodynamic forces axis system can be placed in any position that the user desires, which requires supplemental computing power. The user can define relationships for calculating the aerodynamic coefficient in relation to the angle between the axis and the resulting aerodynamic force in tables. The aerodynamic model has the following parameters: air density, front area, wheel base, and the variation of the aerodynamic coefficient as a function of skid angle. The model computes the coordinate changes if necessary.

The variation of the aerodynamic coefficient as a function of the angle to the x axis can be seen in Figure 13.7.2.

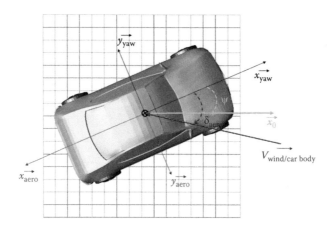

Figure 13.7.1 Amesim aerodynamic model.

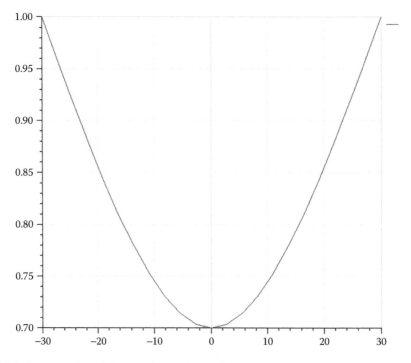

Figure 13.7.2 Amesim plot of the aerodynamic coefficient variation.

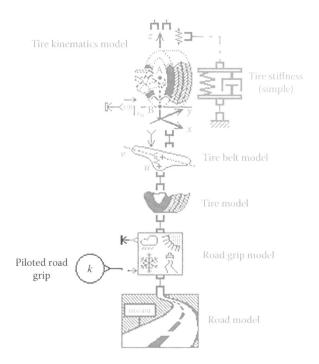

Figure 13.7.3 Tire–road interface for a single wheel.

This way of defining coefficients is very useful if it is needed to insert measurement data from a real vehicle into the simulation model. The Amesim model allows modification to two categories of road parameters: (1) grip coefficients and (2) road coordinate system. The grip between the tire and the road can be changed in the block called Road grip model. This block allows complex modeling of grip changes as a function of the absolute position of the wheel centre, even by using value tables. In the simplest case, the grip coefficient is assumed to be a constant value [7]. The road is considered a plane in space, depending on the absolute position and velocity of a wheel's centre. In order to connect the road model with the rest of the system, the calculation of three parameters is needed: (1) the road plane position, (2) road height, and (3) road slope. The predefined model used by Amesim is presented in Figure 13.7.3.

13.8 Determining the vehicle speed

In order to compare the simulated results to the real ones, the vehicle model has been included in a control loop using a PI controller (Figure 13.8.1). In the initial model, the speed was measured with sensors that are based on coordinate transforms, which did not allow negative speeds. In case the speed became negative, the sensor would just ignore the sign, meaning the respective value could not be used for control. Solving this problem can be done in two ways: either by determining the corresponding sign of the speed and adding it to the speed calculated by the sensor or by deriving the absolute position of the vehicle, which is calculated automatically. The second options pose significant accuracy issues, so the first option was chosen. The sign of the speed results

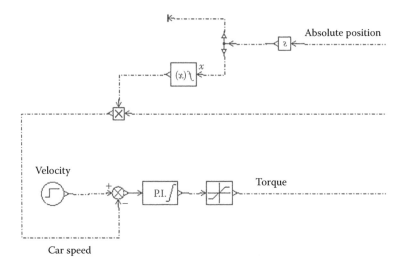

Figure 13.8.1 Speed control architecture for the simulation.

from the derivative of the position: if the vehicle has a negative speed (going in reverse), the absolute position decreases, so the derivative is negative. In the opposite case, if the position is increasing and the derivative is positive to calculate the sign, then the following function is used:

$$f(x) = \frac{\left|x + 1*10^{-9}\right|}{x + 1*10^{-9}}$$

When x is close to 0, problems may appear due to the typical numerical solvers for equations, which would cause $f(x)$ to reach extremely large values. Therefore, an arbitrary value has been added to avoid $(0/0)$ situations around the critical position.

13.9   Results obtained using a model with an ideal power source

For the design and testing of vehicles, standardized speed cycles are used. Even though they are designed for internal combustion engines, they offer a good enough representation of the usual speed that a vehicle travels under the assumed road conditions. In this case, the New European Driving Cycle (NEDC) would be used, which consists of one urban cycle and four extra-urban cycles.

As shown in Figure 13.9.1, the average error in the given consideration is very small. However, given the long timeframe in which simulation is performed, detailed performance will also be analyzed (Figure 13.9.2).

It can be noticed that even though the achieved speed values are close to the reference, there is a delay due to the vehicle having significant inertia which the PI controller cannot compensate. One solution would be to use a more complicated control architecture, with dedicated compensators, but the complexity of the vehicle model makes that difficult. A nonlinear fuzzy controller would be more appropriate but would be difficult to implement in Amesim. Therefore, the above-mentioned response will be considered adequate for now.

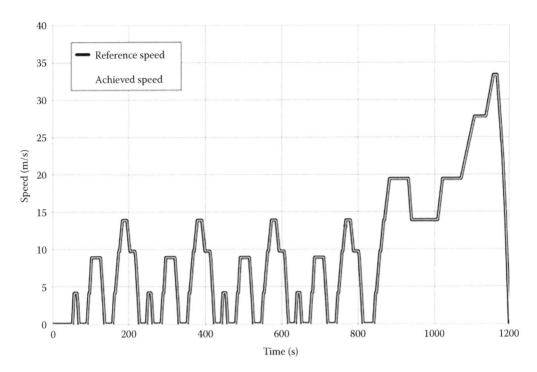

Figure 13.9.1 NEDC reference speed and speed achieved by the PI control.

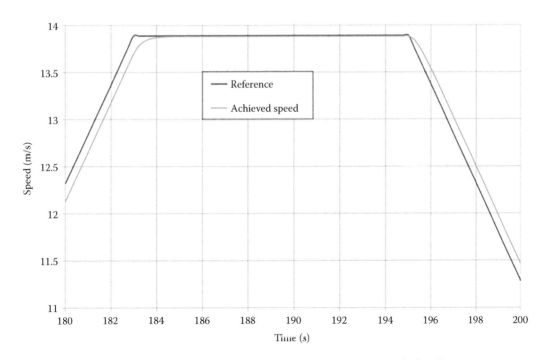

Figure 13.9.2 NEDC reference speed and speed achieved by the PI control (detail).

13.10   Results obtained using a model with a nonideal power source

One of the main advantages of Amesim is the ease of prototyping for determining baseline values. The mechanical vehicle model presented previously can be coupled with electrical elements, enabling the possibility of advanced analysis. Simulation results can be compared to results obtained in the lab, allowing the validation of both elements. From the models offered by the Amesim producers to demonstrate the capabilities of the program, one has been chosen that allows the sizing of components from an electrical vehicle. This model uses detailed modeling for the electrical motor, vector control, vehicle controller but also a driver model to simulate typical human reactions to changes in vehicle parameters during driving. The model is part of an example for designing an electrical vehicle transmission that can be split into five steps: preliminary selection of components, detailed transmission model, comfort analysis, transmission cooling, and battery management.

The first two steps are important for this section, because they can be compared with the results obtained from real electrical vehicles. The model (Figure 13.10.1) consists of a virtual driver, an electrical vehicle control unit, a torque control module, an inverter, a battery, and an electric motor. The vehicle model will be replaced with the one presented previously, which offers greater detail.

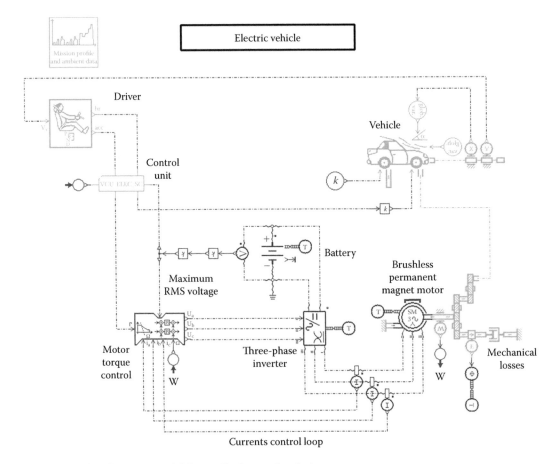

Figure 13.10.1 Amesim model for preliminary simulation.

The virtual driver is a block simulating the human operator. This block calculates the acceleration and brake signals, also taking typical human response times into account. The vehicle control unit transforms the acceleration signal from the virtual driver into a torque command for the electric motor, whereas the brake signal is split between electrical brake torque and mechanical brake torque. The ECU uses a vector control strategy, adapted to the electrical motor used. The motor is a synchronous motor with permanent magnets and apparent poles, using the assumption that the magnetic circuit is linear. The inverter is an average values model, whose goal is determining the power losses without sacrificing dynamic response. The battery is defined in data tables. The model is running the previously presented NEDC cycle. The structure of the vehicle controller (Figure 13.10.2) contains, among other functions, a model of vector control to automatically determine the torque, voltage, and current limits. It also ensures the economical usage of the vehicle, for example, choosing mechanical braking at low speeds instead of the electrically generated torque that would be less effective.

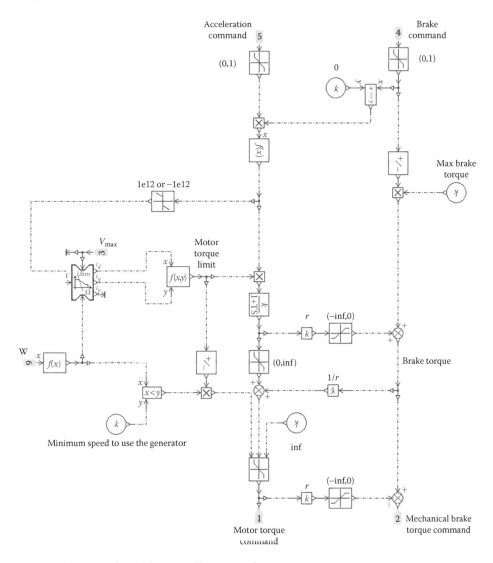

Figure 13.10.2 Electrical vehicle controller internal structure.

The controller shown in Figure 13.10.3 simulates a classical vector command structure. As inputs, it uses a torque or voltage reference, with a maximum phase voltage and a current limit. Unlike many existing models for vector command converters, this one provides a model closer to reality, accounting for the power characteristics of a given motor. To control the currents on the two axes, I_d and I_q, two independent PI controllers are used. These controllers rely on eliminating zeros to make system behavior equivalent to a first order one. The schematic from the Amesim manual can be seen in Figures 13.10.4 and 13.10.5.

Practically, for every situation a different model is used, which helps to insure a behavior as close as possible to a real controller. The inverter uses the average values for current and voltage to approximate the losses through conduction and commutation. This model

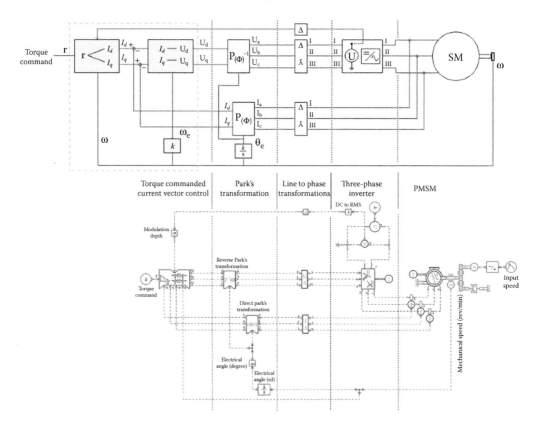

Figure 13.10.3 Vector command controller in Amesim.

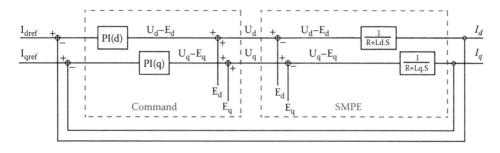

Figure 13.10.4 Controller schematic in Amesim manual.

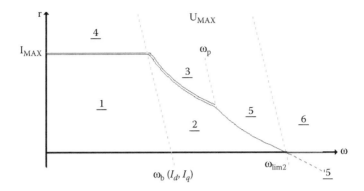

Figure 13.10.5 Controller function in Amesim manual.

is appropriate when the inverter dynamic is much faster than the rest of the system. The characteristics of the power electronics are considered linear. The battery is also an average values model. The battery is considered as a variable voltage source whose internal resistance depends on the charge status. Two tables are being used: (1) output voltage as a function of charge status and (2) internal resistance as a function of charge status. Power losses are being considered, which allows a more exact calculation of effectiveness compared to only using electrical parameters. The brushless motor model includes energy calculation.

13.11   Simulation results for the complete vehicle model in Amesim

The NEDC profile is followed with a comparable precision, as shown in Figure 13.11.1.

At a more detailed analysis however, it can be seen that the delay is gone and has been replaced by oscillations (Figure 13.11.2).

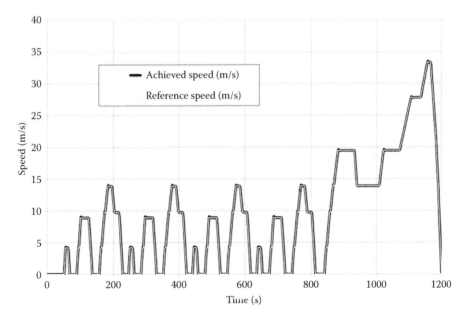

Figure 13.11.1 NEDC cycle reference speed and speed achieved by the complex model.

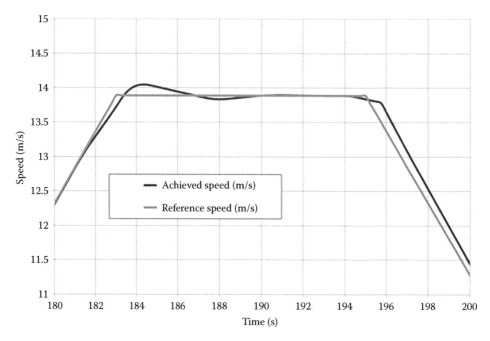

Figure 13.11.2 NEDC cycle reference speed and speed achieved by the complex model (detail).

The error can be measured for both cases and the results can be compared in Figure 13.11.3.

The maximum error is 0.5 m/s, the two results being quite similar, the complex model being only slightly better. Due to these results, it can be concluded that both models offer similar performance, therefore using the complex model for a simulation is not necessary, unless this is the objective for the optimization. The angular speed of the motor is also very close for both models (Figures 13.11.4 and 13.11.5).

A bigger angular speed variation for the complex model can be observed. However, this does not influence the structural resistance or comfort of the vehicle because it has 25 rpm amplitude and a period of about 13 s. The motor torque will be different between the two models due to the different control strategies. The PI controller produces a linear

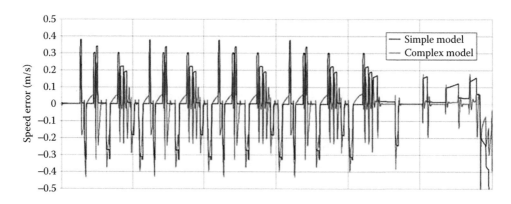

Figure 13.11.3 Error comparison (complete cycle).

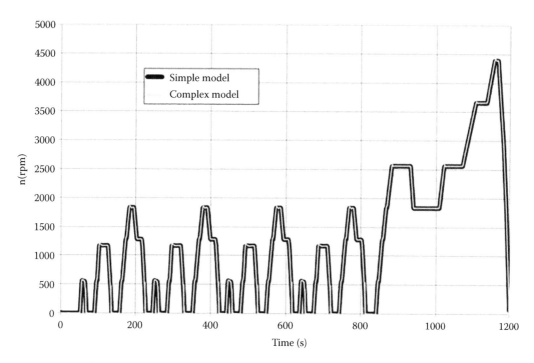

Figure 13.11.4 Motor angular speed comparison.

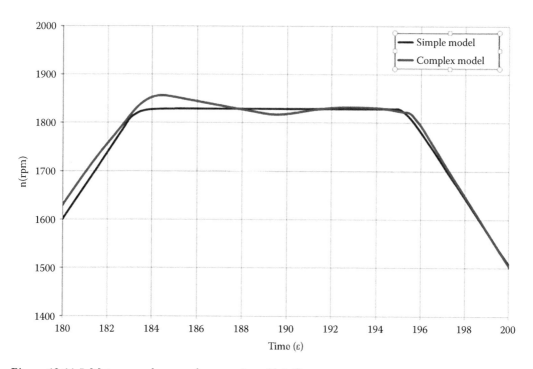

Figure 13.11.5 Motor angular speed comparison (detail).

response, whereas the control strategy of the complex model and the structure including power losses insert several nonlinear elements, which can be seen in the motor torque (Figures 13.11.6 and 13.11.7).

The command strategy influences consumed power. In order to estimate the losses in the electrical transmission, the models allow the measuring of the thermal flow at the exterior of various components (Figures 13.11.8 through 13.11.13).

The Joule losses for the battery have a maximum value around 800 W and the losses for the gear speed reducer are approximately the same.

In order to determine transmission efficiency, the input and output energies need to be calculated. The final results are two values of interest: (1) the efficiency of the transmission and (2) the efficiency of the energy recovery.

The above-mentioned two graphics are the result of integrating the power in respect with time and therefore they need to be interpreted point-by-point. In order to obtain a greater precision, a very large number of measurements would need to be performed. However, the efficiency obtained is very good, 90% at the end of a NEDC cycle, greater than the value for

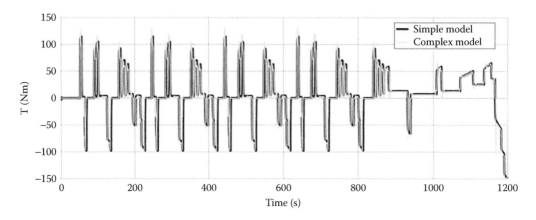

Figure 13.11.6 Motor torque comparison.

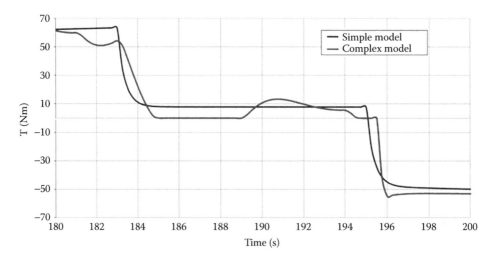

Figure 13.11.7 Motor torque comparison (detail).

an internal combustion engine vehicle. The consumed energy in the whole cycle is 3.34 MJ, and the recovered energy is 1.344 MJ. These values are obtained considering the assumption that energy conversion losses only occur in commutation and the battery accepts the energy directly, regardless of other parameters. On average, the system recovers and stores in the battery about 40% of the consumed energy in a NEDC cycle. Obviously, if other typical vehicle consumers, such as air conditioning, lights, cooling system, and brake pump, would be considered or the control strategy would be different, this value would decrease [8].

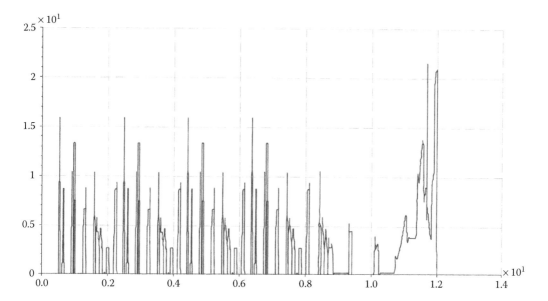

Figure 13.11.8 Electric motor Joule losses.

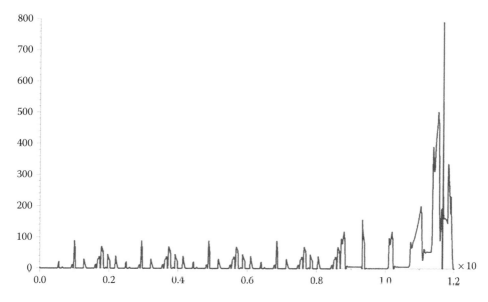

Figure 13.11.9 Battery Joule losses.

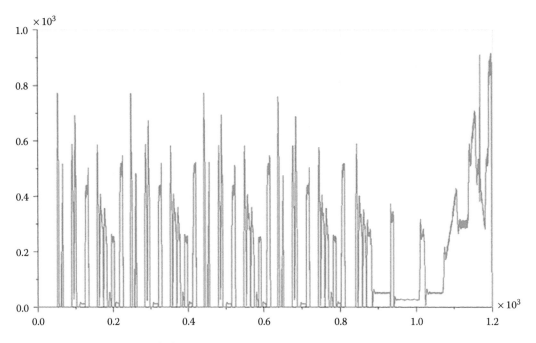

Figure 13.11.10 Inverter Joule losses.

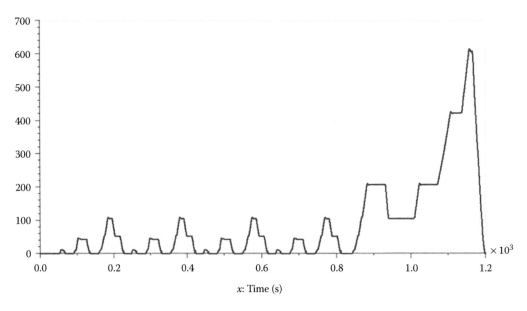

x: Time (s)

Figure 13.11.11 Friction losses for the gear reducer.

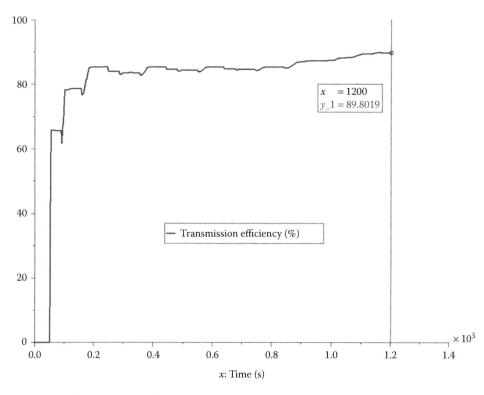

Figure 13.11.12 Transmission efficiency.

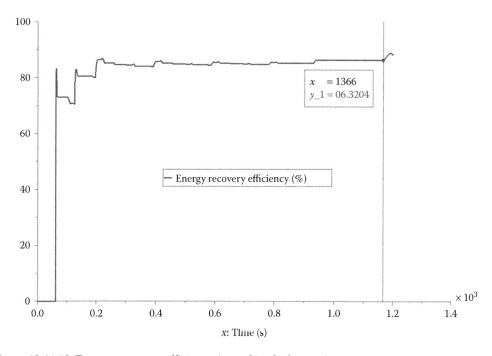

Figure 13.11.13 Energy recovery efficiency (stored in the battery).

13.12   Preparing the Amesim models for Real-Time simulation

The first step in preparing the models for Real-Time simulation consists of performing some preliminary checks. Then, after adding several specific blocks, the actual conversion will be performed automatically. Unfortunately, the integrators commonly used in the Real-Time systems do not allow solving systems of equations with implicit variables. To avoid this issue, most predefined components containing one or more implicit variable offer alternative implementation that avoids these issues. In order to check the number of implicit variables in a system, it is enough to check that the report Amesim is automatically generated before running the simulation (Figure 13.12.1).

A first system analysis tool offered by Amesim is the list of times required by the numerical integration processes. For variable step integrators there are several indicators, such as number of steps for orders 1, 2, 3…12, number of processed discontinuities, number of differential steps, and so on. For fixed step integrators, the situation is much simpler, and the only condition for the system to be suitable for Real-Time simulation is that the integration time is smaller than the real measured time. This condition is only valid for Real-Time systems that have a high rate of user interaction or fixed temporal constraints. For any Real-Time system however, discontinuities in the integration time graph indicate problems that can potentially affect the system's stability during the Real-Time simulation. These discontinuities can be noticed when using variable step integrators for the fixed step integrators, the graph being a straight line (Figure 13.12.2). The time interval for measuring the values must also be taken into account; if the interval is too big, the measurements can simply *jump over* some discontinuities. If there are any integration issues, such as using a dedicated fixed-step integration model with a variable step integrator, the integration time will become greater than the real measured time. In general, the Real-Time for a system is given by a straight line starting from the origin of the coordinate system, with a slope representing the ratio between the systems' Real-Time and the actual time (measured on a clock). In Figures 13.12.2 and 13.12.3, the real time is considered to be equal to the clock time, so the slope of the green line is 1.

The model can be run in real time for the assumption detailed earlier (system Real-Time is the clock time). The average integration step is solved in $200*10^{-6}$ s, which indicates good performance, the integration step being 1 ms. Another way to estimate performance is to split the largest value of the integration time to the Real-Time value. By this criterion, the simulation is 2.66 times faster than the assumed Real-Time. However, speed is not a guarantee of robustness. Other methods will be used to analyze robustness and precision, specifically the state count and linear analysis of the system. The state count facility (Figure 13.12.4) displays the list of all state variables of the system, alongside a number indicating how active the respective state is in the integrating process. This value is defined as the number of steps for which the error of the respective state variable is greater than the integrating process tolerance. If one of the states is being *controlled* very often, a variable step integrator will automatically decrease the integration step size and therefore will increase the time

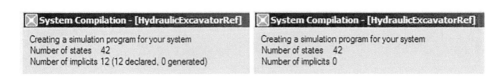

Figure 13.12.1 Amesim variable type report.

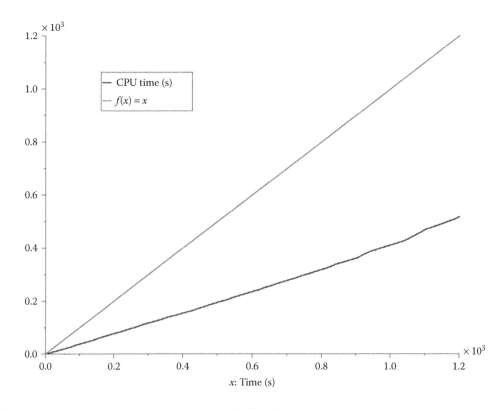

Figure 13.12.2 Amesim integration time graph (fixed step).

needed to calculate the values. Together with the linear analysis, this can offer an indication of which model needs to be modified to increase the performance of the system.

From the results of Figure 13.12.4 that a thermal flow sensor is generating most issues, the chassis model is also significantly contributing. If the first one can be eliminated, the body obviously cannot; so interpreting the data needs to also account for each component's role in the system. After performing the previous steps, a deeper analysis of system dynamics can be performed, which has the goal of determining maximum integration time for which the results converge.

By linearizing the equations on certain time intervals, the eigenvalues can be identified and, after interpreting them, the system's frequency response can be identified. The times for linearization must be defined considering the system's state at a given moment: It is ideal to choose a value for each balanced state and a value for each transient state. Linearizing systems can only be done by using variable steps integrators. If the system does not allow using variable step integrators, the state count remains the only tool to analyze the system's behavior for Real-Time simulation [9]. The theoretical criteria for analyzing the eigenvalues are:

- $f_{int} \geq [(2\pi f_i)^2 / 2R_i]$, if complex eigenvalues correspond to the module i (i undamped)
- $f_{int} \geq [R_i / 2]$, if real eigenvalues correspond to the module i (i completely damped)
- $f_{int} \geq -R_i$ to integrate the complete dynamic without oscillations

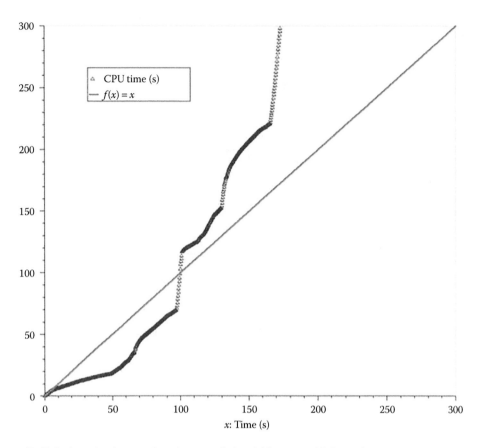

Figure 13.12.3 Amesim integration time graph (variable step with issues).

Summary of which state variables controlled the integration step size:

State No	Controlled	Submodel	Variable	Unit	▲
54	40484	powersensor_thermal_1 [THPOT1-1]	positive energy	J	
15	28688	Chassis15dof_32ports [VDCAR15DOF1-1]	Front left axle (11) : relative rotary velocity (wheel /spindle) - f...	degree/s	
19	21381	Chassis15dof_32ports [VDCAR15DOF1-1]	Front right axle (12) : relative rotary velocity (wheel/spindle) - fr...	degree/s	
27	17006	Chassis15dof_32ports [VDCAR15DOF1-1]	Rear right axle (22) : relative rotary velocity (wheel/spindle) - re...	degree/s	
41	15530	elect05 [DIF1-1]	dummy state variable	1/s	
35	11663	tirecontact_4_2_3 [VDSLIP001-3]	longitudinal deflection of the carcass	mm	
23	11339	Chassis15dof_32ports [VDCAR15DOF1-1]	Rear left axle (21) : relative rotary velocity (wheel/spindle) - re...	degree/s	
37	11051	tirecontact_4_2_4 [VDSLIP001-4]	longitudinal deflection of the carcass	mm	☰
31	4907	tirecontact_4_2 [VDSLIP001-1]	longitudinal deflection of the carcass	mm	
33	4852	tirecontact_4_2_2 [VDSLIP001-2]	longitudinal deflection of the carcass	mm	
34	1439	tirecontact_4_2_2 [VDSLIP001-2]	lateral deflection of the carcass	mm	
32	1286	tirecontact_4_2 [VDSLIP001-1]	lateral deflection of the carcass	mm	
65	1130	powersensor_elect [EBPOT1-1]	negative energy	J	
47	680	emd_SynchronousMachine [EMDSMPE01-1]	stator current on Park's q axis	A	
8	520	Chassis15dof_32ports [VDCAR15DOF1-1]	Port 20 : carbody rotary velocity (R1 vs R0) - expressed in carb...	degree/s	
50	375	VCU_ELEC_SC.elect03 [LAG1-2]	output from first order lag		
46	357	emd_SynchronousMachine [EMDSMPE01-1]	stator current on Park's d axis	A	
42	144	drv_driver_c2 [DRVDRVA00A-1]	dummy state variable	m/s	
58	87	powersensor_thermal_3 [THPOT1-3]	positive energy	J	
36	41	tirecontact_4_2_3 [VDSLIP001-3]	lateral deflection of the carcass	mm	
38	23	tirecontact_4_2_4 [VDSLIP001-4]	lateral deflection of the carcass	mm	
63	9	powersensor_rotary_1 [PTR10-1]	negative energy	J	
43	9	drv_driver_c2 [DRVDRVA00A-1]	integral part of the acceleration control loop	m	
64	0	powersensor_elect [EBPOT1-1]	positive energy	J	
62	0	powersensor_rotary_1 [PTR10-1]	positive energy	J	

Figure 13.12.4 State count.

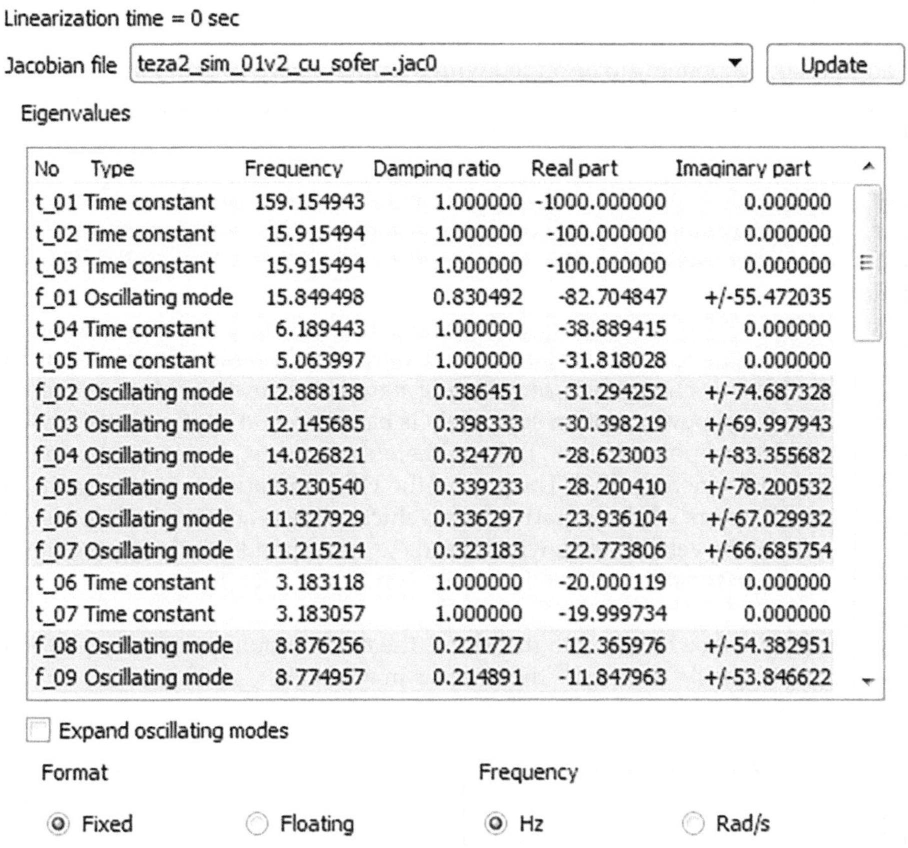

Linearization time = 0 sec

Jacobian file | teza2_sim_01v2_cu_sofer_.jac0 | ▼ | Update |

Eigenvalues

No	Type	Frequency	Damping ratio	Real part	Imaginary part
t_01	Time constant	159.154943	1.000000	-1000.000000	0.000000
t_02	Time constant	15.915494	1.000000	-100.000000	0.000000
t_03	Time constant	15.915494	1.000000	-100.000000	0.000000
f_01	Oscillating mode	15.849498	0.830492	-82.704847	+/-55.472035
t_04	Time constant	6.189443	1.000000	-38.889415	0.000000
t_05	Time constant	5.063997	1.000000	-31.818028	0.000000
f_02	Oscillating mode	12.888138	0.386451	-31.294252	+/-74.687328
f_03	Oscillating mode	12.145685	0.398333	-30.398219	+/-69.997943
f_04	Oscillating mode	14.026821	0.324770	-28.623003	+/-83.355682
f_05	Oscillating mode	13.230540	0.339233	-28.200410	+/-78.200532
f_06	Oscillating mode	11.327929	0.336297	-23.936104	+/-67.029932
f_07	Oscillating mode	11.215214	0.323183	-22.773806	+/-66.685754
t_06	Time constant	3.183118	1.000000	-20.000119	0.000000
t_07	Time constant	3.183057	1.000000	-19.999734	0.000000
f_08	Oscillating mode	8.876256	0.221727	-12.365976	+/-54.382951
f_09	Oscillating mode	8.774957	0.214891	-11.847963	+/-53.846622

☐ Expand oscillating modes

Format

◉ Fixed   ○ Floating

Frequency

◉ Hz   ○ Rad/s

Figure 13.12.5 Eigenvalues for $t = 0$ for the complex model.

The above criteria are valid for an Euler type algorithm, the higher order ones being more robust. The results for the model used can be seen in Figure 13.12.5 for $t = 0$ (least favorable case).

It can be observed that the most dangerous oscillation mode has a 60 Hz frequency, whereas the first damped mode with imaginary part 0 imposes a 1000 Hz frequency or a step time of 1 ms. Beyond this, instabilities can occur, leading to loss of control over the Real-Time simulation system. However, the above-mentioned rule is only valid for the Euler method, for the other methods no similar criteria being available. It can be assumed however that, as long as the first-order method is stable, the higher order ones (Runge–Kutta 2,3, or 4) will also be stable, so the maximum simulation step can be carefully increased.

13.13   *Hardware-in-the-Loop test stand hardware structure*

The HiL simulation cannot be performed unless there is at least one hardware element. As most car producers have launched or desire to launch electrical vehicles, the advanced analysis of new types of electrical motors and also auxiliary equipment such as power source (battery) have been considered useful. However, due to the difficulties in finding the needed resources to simulate the entire energy chain, the authors have decided to focus on a solution that only involved the electrical motor. This does not mean that the other components are being ignored. On the contrary, the test stand and the software allow additional components to be added with minimal effort. In order to test the integration of an electrical propulsion system in a vehicle, it is needed to know its electrical and mechanical parameters. For high-performance applications, it is not enough to consider the electrical motor, its power and command source (in this case an inverter) need to be considered. The high dynamic required for functioning inside a vehicle forces the selection of a high-performance command solution, more exactly a vector command inverter, or DTC. The inverter used in the HiL test stand has a low power but similar control characteristics to larger models. Brushless electric motors are easy to scale, so the only difference between two models of the same series has been assumed to be consumed power and inertia. If the type of motor needs to be changed, only rebuilding the electrical connections to the new motor and changing a few scaling parameters in the software are needed. Treating the electrical power system separately is based on real applications, where, from the car manufacturers' point of view, the parameters of interest concern performance and not specific internal functions [7,8]. The goal of the HiL simulation is studying the parameters of electrical motors while operating on a vehicle. This is achieved while also following the behavior of the vehicle itself, which needs to correspond to existing regulations. In order for this goal to be achieved, the following conditions need to be fulfilled:

- The torque must be identical to the torque the motor would provide on the vehicle.
- The rotation speed needs to be the same as in a vehicle.
- The scaling must be done in such a way as to not change the dynamic performance and allow the date to be interpreted linearly.
- Low command delay (fast response).

Testing electrical motors for vehicles is usually done using another motor coupled to the same axle, which operates in opposition to the tested motor. If all operating conditions need to be simulated, the command unit needs to allow the operation in all four quadrants, which is not very common. In order to allow both imposed torque and imposed speed, these need to be used as references for the motors on the test stand. As an electrical motor cannot be simultaneously driven by a torque and speed reference, these will be split, one for each motor. As the two motors are mechanically connected on the same axle, each will be subjected to the same mechanical effort as in the simulated vehicle. This is only true as long as the response speed of the converter is fast enough to allow it to follow the reference with insignificant errors. The principle schematic of the HiL test stand can be seen in Figure 13.13.1.

In order to better understand the operation of the test stand using the diagram shown in Figure 13.13.2, consider first a vehicle equipped with an internal combustion engine. If the engine is removed beginning from the output axle, we obtain the simulated model. If we look at the axle where the engine used to be connected as an input/output port, the following succession of events takes place:

- The drivers wish to accelerate, pushes the corresponding pedal.
- ECU notices this and in order to increase the vehicle speed increases the electrical motor torque reference.
- The motor control unit receives the new reference and through command signals creates the conditions for the motor to increase its torque, operation that cannot be achieved instantaneously.

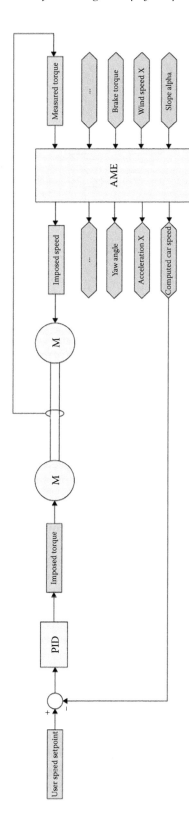

Figure 13.13.1 Test stand command structure.

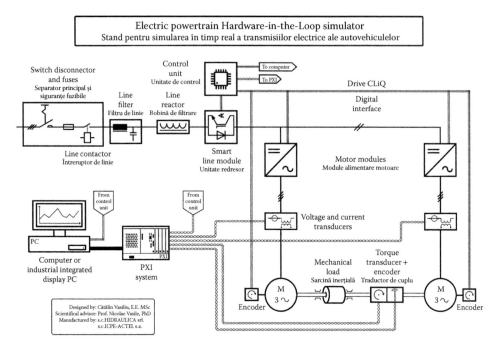

Figure 13.13.2 Simplified test stand structure.

- The increased torque of the motor is transmitted through the entire transmission chain to the wheels, which, depending on conditions (road, wind, slope, grip etc.), will accelerate the vehicle in a certain time.
- The speed of the wheels is connected to the speed of the motor; after considering elasticity and other mechanical effects, the motor is forced to spin at a speed determined by the event chain described earlier.

When reaching the desired speed, ECU acts to maintain a constant speed for the vehicle, changing the engine torque to smaller values.

If a speed transducer were to be placed on the axle remaining in the vehicle, then the measured speed could be transmitted to the motor through another rotating machine. These are the roles of the two motors used in the test stand: (1) one is acting as the vehicle motor and (2) the other ensures that the motor load corresponds to the simulation results. The operating modes are determined dynamically by the power converter. In order to scale the system correctly, apart from the torque, it is needed to reproduce the inertia of the full scaled motor. In order to do this, the inertia of a 1:1 scale motor has been calculated and the difference in inertia between the large motor and the motor used on the HiL test stand has been included on the axle as a disc. Therefore, the motor accelerates as a much larger one [12].

The converter used has a special structure. In order to increase the energy efficiency for multiple axis systems, many producers have created an electrical energy chain consisting of a transformer and a DC–AC converter (inverter) for each motor. Therefore, the system can be improved without completely changing the power components. Moreover, a common current bus helps with energy recovery. Motors working as brakes or generators can power those working in quadrants 1 and 3, the system only consuming energy to cover the mechanical and electrical losses [1,13].

The inverters are commanded using analog signals. This is the only solution that allows Real-Time simulation at the required speed. There are several digital command protocols that are stated to function in Real-Time, but all are proprietary, have not been sufficiently tested in these conditions, and the equipment producers have not explicitly implemented any of them. The analog signal scan be easily measured and generated at much higher rate than the speed of the simulation and therefore has been considered as the optimal solution.

For Real-Time control, Intel dual-core T8000-based PXI computers produced by National Instruments are used. These general-use controllers contain data acquisition boards, including 16 inputs and 2 outputs, enough for the current application. In order to adequately distribute the computing effort and to make troubleshooting easier, three units have been used: (1) one contains and simulates the vehicle model, (2) one controls the motor (acting as an ECU analog), and (3) the third measures the electrical parameters of the system. The three PXI units are connected to a host Windows PC that provides the GUI and also offers the possibility to communicate the results through a network (local or Internet). These host PCs were required because the Real-Time PXI computers are headless: They only contain the ETS PharLap Real-Time operating system and are unable to provide a graphical interface on their own. The data transfer between the PXI machines and the host computer is done through a standard Ethernet connection.

13.14   Hardware-in-the-Loop test stand software structure

The software that runs the HiL test stand has been developed using the National Instruments LabVIEW programming environment and a detailed analysis of this code is beyond the scope of this paper. The integration between the Amesim model and the LabVIEW code is of interest. The algorithm used by Amesim to transform model files into files that are compatible with Real-Time simulation software relies upon aggregating all parameters in a single file. A general structure of the software components can be seen in Figure 13.14.1.

The Real-Time model contains both predefined models (from the Amesim Libraries) and user-created models. The user can modify only a part of these, the rest being protected.

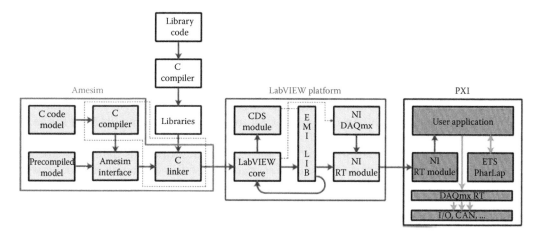

Figure 13.14.1 Real-Time simulation platform software structure.

The compiling is done with the help of a graphical interface but it can also be done manually using a make file type file and an external C compiler such as GCC or the one from Microsoft Visual Studio. The version of the compiler is very important, because it might result in difficulty to solve the incompatibilities that cannot be detected before attempting to load the model into the PXI machine. Following the compiling process, the result is a.dll file that can be run on x86-compatible machines. This file contains all the needed equations for running the transformed mathematical model. A.dll file is a collection of software functions that can be dynamically called from a compatible program. In practice, certain code segments are only run when the calling program (in this case the LabVIEW program running on the PXI) requests it. As a dynamic link library, the functions can be directly accessed by any program, unlike a static link library, where the functions would be inserted in the calling program when they are linked.

The functions which the.dll file contains correspond to the Amesim ones, being a direct transform of these to allow the compatibility with the External Model Interface module from LabVIEW (Figure 13.14.2). A specialized LabVIEW structure ensures that the functions in the.dll file are called in the correct order, and therefore the model functions properly.

In order to support those that are familiar with the classic simulation paradigm used by MATLAB and SIMULINK, for example, National Instruments has developed a toolkit for simulation based on mathematical equations and for dynamic system control design, which is called Control Design and Simulation Module (CDSIM). The blocks from this toolbox cannot be placed directly into the virtual instrumentation (VI) diagram; a special structure is needed, where simulation-specific parameters, such as integrator type, simulation time, time step, and synchronization with external sources, can be defined. Simulation subsystems can be inserted in any VI diagram. The general parameters can be defined, such as the loop ones, through a dialog box.

LabVIEW relies on predefined function libraries stored in.dll files, such as the Amesim model. Therefore, a very important facility is the ability to import C code stored in these libraries. Importing the Amesim model has been done through an example offered by National Instruments, unfortunately without any documentation. A complex analysis was performed on it together with the Amesim model to establish its exact function and any optimization possibilities. Starting with LabVIEW 2009, new ways to import external models have been developed, but the old solution used in this LabVIEW program is still functional. External Model Interface is a set of blocks for loading external models into LabVIEW and simulating them using CDSIM (Figure 13.14.3). The example can be found in the following LabVIEW path: LabVIEW\examples\Control and Simulation\ Simulation\External Model Interface\External Model Interface Node. The creator of the simulation model must use a C/C++ compatible compiler and must utilize the functions offered by the example mentioned earlier to define the model. The example is a model skeleton, which implements the functions in a certain way. The same functions must be implemented in the Amesim model, taking care that their results have the same meaning.

On a more in-depth analysis, it can be noticed that the functions defined in the .dll files are just *wrapper* functions for the usual functions used in a simulation model, with some specific LabVIEW structure elements added. It is therefore expected that, if a model works correctly in Amesim, the same thing will happen in LabVIEW. Starting from the example VI and eliminating the functions that are void in the Amesim model, better performance can be achieved. Also, certain functions that are called only for initialization and end of execution can be separated, further optimizing the code.

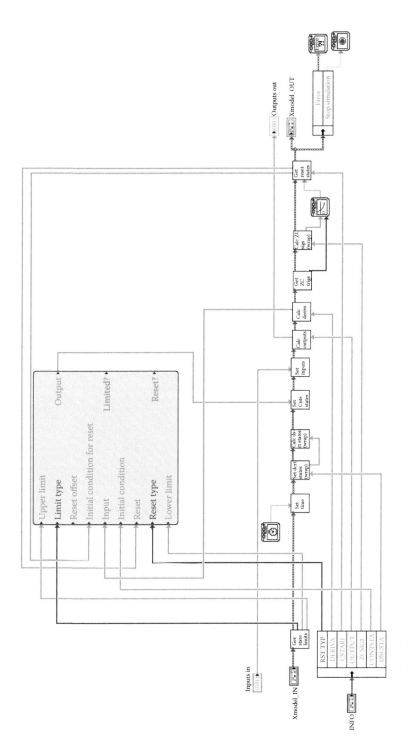

Figure 13.14.2 LabVIEW EMI internal structure.

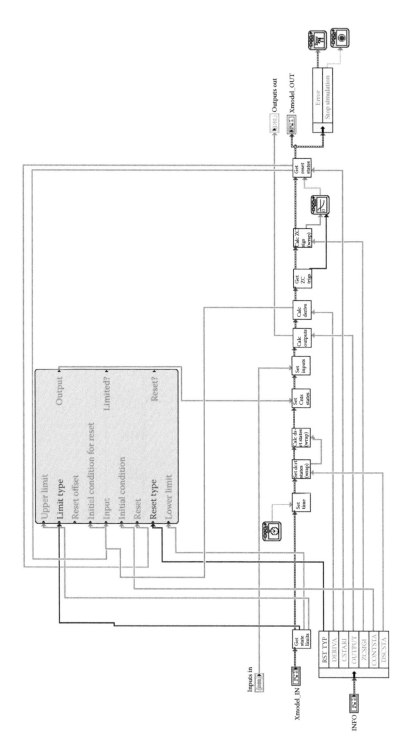

Figure 13.14.3 LabVIEW simulation model import core.

13.15   The graphical interface

The graphical interface was created to assist the user in performing various operations on the test stand. A well-designed graphical interface eliminates ambiguity, offers easy access to important device functions and contributes to reducing the time needed to obtain results. Due to the large amount of data present in the system at a given time, the voltage and current panel is separated by the main simulation panel. In this way, the data acquisition for these measurements can be performed in a stand-alone VI. In order to completely characterize a three-phase voltage and current system, it is necessary to view the wave forms, increase or decrease the displayed part of the wave form, perform harmonics analysis, or display the power on each phase [1,13]. In this case, satisfying all the requirements involved the use of 12 signals. For this reason and considering the form factor of the LCD screen (4:3), the wave forms were presented using four displays in the upper part of the available space (Figure 13.15.1). The buttons and the controls are placed in the lower side of the screen so, when selecting them, all the graphs remain visible. The user can select between the two available modes: WAV and RMS, using the MODE button. An LED shows the mode in use

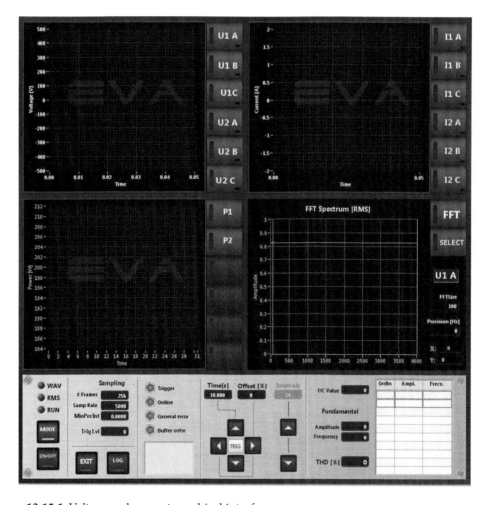

Figure 13.15.1 Voltage and current graphical interfaces.

at a given time. Next to it, system information is presented, such as sampling speed, frame size, or trigger level. The buttons for starting the system, stopping it, and saving the data to disk are below the numerical indicators. To the right of these indicators are LEDs showing that an instrument is active, the activation of the trigger or various errors. Further, four buttons allow the user to change the displayed time window size or the initial offset. The number of intervals based on which the RMS value is calculated can also be changed. The last portion contains information about the fast Fourier transform (FFT) analysis, only present if the FFT analysis is on. The graphs on the first row are dedicated to the waveforms of voltage and current. Through the buttons on the left side of each graph, certain signals can be activated or deactivated. The color of the signals on the graph is reflected also on the button, making them easier to identify. The second row has the graph dedicated to the system powers, calculated like the RMS, for a certain number of intervals. Next to it sits the window dedicated to the Fourier analysis. The FFT button activates it, and the SELECT button chooses one of the 12 signals of the system for analysis. The SELECT button also acts on the trigger, the chosen waveform for the trigger being the same waveform chosen for the FFT transform. The size of the FFT analysis can be changed in the FFT SIZE box, and below it, the precision in Hz is displayed. In order to make identifying certain frequencies easier, a free cursor on the graph is available, whose coordinates are displayed in the X and Y boxes. When activating FFT, the table is automatically populated with the harmonics' values.

The simulation control interface (Figure 13.15.2) was designed to look like a vehicle instrument panel. The lower part provides information about vehicle speed, engine

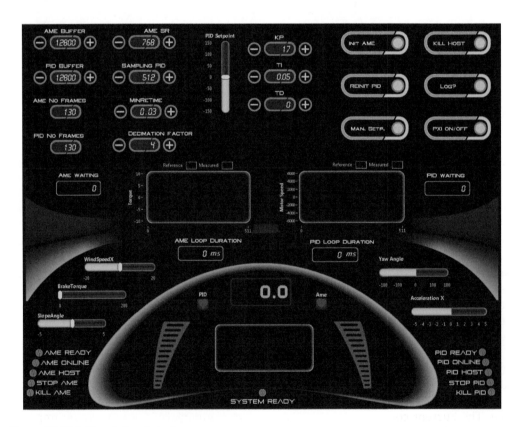

Figure 13.15.2 Simulation interface.

rpm, a graph plotting the reference and achieved speed and indicators for auxiliary variables, steering angle, and acceleration. The left side hosts the controls for the mechanical brakes, wind speed, and slope. The system state is displayed through LEDs placed on the lower left side (for the simulation PXI, called here AME) and lower right side (for the controller PXI, called here PID). Above, two graphs show the torques and speeds, both reference and achieved. Additional information regarding loop duration and number of frames in queue are offered for both AME and PID. The top side is reserved to setting initial parameters. On the left and center the user can set values for the speed in manual mode and several internal variables of the system (PID parameters, sampling, buffer etc.) whereas the right side hosts the system controls that allow the startup and shutdown of the entire system, logging data to disk and changing between automatic and manual modes.

13.16   Simulation results

The first set of tests, presented in Figures 13.16.1 through 13.16.5 studies the electric system's capacity to reach and maintain the reference values for torque and speed. For this, the two motors have been tested separately as well as together. Some of these tests were performed outside the test stand, as the motors cannot be decoupled from each other on the stand. Reaching the reference value happens in about 0.5 s (the natural inertia of the system is large), without any significant overshoot. The braking is done in an equivalent time. As the system is not equipped with any braking resistance, the process' efficiency varies with the local conditions (possibility to recover some energy, temperature, speed etc.). Comparing the time it takes for the motor to accelerate to the time it takes for the vehicle to accelerate, it is easy to notice that there is a significant enough difference so the dynamic behavior of the motor does not affect the behavior of the vehicle. For a 100% speed step, the motor response is also fast, but a torque limitation can be noticed close to the maximum speed. This time the response is overdamped, so an adaptation of the PID controller parameters is needed. This however does not negatively influence the usage of the motor and controller on a vehicle, as around maximum speed, a slightly reduced torque is often required. Testing the torque response is slightly more difficult, taking into account that the mechanical system cannot be disconnected and reconfiguring it is a difficult task. Similar to previous tests, the response is influenced by the natural inertia of the system. When subjected to a torque step, the motor will accelerate until reaching the maximum speed, 6000 rpm, after which the torque will be limited to keep this value constant. Due to the functional principles, the torque response of electric motors can be considered instantaneous, depending only on the current supplied into the circuit by the inverters. Time constants for inverter controllers are several milliseconds, which allows the logical assumption that they can be neglected. The dynamic response can be seen in the tests, accounting for the fact that the torque is the one estimated by the inverter, not the one measured by the transducer. This variant was chosen because it is simpler to implement but does not suffer from large inaccuracies.

The action of the inverter overload protection can be observed in Figure 13.16.6. For a torque variation between 0 and 70% (green line), accounting for the 50% reference speed (3000 rpm, nominal motor speed), a noticeable speed drop can be noticed as a result to current limitation and implicitly torque limitation of the motor with a speed reference. The current consumed by the two motors cannot exceed a limit value imposed by the parameters of the electrical circuit. Both motors operating at 200% of nominal parameters exceed this limit value and therefore the system performance is reduced. The converter provides 6.8 kW power, but the current is limited to 8 A. The motors have a nominal consumption of 2.75 A, which yields a maximum 100% overload capacity for the system (Figures 13.16.7 and 13.16.8). If the torque is kept at 50% of the maximum value and a 6000 rpm speed is requested, it can be seen that the achieved speed is limited to 1000 rpm. When eliminating the torque reference, the speed quickly increases to the requested value.

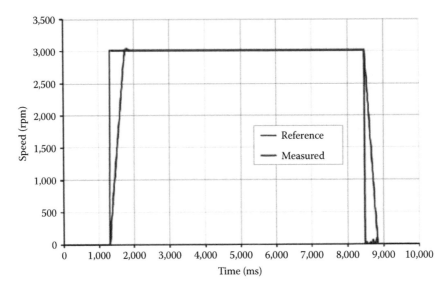

Figure 13.16.1 Motor response to a step signal of 50% of maximum speed (3000 rpm).

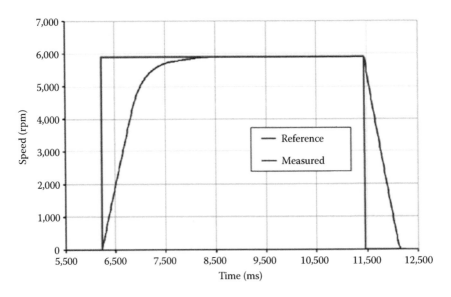

Figure 13.16.2 Motor response to a step signal of 100% of maximum speed (6000 rpm).

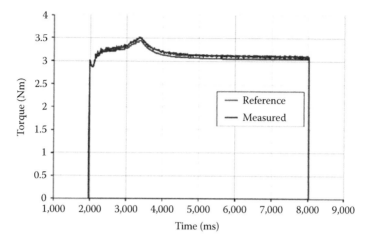

Figure 13.16.3 Motor response to a torque step signal of 30% of maximum value.

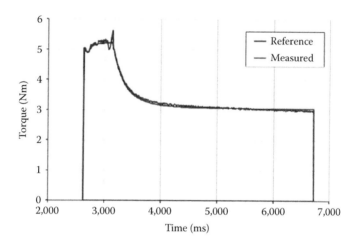

Figure 13.16.4 Motor response to a torque step signal of 50% of maximum value.

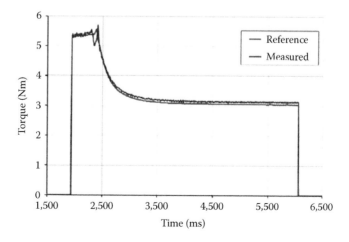

Figure 13.16.5 Motor response to a torque step signal of 90% of maximum value.

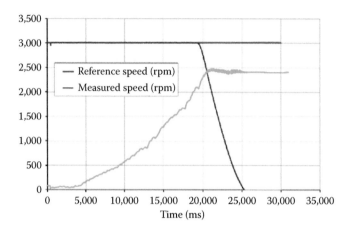

Figure 13.16.6 System response with load, speed = 50%, and torque = 0%–70%.

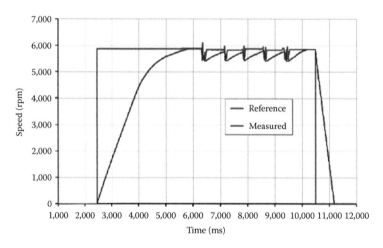

Figure 13.16.7 System response for a 100% speed step and torque = 30%.

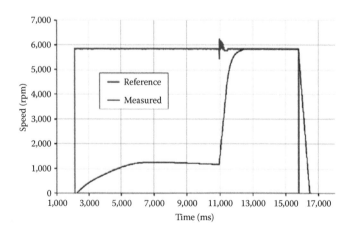

Figure 13.16.8 System response for a 100% speed step and torque = 50%.

Analyzing the data coming from the Real-Time system is difficult, as the sampling rate is quick and the time interval is very long (1200 s). In order to manipulate the recorded data effectively, a specialized software called DIADEM produced by National Instruments has been used. After compressing, the aforementioned data file exceeds 1GB. The data are stored in raw form, as points. As they were all recorded in real time, the time distance between two consecutive points is always equal to the sampling time. Based on this, any series of points can be reconstructed in a real waveform.

Synchronizing multiple PXI machines can be done with dedicated equipment, but this is quite expensive. A second option is using a dedicated port, but this method has much lower precision. A third option is introducing markers in the recorded data. In this case, the time constants of the physical systems are big (seconds), so the waveforms can be manually synchronized and aligned. DIADEM allows the user to perform multiple useful post processing tasks. All of these can be easily performed on an entire data channel, without the need for scripts or individual points. For example, the software can scale a channel automatically, determine its RMS value and filter it through different algorithms [14].

The most important aspect of the simulation is how the vehicle follows the speed reference (Figure 13.16.9). The reference is followed precisely, with deviations no higher than 0.1 km/h. The required torque can be seen in Figure 13.16.10.

The difference between the two values is likely due to noise affecting the torque reference, subsequently filtered. On the host machine screen this difference was not recorded. The speed reference for the electric motor is also followed (Figures 13.16.11 and 13.16.12). The graph confirms the hypothesis that the speed controller of the inverter can perfectly cover the space defined by the reference signal. This way, even if the motor speed is not used to calculate the vehicle speed, the tested motor is subjected to the same forces as in a real situation. Comparing the vehicle speed to the motor speed, similar shapes can be noticed, which allows the conclusion that the elasticity of the transmission has no visible influence in this case.

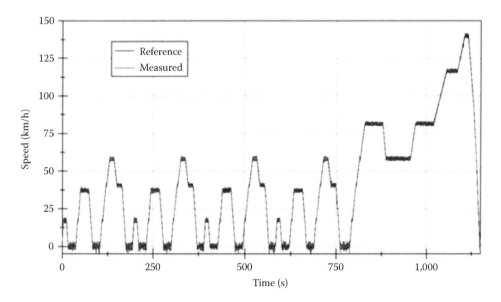

Figure 13.16.9 Speed profile tracking performance.

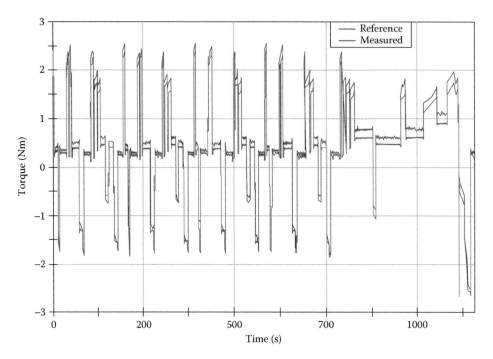

Figure 13.16.10 Reference and measured torque.

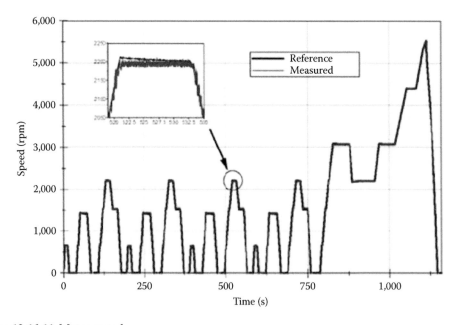

Figure 13.16.11 Motor speed.

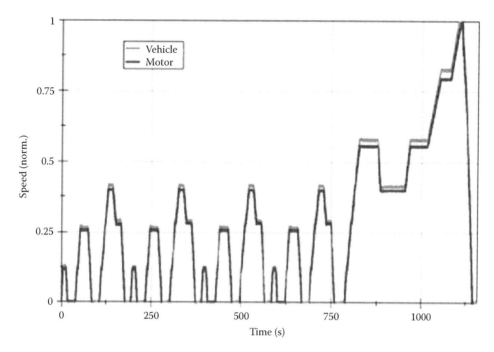

Figure 13.16.12 Vehicle speed and motor speed.

The speed–torque diagram of the analyzed engine can be seen in Figure 13.16.13. The values were normalized according to the maximum recorded value. The torque measured by the transducer is proportional to the torque difference between its two ends (Figures 13.16.14 to 13.16.16). When its sign is negative, it can be said that the axle switches from motor to generator. The speed reference to one of the motors is always positive (in the test the vehicle is always moving forward). Considering this, when braking, the motor with the torque reference becomes a generator, its torque being opposite to the axle movement. The RMS value of the current on a phase cannot have negative values, therefore in the previous graphs the two motors seem to be operating in the same mode. It can be noticed that one of them has a larger torque, indicated by a larger current value. This is the motor with the speed reference, which need to resist the other motor to keep the axle moving.

A more detailed look on the correlation between the measured values is shown in Figure 13.16.17. Figure 13.16.18 shows the current of the motor with the torque reference and the voltage of the motor with the speed reference together with other mechanical signals of interest. The sign of the torque measured by the transducer has been chosen so it has the same sign as the vehicle speed when the vehicle is going forward.

It can be seen that the voltages have reverse amplitudes compared to the currents. This is possible due to the vector control, the vehicle speed being controlled independently of the torque. The power of the two motors, calculated from the measured current and voltage, can be seen in Figures 13.16.19 and 13.16.20.

Comparing the electrical and mechanical power, the losses in the transmission can be calculated (Figure 13.16.21).

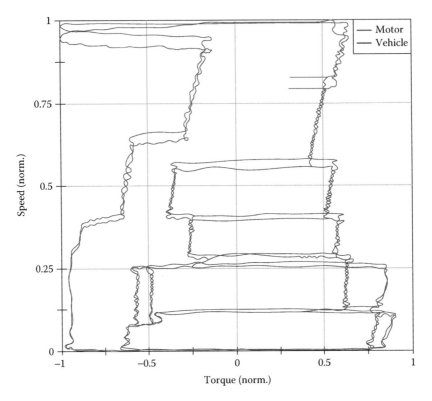

Figure 13.16.13 Speed–torque diagram.

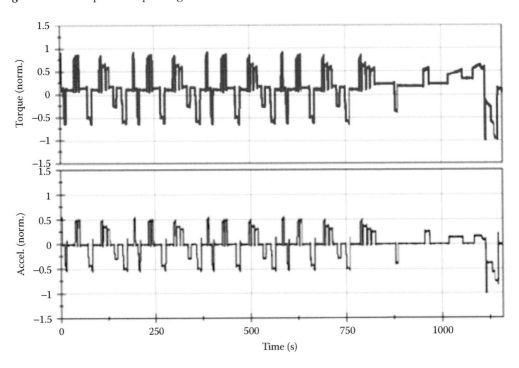

Figure 13.16.14 Comparison between torque and vehicle acceleration.

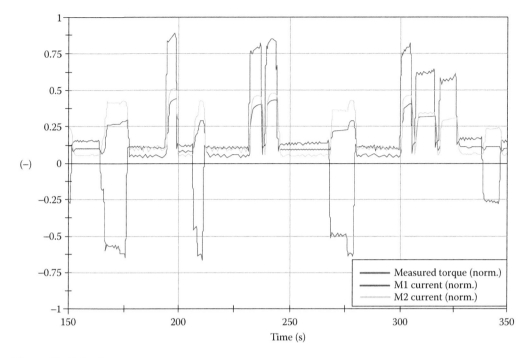

Figure 13.16.15 System torques.

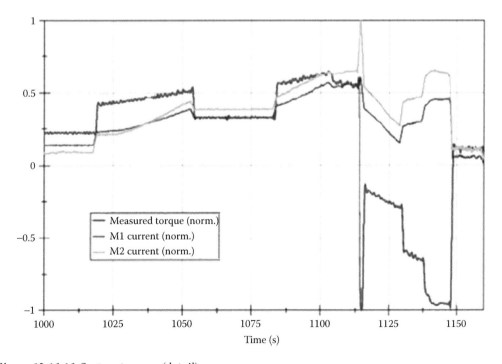

Figure 13.16.16 System torques (detail).

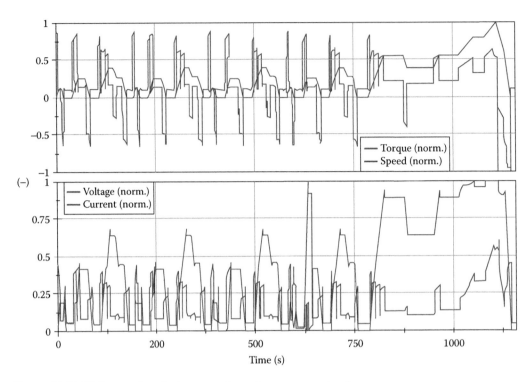

Figure 13.16.17 Comparison between main system measured values.

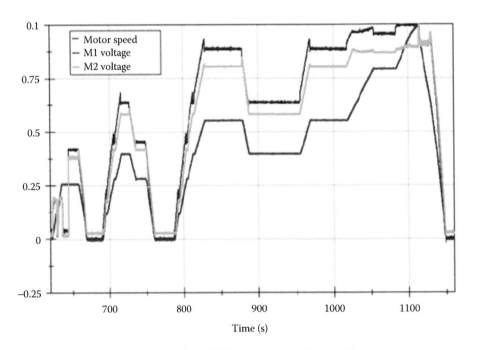

Figure 13.16.18 Comparison between the vehicle speed and voltage of the motors.

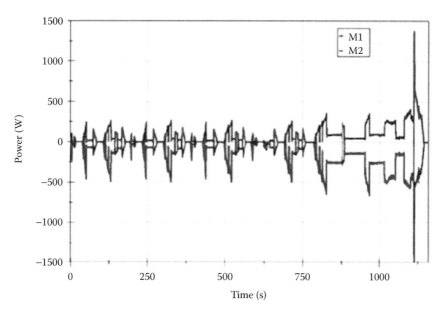

Figure 13.16.19 Power of the two motors.

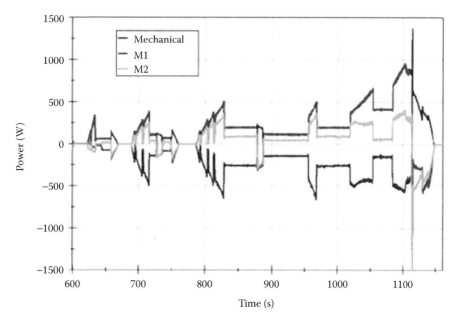

Figure 13.16.20 Power balance of the system (detail).

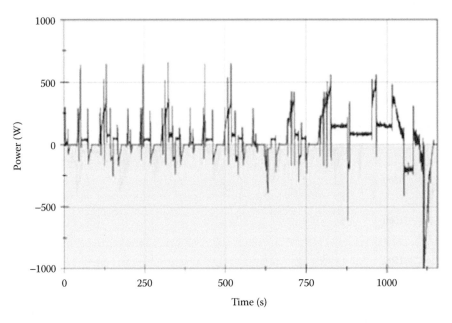

Figure 13.16.21 Transmission losses.

The negative losses represent the recovered energy during braking. It can be noticed that this is not very significant. By integrating the power losses, the total lost energy in a cycle can be determined. On average, the total energy loss in a whole cycle is 25 kWs, approximately 7 Wh.

13.17 Conclusion

This last part of the book tries to reveal the utility of the Amesim software in solving multiphysics problems. The high level of modeling the physical phenomena is very important for the quality of the whole HiL process. The time and money saved by this way of rapid fine tuning of the new products lead to a continuous development of the field. At the same time, the complexity of the hardware and software needed to build realistic simulation in any field requires strong interdisciplinary research teams, which can be *assembled* by the strong corporations aided by high-level research universities. The education of the master students in the field, by companies such as Siemens PLM Software [15] and dSPACE [16], will accelerate the use of RTS and HiL technologies in any common industry.

Bibliography

1. Vasiliu C., *Real Time Simulation of the Electric Automotive Transmissions*, PhD Thesis, University Politehnica of Bucharest, Bucharest, Romania, 2011.
2. Lebrun M., Claude R., How to create good models without writing a single line of code. *5th Scandinavian International Conference on Fluid Power*, Linköping, Sweden, 1997.
3. LMS International, *AMESim Reference Manual, Rev. 9*, 2009.
4. National Instruments Corporation, *LabVIEW User Manual*, 2009.
5. Petersheim M.D., Brennan S.N., Scaling of hybrid electric vehicle powertrain components for hardware-in-the-loop simulation. *17th IEEE International Conference on Control Applications*, Texas, 2008.

6. Brossard J.P., *Dynamique du véhicule: Modélisation des systemes complexes. Sciences Appliquees,* INSA Lyon – PPUR, Villeurbanne, France, 2006.

7. Andreescu C., *Wheel Automotive Dynamics, Vol. 1,* (in Romanian), Politehnica Press, Bucharest, Romania, 2010.

8. Elmqvist H., Mattsson S.H., Olsson J. et al., *Realtime Simulation of Detailed Vehicle and Powertrain Dynamics,* SAE Technical Paper, 2004.

9. Betz R., *Introduction to Real time Systems—ELEC371 Class Notes,* Electrical and Computer Engineering Department, Newcastle University, Australia, 2000.

10. Alles S., Swick C.A., Hoffman M.E., Mahmud S.M., Lin F., The hardware design of a real-time HITL for traction assist simulation. *IEEE Transactions on Vehicular Technology,* 3(44), 668–682, 1995.

11. Apsley J.M., Varrone E., Schofield N., Hardware-in-the-loop evaluation of electric vehicle drives. *5th IET International Conference on Power Electronics Machines and Drives,* Brighton, UK, 2005.

12. Hanselman D.C., *Brushless Permanent Magnet Motor Design,* 2nd ed., Magna Physics Publishing, Lebanon, OH, 2006.

13. Vasiliu C., *Method and Devices for RTS of an Electric Car Transmission,* Romanian Patent Demand No. A2010/01079, 2010.

14. Clark C.L., *LabVIEW Digital Signal Processing,* McGraw-Hill Education, New Delhi, India, 2005.

15. http://www.plm.automation.siemens.com/en_us/about_us/index.shtml.

16. https://www.dspace.com/en/inc/home/medien.cfm.

Index

Printed and bound by CPI Group (UK) Ltd, Croydon, CR0 4YY

24/10/2024

01778290-0015